Environmental Anthropology

a reader

Benjamin G. Blount

SIMON & SCHUSTER CUSTOM PUBLISHING

Printed in the United States of America

10 9 8 7 6 5 4 3 2 1

ISBN 0-536-00253-3

BA 97167

SIMON & SCHUSTER CUSTOM PUBLISHING
160 Gould Street/Needham Heights, MA 02194
Simon & Schuster Education Group

COPYRIGHT ACKNOWLEDGMENTS

Grateful acknowledgment is made to the following sources for permission to reprint material copyrighted or controlled by them:

"Living the Land Ethic: Aldo Leopold's Vision," by Teresa Opheim, reprinted from *EPA Journal*, vol. 18, no. 4, Sept./Oct. 1992, by permission of the U.S. Environmental Protection Agency.

Quotations, by Aldo Leopold, reprinted from *A Sand County Almanac*, 1949, by permission of Oxford University Press.

Quotations from "Engineering and Conservation," by Aldo Leopold, reprinted from *The Mother of the River of God and Other Essays*, edited by Susan L Flader and J. Baird Callicott, 1991, University of Wisconsin Press.

"A Call for Sustainability," by Russell E. Train, reprinted from *EPA Journal*, vol. 18, no. 4, Sept./Oct. 1992, by permission of the U.S. Environmental Protection Agency.

"Dilemmas in Ethics: Towards a Sustainable Society," by Poul Harremoës, reprinted from *Ambio*, vol. 25, no. 6, Sept. 1996, Royal Swedish Academy of Sciences.

"The Ecological Lessons of the Past: An Anthropology of Environmental Decline," by Timothy C. Weiskel, reprinted from *The Ecologist*, vol. 19, no. 3, 1989.

"Economic Interdependence and the World Food Summit," by Amartya Sen, reprinted from *Journal of Society for International Development*, vol. 4, 1996, Blackwell Publishers.

"Can the Growing Human Population Feed Itself?" by John Bongaarts, reprinted by permission from *Scientific American*, March 1994, copyright © 1994 by Scientific American, Inc. All rights reserved.

"Where the Poor Live: Are the Assumptions Correct?" by Robert W. Kates and Viola Haarmann, reprinted from *Environment,* vol. 34, no. 4, May 1992, Heldref Publications, Inc.

"Back to Eden,"by Evan Eisenberg, reprinted from *The Atlantic Monthly,* November 1989.

"Making Peace with the Earth: Indigenous Agriculture and the Green Revolution," by Deane Curtin, reprinted from *Environmental Ethics,* vol. 17, no. 1, Spring 1995, Environmental Philosophy, Inc.

"Biotechnology to the Rescue? Twelve Reasons Why Biotechnology is Incompatible with Sustainable Agriculture," by Jack Kloppenburg, Jr. and Beth Burrows, reprinted from *The Ecologist*, vol. 26, no. 2, March/April 1996.

"Environmental Ethics and Tropical Rain Forests: Should Greens Have Standing?" by Alastair S. Gunn, reprinted from *Environmental Ethics*, vol. 16, no. 1, Spring 1994, Environmental Philosophy, Inc.

The following are reprinted from *Human Organization*, by permission of the Society for Applied Anthropology:"Land-Use Strategies in the Amazon Frontier: Farm-Level Evidence from Ecuador," by Francisco J. Pichón, vol. 55, no. 4, 1996, pp. 416-424; and "Grassland Enclosures: Catalyst of Land Degradation in Inner Mongolia," by Dee Mack Williams, vol. 55, no. 3, Fall 1996, pp. 307-313.

"Agroforestry," by Sara J. Scherr, reprinted from *National Geographic Research and Exploration*, vol. 10, no. 2, 1994, National Geographic Society.

"Deforestation in Tanzania: Beyond Simplistic Generalizations," by Solon L. Barraclough and Krishna B. Ghimire, reprinted from *The Ecologist*, vol. 26, no. 3, May/June 1996.

"The Ecological Significance of Biological Diversity," by Martin Holdgate, reprinted from *Ambio*, vol. 25, no. 6, September 1996, Royal Swedish Academy of Sciences.

"Using Biodiversity as a Justification for Nature Protection in the US," by R. Edward Grumbine, reprinted by permission from *Wild Earth*, Winter 1996/97, P.O. Box 455, Richmond, VT 05477.

"Burning the Library of Amazonia," by Richard Evans Schultes, reprinted by permission of *The Sciences*, vol. 34, no. 2, March/April 1994. Individual subscriptions are $21 per year in the U.S. Write to: The Sciences, 2 East 63rd Street, New York, NY 10021.

"Biodiversity in Southern Africa: An Overview," by Joseph P. Dudley, reprinted by permission from *Wild Earth*, Fall 1996, P.O. Box 455, Richmond, VT 05477, and the author. The author's support at the time of writing was provided under grants from the National Science Foundation Program in Conservation and Restoration Biology and the Center for Field Research—Earthwatch. The author is currently a Ph.D. candidate in Wildlife Ecology at the University of Alaska, Fairbanks.

Map: "Vegetation Zones of Africa," International Union for Conservation of Nature and Natural Resources.

"Culture and Biodiversity in the Andes," by Grimaldo Rengifo Vásquez, reprinted from *Journal of Society for International Development*, 1996, Blackwell Publishers.

"Plant Management Among the Nahua and the Mixtec in the Balsas River Basin, Mexico," by Alejandro Casas, María del Carmen Vázquez, Juan Luis Viveros, and Javier Caballero, reprinted from *Human Ecology*, vol. 24, no. 4, 1996, by permission of Plenum Publishing Corporation.

"The Seeds of Our Future," by Vandana Shiva, reprinted from *Journal of Society for International Development*, vol. 4, 1996, Blackwell Publishers.

"A Fishy Tale," by Yva Momatiuk and John Eastcott, reprinted from *Geographical*, vol. LXVI, no. 11, 1994, American Geographical Society.

"Logging versus Fisheries in the Philippines," by Gregor Hodgson and John A. Dixon, reprinted from *The Ecologist*, vol. 19, no. 4, 1989.

"Collective Action for Common Property Resource Rejuvenation," by John Kurien, reprinted from *Human Organization*, vol. 54, no. 2, 1995, pp. 160-168, by permission of the Society for Applied Anthropology.

"Reversal of Fortune," by Denise Heywood, reprinted from *Geographical*, vol. LXVI, no. 11, 1994, American Geographical Society.

"Catastrophic Change in Species-Rich Freshwater Ecosystems," by Les Kaufman, reprinted from *BioScience*, vol. 42, no. 11, 1992, American Institute of Biological Sciences.

"Percentage of Commercial Catch in Kenya," "Demise of the Native Fishes," and "Total Landings for Haplochromines," Kenya Marine and Fisheries Research Institute.

"Resolutions of the Workshop on People, Fisheries, Biodiversity, and the Future of Lake Victoria," reprinted from a workshop, August 17-20, 1992, in Jinja, Uganda, sponsored by the National Science Foundation.

"Eco-Forestry or Protected Status?" by Ken Wu, reprinted by permission of *Wild Earth*, Winter 1996/97, P.O. Box 455, Richmond, VT 05477, and the author. Ken Wu is a graduate of the University of British Columbia's Ecology program. He is a wilderness activist and a correspondent for *Wild Earth.*

"Taming Tourism," by Sue Wheat, reprinted from *Geographical*, vol. LXVI, no. 4, 1994, American Geographical Society.

"Between the Devil and the Deep Blue Sea: Development Dilemmas in Mozambique," by Antonio Massinga, reprinted from *The Ecologist*, vol. 26, no. 2, March/April 1996.

"Contribution of Conservation to Sustainable Living through Health Promotion," by Manuel Cesario, reprinted from *Ambio*, vol. 25, no. 1, Feb. 1996, Royal Swedish Academy of Sciences.

"Socio-Economic Challenges for Sustainable Tourism in the Mediterranean Basin," by Andreas Ashiotis, reprinted from *Naturopa*, no. 78, 1995, by permission of the Council of Europe.

"Outcomes of a Community Controlled Wildlife Utilization Program in a Zambezi Valley Community," by Gordon E. Matzke and Nontokozo Nabane, reprinted from *Human Ecology*, vol. 24, no. 1, 1996, by permission of Plenum Publishing Corporation.

TABLE OF CONTENTS

PREFACE

This volume of readings grew out of a practical need to have available a reader in environmental anthropology. That need arose from several sources. One source is highly particular. A number of years ago, the University of Georgia established a new requirement of its undergraduates, an environmental literacy requirement. All students entering the University as undergraduates must satisfy that requirement prior to graduation, the expectation being that they will as a consequence have some degree of literacy on environmental concerns and issues. The requirement is satisfied through successfully completing courses that are designated as appropriate in terms of environmental content. The introductory course to anthropology is one of those courses, in fact, one that is selected often by students. Introduction to anthropology was a natural choice to satisfy environmental literacy, since the content of the field is heavily dependent on environmental concerns. Human origins, to take only one obvious topic, cannot be adequately addressed and discussed without bringing environment, environmental change, and natural selection and adaptation into the picture.

Incorporation of environmental topics into course lectures was more a matter of highlighting the environment than of adding new material. Readings on environmental anthropology was another matter. No reader dedicated to that topic existed, not for the whole of anthropology. Instructors for the introductory course have adopted a variety of readings in an effort to meet the needs and demands, from introductory textbooks in ecology and in geography to readers in applied anthropology. Packets of photocopied material has also been pursued. This reader is the culmination of my efforts to make available to students a reader more directly focused on environmental issues as seen by anthropology and addressed by anthropological research. The reader integrates the topics of lectures, which are more focused directly on anthropology, with environmental issues in which anthropology is only one of the participant disciplines.

Selection of essays and articles was difficult, given the rapidly expanding number of ones available. Several criteria drove the selection process. One was to try to cover the major topics, issues, problems, and concerns that a student would encounter in any introductory book on the environment, e.g., land use patterns, agriculture, forestry, fisheries, biodiversity, development, and tourism and natural parks, among others. Another criterion was to select essays or articles that either gave succinct accounts of environmental issues, problems, etc., or in-depth accounts of particular societies or regions of the world. Breadth and depth of coverage, in other words, was a concern. Another criterion was geographical diversity. A mix of essays and articles on the United States and other developed and developing areas of the world was considered essential. The types of concerns that a focus on environment raises are often not unique to any part of the world but are present in different forms in other areas. Conservation of topsoil for agriculture is important, indeed crucial, whether it occurs in Kansas, the Philippines, or Bhutan. How conservation is pursued and the particular issues associated may vary in those places, but the commonality of environmental problems and issues across the world needs to be understood. Lastly, an effort was made to include essays and articles written not just by anthropologists but by other individuals interested in environmental concerns and especially by individuals who are from the countries in which the discussions were based. Thus the authors of an article on ethnobotany in Mexico are Mexicans, and the author of an article of an article on tourism in the Mediterranean Basin is Greek. Some approximation of global coverage is present not just in the geographical area under discussion but in the authors themselves of the essays and articles.

A second source of felt need for the reader went beyond the introductory anthropology course at the University of Georgia. Anthropological attention to environmental issues has been expanding considerably during the past few years, witness the new section of the American Anthropological Association, Anthropology and Environment, and a new interest group on the environment in the Society for Applied Anthropology. A number of graduate programs in anthropology have begun to include environmental/ecological anthropology as a field of concentration and specialization. At the University of Georgia, that is the exclusive focus of the graduate program, where the curriculum is oriented toward environmental and ecological topics and where students are expected to specialize in fields that are directly related or tied to similar concerns. Given all of those developments, the time seemed appropriate for a first-effort at a reader in environmental anthropology.

Gratitude and appreciation is expressed to Laura German, who helped with the compilation of the reader in a variety of ways.

Benjamin G. Blount
8 May 1997

PART 1

Setting the Stage—Introductory Essays and Overviews

Environmentalism is a historical movement of increasingly complex proportions. As more and more attention is focused on the natural environment, the number of issues that relate to the state and health of the environment is expanded. As the focus becomes more close-up, new factors affecting the environment are recognized, new needs for concern and protection are noticed, and new areas for research and education emerge. An example would be recognition of the existence of the colorless and odorless gas, radon, that builds up inside houses and that in high concentrations over time can be cancer inducing. The construction of houses on soil that was contaminated by chemicals, perhaps years in the past, is a more graphic example, or in many areas of the world, of the increased concentration of pesticides in fresh water and the resultant contamination of virtually all living things that use or are associated with the water in any way.

As we learn more and as the range of environmental topics increases, the magnitude of the issues becomes more apparent. Environmental problems are not unidimensional. The "problem" is not just radon, or toxic chemicals in contaminated soil, or pesticides in water but of their impact on the health and even the continued existence of living organisms that come into contact with the agents. The status of the biological organisms becomes part of the concern, and in many, probably most cases, the net outcome of the use of the agents is mixed. A chemical compound may also serve useful purposes, such as making house paint mildew resistant, but in time, the discovery might be made that there are also serious health costs to the use of that particular compound. Health costs then involve a host of other factors. What are the relative costs to individuals, to the industries that produce agents that become environmental controversial, to communities whose health may be affected but whose livelihoods may depend on the very industry in question? Who gets to decide how such issues are resolved?

Politics, local and global, can become involved in any environmental issue, and with that involvement, resources can be put into play to sway outcomes of conflict in particular directions. Jobs may take precedence over health concerns of individuals, at least in the short term. Local ecosystems come under the influence of political economy, and fields of inquiry such as political ecology emerge to study the multidimensional nature of the interrelationships in the human-environmental equation. Ethical questions and dilemmas emerge. Who has the right to make and influence decisions and outcomes? What are the ultimate goals?

The essays and articles in this first section of the reader are addressed to the types of topics, issues, concerns, etc., that are posed above. Despite the diversity of topics and concerns, a number of basic problems and issues can be identified, and discussions about and inquiry into the environment center around them. One of those is the orientation of people toward the land, how they view their relationship to the land and thus ultimately to the products of the land, plant, and animal. Is the land simply to be exploited for what it will maximally produce, or is the land viewed in some other terms, as something not only valuable but even revered or held sacred? The first essay, by Teresa Opheim, provides an insight into the land ethic that one of the pioneers of the environmental movement in the United States held. Aldo Leopold made a simple but profound extension of an idea. He took the ethical notion that individuals in communities are inextricably tied to each other, and that for the good of each and for the whole, they needed to cooperate with each other, and he extended that idea to the natural environment in which the individuals and communities existed. Thus plants, animals, soil, water, the land itself, were all to be treated with respect toward cooperation. The idea was simple, but its consequences are potentially revolutionary, if adopted by a community.

The second essay, by Russell Train, addresses a host of issues that are central to environmentalism. Those include what he identifies as a choice between development that aims to be sustainable, that is to use resources in a way that maximizes the continuity of their availability, and that is exploitative and ultimately destructive. Control of economic growth, energy production and use, and land use policy are all seen as key issues in the United States in relation to the type of development that we currently have and that we need in the future is sustainability is to be taken seriously. He identifies environmental literacy as essential to an understanding of the issues facing contemporary human societies. Not only do we need a more in-depth understanding of the issues, but we need to be informed with a greater appreciation of ethical responsibility for the environment.

The third essay, by Poul Harremoës, addresses the history of what he calls egocentric philosophy, of people viewing themselves in social contexts and terms, and ecocentric philosophy, in which respect for all natural phenomena is accorded. He explores the causes and consequences of the two types of philosophy for sustainable societies, and he follows those out in the context of water, water use, water quality, and water quantity, ultimately placing the discussion in a context of ethical issues. Sustainability is usually talked about in terms of continuity of material resources, such as water, but sustainable societies cannot be simply aggregates of people who just happen to conserve and use their resources wisely. Intellectual issues arise. Information must be shared among such a community, information that anthropologists would call culture, and that must be developed and sustained through time also. Sustainability must also include and contain knowledge.

The focus of the discussion in the fourth essay, by Timothy Weiskel, is on the foundation of all civilizations, their agricultural base, and on the unexpected fragility of that base. The archaeological record is clear that ancient civilizations declined and fell as consequences of the degradation of their ecological bases and thus of their ability to produce sufficient food through agriculture to feed their populations. He extends the complex of problems that can be seen in the demise of ancient civilizations to the contemporary scene and points out that the nature of the problems is the same. A series of destructive processes—called the underdevelopment spiral—involves use of land, water, and forest products in destructive ways and points to similar ecological and environmental collapse. Brief case studies are presented, such as ecological devastation in areas of West Africa.

The next two essays, by Amartya Sen and John Bongaarts, are on the related issues of the world's food supply and human population growth. The United Nations Food and Agricultural Organization (FAO) estimated in 1996 that 800 million of the world's population, approximately one person in seven, is chronically undernourished. Undernourishment, malnutrition, and starvation are all part of the current world scene, and they reflect a variety of interrelated features. Population levels and growth are certainly a central component of that complex, but as the discussion by Bongaarts shows, the connections are not always direct and simplistic. Distribution and availability of food resources is subject to political and economic forces in addition to ecological and environmental ones, and thus overall levels of food production in relation to overall levels of population do not directly equate. When one projects into the future and tries to see if the world can feed itself when the current population doubles, the problems become even murkier and more complex. All of those issues, however, must be seen against the current backdrop of the magnitude of hunger and malnutrition in today's world.

The final essay in this first section, by Robert Kates and Viola Haarmann, asks the question "Where do the poorest people in the world live?" Wealth and poverty are somewhat culturally relative terms, but by almost any terms a substantial portion of the world's people live in circumstances where resources are meager and where efforts to make a semblance of a living have significant impact on the environment. People who are the poorest in terms of food supply and natural resources live in environments that are especially fragile and degradable. Where are those people and those environments? What Kates and Haarmann show is that the poorest countries are the ones that have the most arid lands. The areas in which the productivity of the land is the lowest, principally because of lack of water, are the ones in which the standards of living are the lowest. The central importance of land for food production is tied, naturally, to availability of water for the production of crops.

LIVING THE LAND ETHIC:

Aldo Leopold's Vision

TERESA OPHEIM

"Like winds and sunsets, wild things were taken for granted until progress began to do away with them," Aldo Leopold wrote in the foreword to his book *A Sand County Almanac.* "Now we face the question whether a still higher 'standard of living' is worth its cost in things natural, wild, and free. For us of the minority, the opportunity to see geese is more important than television, and the chance to find a pasque-flower is a right as inalienable as free speech."

Aldo Leopold devoted his life to preserving his expansive version of "inalienable rights." In his time, Leopold made seminal contributions to ecology, wilderness preservation, and wildlife management, among other fields. Today, however, he is known chiefly for his authorship of *A Sand County Almanac.* In *Almanac,* Leopold wrote of a woodcock's dance, the corky bark of a burr oak, and other observations typical of the genre of nature writing. He also discussed conservation issues, including, in his words, "some of the episodes in my life that taught me, gradually and sometimes painfully, that the company is out of step." *A Sand County Almanac's* most enduring passages, however, are Leopold's presentation of his now-famous "land ethic":

> "All ethics so far evolved rest upon a single premise: that the individual is a member of a community of interdependent parts. His instincts prompt him to compete for his place in the community, but his ethics prompt him also to co-operate (perhaps in order that there may be a place to compete for).
>
> "The land ethic simply enlarges the boundaries of the community to include soils, waters, plants, and animals, or collectively: the land.
>
> "This sounds simple: do we not already sing our love for and obligation to the land of the free and the home of the brave? Yes, but just what and whom do we love? Certainly not the soil, which we are sending helter-skelter downriver. Certainly not the waters, which we assume have no function

except to turn turbines, float barges, and carry off sewage. Certainly not the plants, of which we exterminate whole communities without batting an eye. Certainly not the animals, of which we have already extirpated many of the largest and most beautiful species. A land ethic of course cannot prevent the alteration, management, and use of these 'resources,' but it does affirm their right to continued existence, and, at least in spots, their continued existence in a natural state.

"In short, a land ethic changes the role of *Homo sapiens* from conqueror of the land-community to plain member and citizen of it. It implies respect for his fellow-members, and also respect for the community as such."

"A Pinchot-like Intent"

A Sand County Almanac has been called a subversive book, one that questions the deepest values of our civilization. Leopold, however, did not live his life as a subversive man. Although he had the aesthetic and ethical sensitivity of a Romantic, he was pragmatic and spent his days working for change through committees, governmental reports, and articles directed toward garden club members, hunters, and farmers. He had, according to his biographer, Curt Meine, "both a Muir-like appreciation of nature and a Pinchot-like intent to use nature wisely."

Aldo Leopold was born in 1887 in the river town of Burlington, Iowa, to a German-American family of avid naturalists and hunters. The Leopolds lived in a home atop the bluffs of the Mississippi River. Although the location offered spectacular wildlife displays along the river's flyway, the wildlife was diminishing rapidly by the time of Aldo's birth. Leopold left the Midwest in 1904 to attend the college preparatory Lawrenceville School in New Jersey. He entered Yale University in 1905 and graduated with a Master of Forestry from the Yale Forest School in 1909.

After graduation, Leopold joined the U.S. Forest Service and headed to the Arizona and New Mexico territories for his first assignment. He quickly rose to become supervisor of the Carson National Forest in northern New Mexico, a position he coveted. After he was striken with acute nephritis a few years later, though, the physical rigors of a supervisor's job were no longer an option for Leopold, and he spent the rest of his life working at less physically demanding desk jobs and on consulting projects.

Leopold wasn't long in the Southwest before he began to notice a development pattern of waste and abandon. He lamented the serious erosion he routinely saw, and he came to challenge the Forest Service's assumption that heavy grazing by domestic stock was necessary to hold down grass that could carry fires. Leopold noted that, before white settlers arrived in the region, Southwestern watersheds had maintained their integrity despite centuries of periodic wildfire.

The young Leopold's thinking on watersheds was ahead of his time, but in other ways he reflected the conservation status quo. He was instrumental in a campaign to rid Southwestern public lands of "varmints" such as wolves, foxes, and bears, predatory animals that Leopold said "are continuing to eat the cream off the stock grower's profits."

Leopold grew to love the Southwest deeply and knew that, before Anglo-Americans pushed their way into every continental cranny, a governmental program would be needed to save any remaining areas of wilderness. Leopold wrote of the West and the nation's expansion: "For three centuries that environment has determined the character of our development; it may, in fact, be said that, coupled with the character of our racial stocks, it is the very stuff America is made of. Shall we now exterminate this thing that made us American?"

In 1922, Leopold drafted a wilderness area plan that became the basis for an administrative designation of the Gila National Forest as a wilderness area in 1924. His work set the pattern for what eventually became a nationwide system of wilderness areas. Leopold continued writing about the importance of wilderness areas his entire life, and later, in 1935, he helped found The Wilderness Society.

Ownership and Obligation

In 1924, Leopold accepted a transfer to the U.S. Forest Products Laboratory in Madison, Wisconsin, where he worked for four years before leaving to become a consultant on forestry and game management issues. A life-long hunter (for which he has been criticized by Rachel Carson and others), Leopold then surveyed wildlife habitat and game restoration policy in many Midwest states and helped inaugurate wildlife research projects at a variety of universities. He wrote *Game Management,* still a classic textbook for its field, and spent the last 15 years of his life as chairman of game and wildlife management studies at the University of Wisconsin. According to historian Susan Flader, "Today he is acknowledged by many in the field as the 'father' of the profession of wildlife management in America."

In Wisconsin, Leopold also focused on soil erosion and other issues of what he called "biotic farming." He wrote, "Soil is the fundamental resource, and its loss the most serious of all losses. . . . The day will come when the ownership of land will carry with it the obligation to so use and protect it with respect to erosion that it is not a menace to other landowners and the public."

Leopold's extensive fieldwork and his acute powers of observation on such a wide variety of issues—game depletion, soil erosion, wilderness preserves—reflected his understanding of the science of ecology at a time when that science was just coming of age. Leopold realized that, at its basic, the land was a complex organism that functioned through the interaction of its components. His keen understanding of the complexity of the natural system eventually led him to alter his thinking on predators. Decades after leading a campaign to eliminate predators, Leopold then led the call for their restoration to American public lands. His reversal was eloquently described in his essay "Thinking Like a Mountain," which described his interaction with a wolf he and others had killed:

"We reached the old wolf in time to watch a fierce green fire dying in her eyes. I realized then, and have known ever since, that there was something new to me in those eyes—something known only to her and to the mountain. I was young then, and full of trigger-itch; I thought that because fewer wolves meant more deer, that no wolves would mean hunters' paradise. But after seeing the green fire die, I sensed that neither the wolf nor the mountain agreed with such a view. . . ."

The Shack

Leopold's professional work provided the seeds for his theories about a human relationship with the land, but it was on his farm in Central Wisconsin that Leopold refined those theories and practiced his own personal compact with the land.

In 1935, Leopold bought some land in Wisconsin's Sand Counties Region. The only building on the abandoned Wisconsin River farm was a dilapidated chicken house-turned-cowshed with knee-deep manure on the floor. The farm had bare, blowing sands, it had been stripped of much of its timber, and its marsh had burned. For those who must go to towering mountains or crashing seas to find displays of nature they consider spectacular, the Sand Counties land had little to offer. For Leopold, however, the aesthetic appeal of country had nothing to do with its scenic qualities; he was interested in the land's evolutionary heritage and ecological processes. He sought to restore his land to ecological integrity, a goal he knew he would not live to fulfill completely.

Leopold's first recorded act as a landowner was to plant a food patch for wildlife. He, his wife, Estella, and their five children (all of the Leopold children became well-respected scientists) built birdhouses for martins, screech owls, and wood ducks and planted prairie grasses, wildflowers, and shrubs. Through the years, they planted thousands of pines—up to 5,000 to 6,000 a year—and Leopold devoted a good deal of effort trying to keep his trees healthy and free from the deer, drought, and fires that killed so many of them.

At The Shack, Leopold continued his life-long record keeping. His detailed journal includes entries such as the one for May 31, 1945, that only an insomniac such as Leopold could have observed:

Weather Warm, calm, hazy at daybreak, 58 degrees at 3 AM. Changed suddenly to cold NE wind at 7 AM, with a long streak of cloud marking the NE front. 54 degrees at 8 AM, cloudy. 41 degrees 6 PM.

Daybreak Song A favorable morning for early song:
2:45 song sparrow
3:00" "
3:05 field sparrow
3:07 song sparrow
3:09" ", field sparrow
3:11 field sparrow
3:15 Yellowthroat, field spar.

3:18 Crested Fly field spar.
Indigo (after which all cut loose). . . .

On April 21, 1948, Aldo Leopold died of a heart attack while helping a neighbor fight a grass fire. Just a few days before, Oxford University Press had notified Leopold that *A Sand County Almanac* would be published. Albert Hochbaum, a former student and colleague of Leopold's, said in his eulogy for him: "Few men loved the land so deeply as he loved America; few who have loved the land have examined it so carefully; and few who have examined the land have been so articulate in detailing their discoveries."

Leopold once wrote that "No important change in human conduct is ever accomplished without an internal change in our intellectual emphases, our loyalties, our affections, and our convictions." Given the momentous way in which he was asking Americans to redefine their relationship with the land, the practical Leopold fully expected the necessary adjustments to take lifetimes. Said Leopold: "I have no illusions about the speed or accuracy with which an ecological conscience can become functional. It has required 19 centuries to define decent man-to-man conduct and the process is only half done; it may take as long to evolve a code of decency for man-to-land conduct." Given the urgency of today's environmental problems, Leopold's vision of a land ethic remains poignantly relevant; his patience for change is not.

(Editor's Note: Quotations from A Sand County Almanac *(Oxford University Press, 1949) are reprinted with permission; quotations from the essay "Engineering and Conservation" are from* The Mother of the River of God and Other Essays by Aldo Leopold. *Edited by Susan L. Flader and J. Baird Callicott (Madison: The University of Wisconsin Press) Copyright 1991, the Aldo Leopold Shack Foundation; reprinted by permission of the publisher.)*

A CALL FOR SUSTAINABILITY:

To Ensure Our Future Survival, Major Changes are Needed Now

RUSSELL E. TRAIN

The coming together of more than 170 nations under the auspices of the Earth Summit was, if nothing else, the first *global* acknowledgement that environmental quality and economic health are inextricably linked—that the economic well-being of the Earth's peoples depends directly on the continued health of its natural resources.

This synthesis of environment and economics—and put forward in *Agenda 21*, the lengthy charter for the future adopted by the conference plenery—known as sustainable development, was only advanced, not discovered, by the diplomats in Rio. I suspect that if one were to search the literature, one would find references to the basic relationship hundreds, if not thousands, of years ago. I do know that 85 years ago, in the annual message to Congress which has since become known as the State of the Union address, President Theodore Roosevelt said, "To waste our natural resources, to skin and exhaust the land instead of using it as to increase its usefulness, will result in undermining in the days of our children the very prosperity which we ought by right to hand down to them amplified and developed."

The choice between crisis and sustainable development is one our nation shares with the rest of the world, and the only way to address it is through international cooperation and through U.S. commitment to leadership at home and abroad.

As the world's single largest economy, the largest user of natural resources, the largest producer and consumer of energy, and the largest producer of carbon dioxide pollution, the United States has not just a special responsibility to exercise world leadership but a particularly high stake in meeting the environmental challenges of the future.

I am convinced that the natural processes that support life on Earth are in serious jeopardy and that by acting now—or not acting—our country is choosing between two radically different futures. If the United States continues down its current path, merely reacting to and trying to repair environmental injuries, then the nation's natural resources, economy, and way of life will deteriorate. However, if our country pioneers new technologies, realigns government policies, makes bold economic changes, and embraces a new ethic of environmentally responsible behavior, we can expect the coming years to bring a higher quality of life, a healthier environment, and a vibrant economy.

The time is now for new strategies to address the environmental challenges of the future. The National Commission on the Environment (see box on page 9) spent more than 18 months deliberating and debating ways to address the overwhelming environmental problems we face. Let me share some thoughts of mine that arose from the commission's work.

The Picture Today

Over the past 20 years, an impressive array of federal, state, and local pollution control and resource management programs, both public and private, have been instituted in the United States. Total U.S. expenditures on environmental protection now average more than 2 percent of gross national product per year.

The United States had the foresight to begin adopting stringent environmental laws and regulations more than two decades ago and to make sizable economic investments in pollution control and energy efficiency. As a result, this country does not have to contend with landscapes as blighted, air and water as polluted, soils as poisoned, or public health as ravaged as those of Central and Eastern Europe. The measurable environmental progress made by the United States should be a source of national pride.

Still, our country's environmental achievements allow no room for complacency. Despite numerous victories, the United States is losing the battle:

- Global environmental problems to which we make no small contribution—climate change, loss of biodiversity, stratospheric ozone depletion, for instance—are placing both human and natural systems at grave risk.
- The air in U.S. cities threatens to deteriorate further as improvements in auto emissions controls are overwhelmed by the sheer numbers of cars and miles driven and by congestion. Meanwhile, indoor air pollution is largely ignored.
- Disposal and cleanup of the vast amounts of waste generated each year pose ever greater difficulties and consume an increasing proportion of the limited funds available for environmental protection.

Indirect sources of pollution, such as urban and agricultural runoff, continue nearly unabated.

- Encroaching land development is displacing and undermining critical ecosystems, such as wetlands, and threatens rural landscapes, natural areas, and biological diversity.
- Large areas of national forest and other public lands and resources are not managed sustainably.
- Farmlands are suffering from the loss of soil and excessive use of chemicals.
- Aquifers, a major source of water supply, are being consumed and contaminated at an alarming rate in many areas of the nation.
- Overfishing is seriously depleting our most important commercial fisheries.
- In many U.S. inner cities, the physical environment has the look of a wasteland.

While this litany of environmental ills, familiar-sounding and by no means complete, is a product of *today's* level of economic activity and human population, consider tomorrow. Over the next 50 years—within the lifetimes of many of us and of all of our children—economic activity in the United States is projected to *quadruple* and global population to at least *double*. If growth of this magnitude occurs with today's industrial processes, agricultural methods, and consumer practices, the result could be both environmentally and economically disastrous.

Forecasts based on linear projections are often wrong. In the case of environmental conditions, such projections may be too optimistic. Ample evidence suggests that, unless we act decisively, the price will be serious—in some cases, irreversible—environmental damage. Clearly, it would be the height of folly for the nation to sit back and simply hope that the future will be greened by an invisible hand. Excuses for inaction, such as budget deficits and opposition to taxes, abound. Yet the continuing pursuit of politics as usual will almost certainly guarantee failure.

There must be an end to ambivalence about both the importance of environmental policy and our environmental policy priorities. The United States must have a long-term strategy for pursuing the goal of sustainable development. Such a balanced strategy may anticipate or avoid severe local and regional economic dislocations or stimulate adjustment assistance and job retraining.

Economic Growth

Economic and environmental well-being are mutually reinforcing goals that must be pursued simultaneously if either one is to be reached. Economic growth will create its own ruin if it continues to undermine the healthy functioning of Earth's natural systems or to exhaust natural resources. By the same token, healthy economies are most likely to provide the necessary wherewithal for investments in environmental protection. For this reason, one of the principal objectives of environmental policy must be to ensure a decent standard of living for all.

Sustainable development innovations will themselves bring major economic benefits. The economic advantage of efficiently using materials and energy is obvious, and the domestic production and use of environmentally sound technologies will reap profits both for the U.S. firms that sell them and for those who use them.

The most efficient way to achieve environmental progress is to harness market forces. Here, the role of public policy is to send the right signals to the economy—"getting the prices right" and making the marketplace work *for*, instead of *against*, environmental protection. Available tools include social-cost pricing, taxes, and removing or instituting subsidies.

The National Commission on the Environment harbored no illusions that market economics alone will put the United States or the world on the path to sustainable development. Government, private, and personal initiatives are also required. Regrettably, the U.S. statutory and regulatory system is woefully inadequate, cumbersome, and sometimes even perverse. A regime that now emphasizes "end of the pipe" cleanup must be radically reformed into one that encourages pollution prevention. Changing product design or manufacturing processes to minimize or prevent pollution is obviously superior to mandating expensive cleanup after the fact. An environmentally literate public will encourage such efforts by demanding environmentally acceptable products.

If environmental prevention is to prevail over environmental cure, and if the United States is to remain an industrial leader, our country must rapidly develop and deploy a wide array of more efficient and environmentally safe technologies. This need for new technology is particularly acute in transportation and energy generation.

Energy

There must be a fundamental change in how our country produces and uses energy. No single area of activity is so closely interwoven with the environment. Were it not for the world's predominant reliance on fossil fuels for both energy production and transportation, the problems of global warming, acid rain, and urban smog would be relatively minor. A progressive shift away from fossil fuels as quickly as possible in both the energy and transportation sectors is therefore crucial. Because the United States uses in excess of one-fourth of the world's energy—most of it generated by fossil fuels—it must accord this matter the highest priority.

We need dramatically different economic incentives in the energy sector. Coal, oil, and gas prices, for instance, must reflect the environmental costs associated with their combustion. Over the long term, the United States must develop alternative nonpolluting sources of energy, principally from renewable sources.

Meanwhile, the country must develop technologies that use energy more efficiently, thereby consuming less fuel.

In transportation, the long-range need is a shift in auto technology to a nonpolluting source of energy. Electricity—and in the long run, hydrogen—holds promise as a nonpolluting energy carrier. More research on nonpolluting sources is needed. The immediate need is for incentives for more fuel-efficient autos and for fewer miles driven.

In agriculture, in manufacturing processes, in consumer products, and in almost every other sector of the economy, new technologies will give the United States a competitive edge as well as a healthier environment. The worldwide market for such technologies can only continue to grow as the connections between environmental and economic well-being become more apparent. The economic potential of trade in such technologies is no secret: Japan and Germany, among others, have already moved aggressively into this field. If the United States moves up in this technology race, it will be because we have at last understood that we need a technological revolution, not just another technical fix.

Land Use Planning

A key failure in the U.S. effort to address environmental needs has been widespread resistance to land use planning. I am not suggesting that the federal government should impinge on state and local governments in this area, though certainly it should manage its own property and facilities better. State and local governments and other regional groups should undertake land use planning for a variety of reasons: to protect environmentally sensitive areas, including watersheds, aquifers, and wetlands; to maintain biological diversity; to continue productivity of agricultural land; and to protect sites of natural beauty and of historic and cultural value.

The Global Environment

While the United States must put its own environmental house in order, there is no denying it has a huge stake in addressing global environmental problems. As environmentalists Barbara Ward and René Dubos have said, all of us have two countries—"our own and Planet Earth."The threat of global climate change, for example, certainly requires national initiatives in the United States, but the problem really can be addressed only through a common worldwide effort.

Similarly, the destruction of forests will exacerbate global warming and accelerate the loss of species and ecosystems, foreclosing medical, recreational, and trade opportunities for the United States, as well as diminishing the world's shared biological heritage. Economically, the large and growing trade with developing countries will ultimately collapse unless these nations achieve sustainable development. Politically, U.S. national security interests depend increasingly upon achieving a level of international stability that can come only from sustainable development.

The most critical need facing the world is the control of human numbers. Continued global population growth of the current magnitude—1 billion more people every decade—will swamp economic and social progress, as well as efforts to protect the environment. Our country and every other country stands to gain by efforts to stabilize world population and improve living standards in developing countries, where 90 percent of projected population growth will occur. The burden placed on the environment is a product of population and consumption. The priorities for the developed countries must focus on switching to sustainable technologies to reduce wasteful consumption; the priorities for the developing countries are to develop sustainably and curb population growth.

The United States must make a major commitment to cooperate with the world community to stabilize global population, recognizing the linkages between birth rates, child survival, economic development, education, and the economic and social status of women. Universal access to effective family planning information, contraceptives, and health care is essential.

Environmental Literacy

An informed citizenry with an ethical commitment to care for the environment is essential to the future. Success with the necessary technological, economic, and governmental changes is predicated on the understanding and support of the American people—individuals and families, government at all levels, and business.

The U.S. society and its schools must pledge themselves to the goal of ecological literacy. U.S. citizens must have the knowledge, practical competence, and moral understanding to cooperate in building a sustainable civilization. The pursuit of environmental literacy will require curricular innovations from kindergarten through college, changes in teacher education programs, and expanded graduate programs.

All of us must develop a greater sense of ethical responsibility for the environment. Environmental ethics are founded on an awareness that humanity is part of nature and that nature's myriad parts are interdependent. In any natural community, the well-being of the individual and of each species is tied to the well-being of the whole. In a world increasingly without environmental borders, nations, like individuals, have a fundamental ethical responsibility to respect nature and to care for the Earth, protecting its life-support systems, biodiversity, and beauty, caring for the needs of other countries and future generations.

It is only within such a framework that sustainable development will be achieved. Religious institutions, schools, businesses, governments, the news media, and, perhaps, above all, families must share in the tasks of achieving it.

Humanity *can* live and prosper in harmonious and sustainable balance with the natural systems of the Earth. Americans have an opportunity to rise to the challenge of environmental leadership as they have to the causes of human liberty, of equality, and of free and open markets. The challenge starts at home.

DILEMMAS IN ETHICS:

Towards a Sustainable Society

POUL HARREMOËS

Sustainable development requires significant changes to philosophical and religious attitudes. The western world needs ecocentric awareness and changing attitudes to mass transport, utilization and release. The developing world will experience pressures on water and the environment from population growth and from demands for better material welfare in mutual conflict.

Introduction: Anthropocentric Versus Ecocentric Philosophy

Mankind has the potential ultimately to reach a solution: a sustainable society, that can continue for generations. Current development is, however, not even close to anything like sustainability; in fact development is rapidly decreasing the prospects for sustainability. This applies to both the industrial and developing countries. A sustainable society cannot be achieved without fundamental changes in our basic thinking, ethical values, moral concepts and religious beliefs. The dilemma, is that most of our fundamental beliefs are in direct conflict with the anticipated changes. That is the challenge.

The environmental debate goes right to the root of our basic thinking. Environmentalists advocate that we have to substitute our anthropocentric thinking with ecocentric thinking. Anthropocentric philosophy considers everything from the point of view of mankind, and the inalienable right to pursue his fortune as he sees fit. The *egocentric* person thinks only of himself in a social context as opposed to an *ecocentric* philosophy, which advocates respect for all nature and all creatures, basic rights.

This issue is at the very heart of philosophy and religious beliefs. European philosophy and Christianity is founded on anthropocentric concepts. God is visualized and symbolized as almighty. In fact, Christianity developed in competition with religions devoted to many gods of all shapes and features, to a concept of one all embracing God. Two basic concepts are identified as being particularly anthropocentric:

- Love from God to man and the mutual love of one's neighbor are basic concepts of Christianity. The latter aspect of the two is an ideal which the Christian world has had some difficulties with. It is fair to say that through 2000 years of history the concept has had a tendency to consider people of other cultures and beliefs as outsiders to be converted and not as neighbors. It is a fair question whether this has changed during this century.
- The other anthropocentric extreme is the concern for life after this. This concern took its most extreme forms during the middle ages. However, philosophically speaking this is the anthropocentric thinking which was the driving core of the approach to life. There was little concern for nature and other creatures as equal partners.

This is seconded in European philosophy by our Greek heritage. This started with the sophistic thinking, which took its starting point in the human being and his ability to think—as opposed to a competing concept of the human being in an all embracing cosmos. From this developed the roots of logic and scientific thinking. In this regard, environmentalists in particular are antagonistic to one of the most prominent European philosophers, René Descartes (1596–1650), for his statement: "*Cogito, ergo sum*". Everything starts with man and his ability to think. All values, all concepts are derived from man.

It is thought provoking that the most basic and scientifically fundamental considerations of the renaissance were devoted to something as "useless" as astronomy. Galileo Galilei (1564–1642) proved that the earth circled the sun and not the other way around and was condemned by the Church. He introduced experiments and applied mathematics, further developed by Isaac Newton (1642–1727), Pierre de Fermat (1601–1665), G. W. Leibniz (1646–1716) and many others to follow. Science became one of the pillars in European philosophy and formed the basis for the industrial revolution of the last century.

In this context, the result was the western concept to conquer the world—not only the world in a geographical sense, but also in the sense of mastering the universe. Man can shape his own destiny without constraints. This anthropocentric attitude is quite understandable in view of what has been achieved. It is a frequently stated misconception that "the good old days" were romantic and pleasant. Diseases, famine, cold, violence—all accounted for a misery which is hard to conceive today. Scientific development and industrialization have been a success from an anthropocentric point of view, and an ecocentric alternative cannot be based on the naive proclamation: "return to nature", as advocated by Rousseau (1712–1778).

While science and industrialization have their roots in classical physics as initially applied to cosmos, the critic of anthropocentric thinking is derived from the atoms. It has become clear that it may not take an accidental comet to disrupt the world. We now have the potential for nuclear self-destruction. This was the overture to questions about pollution, resource depletion, and to the environmental debate as we see it today.

The antipode to anthropocentric thinking is frequently associated with, e.g. the American Indian. In Indian philosophy, man is intermingled with nature and must live in harmony with it. The spirits are the nature in all its forms. This is very close to fetishism (the belief that the spirits are in all objects of nature), which I was taught in school to be the most "primitive" of all religions.

The Asian religions have mixtures of anthropocentric and ecocentric features. Reincarnation and nirvana are basically anthropocentric, while both Hinduism and Buddhism have gods symbolized as animals, and built-in concepts of a holy nature.

Environmentalists advocate conversion to an ecocentric philosophy as opposed to an anthropocentric philosophy. European religion and philosophy is anthropocentric by origin and by status. It would take generations and centuries to change such deep-rooted concepts. It is not just semantic to argue that the solution is to accept the fact and convince people that:

> *It is in the anthropocentric interest of man to incorporate ecocentric aspects in religion and philosophy.*

Descartes axiom: "*Cogito, ergo sum*" cannot be removed from the minds of the European nor the American. It is up to them to recognize their self-interests and act accordingly with respect to the environment.

There are many examples of environmentalists going overboard with utopian ideals. For example, a statement claiming the right of all organisms as equal partners; the practical consequence of which would be human life threatening, but pathogenic bacteria are also organisms, though micro. The Who program for global eradication of the smallpox virus would have to be discontinued. Use of penicillin should be banned. Ecocentric approaches have to be argued within the context of anthropocentric interests, in order to have impact.

Definitions and Interpretations of Sustainability

Sustainability is the new buzzword. The short version of the definition of the Brundtland Commission reads:

> *A development is sustainable, if it meets the needs of the present without compromising the ability of future generations to meet their own needs.*

Note the anthropocentric version of the definition. Since then much thought has been devoted to alternative definitions and interpretations of the term. However, the essence is not the semantics of the definition, but the introduction of a whole new dimension in European philosophy, the implications of which are yet to be seen. In the process, the debate itself and the awareness that it introduces are important factors.

> *There is a severe contradiction between the development of a society and the frequently formulated objectives for that society; sustainability is a striking example.*

The changes that have already taken place in western societies with respect to pollution abatement, are quite encouraging. Purification of sewage and cleaner production techniques are changing industrial production. Environment has become an accepted strategic consideration in industrial planning. It is no more than a decade ago, that industry considered pollution abatement as a burden.

I will use the following interpretation of"sustainability"which has a resource component and a pollution component:

> *Society should use its resources such that the society can continue its mode of operation without exhausting its resources.*

Our present mode of operation illustrates the thermodynamic principle of entropy. When mineral resources are mined and used, the result is a conversion from a concentrated to a dispersed state. It will require energy to reconcentrate the matter, which makes renewable energy basically the ultimate measure of sustainability.

The practical solution is to approach the ideal, which makes sense in societies which have violated the principle of sustainability due to lack of concern in the past. Water and a few other resources are unique in making sustainable operations possible.

> *Society should protect the environment against irreversible damage, including protection of unique species and habitats.*

The conflict is that the very basis for the success of man in a modern society is his ability to change the environment to sustain larger populations. Man has now pursued this development track to the extent where the changes are a threat to man himself.

There are many examples in history where the exploitation of resources and the environment have led to environmental disasters. What is new, is the scale at which this is taking place in the 20th century.

Increasing Entropy: Resource Exploitation and Pollution

In scientific, philosophical terms what happens is that elements in a geologically concentrated form are mined and converted into products which are "consumed", then dispersed in a mixed flow of waste, which again may be dispersed in the environment. Energy is consumed during mining, in production and in waste treatment. The whole process is an increase in entropy. It illustrates that energy is the key factor in the materialistic interpretation of the running of modern society. In principle, all dispersion that results in pollution could be minimized with sufficient energy to reconcentrate the waste back to a resource.

Box 1. Consider a chemical that is used in society: What options with respect to ultimate fate are available?

Options for sustainability: No use, Reuse, Convert, Contain and Disperse

No use: Stop use of an unwanted substance, because the advantages of its use are overshadowed by detrimental effects to the environment. *The historical example is the hard detergents in washing powder, which were banned in Germany in the early 1960s. Other examples are DDT, PCB, etc. The key is to find environmentally sound substitutes.*

Reuse: Decrease the amount reaching the environment by internally reusing the substances. This approach is under development and implementation is just beginning. Cleaner production together with mass balances and life cycle analyses will have a much greater effect on daily life than we can imagine today. *This is a very viable solution, but it has to be realized that there will still be a residue to be considered.*

Convert: When a substance has been introduced it is important to control the route of that substance such that the transport can be identified and the flow treated. Treatment mostly means converting the substance from an obnoxious form to a form, acceptable for further transport by air or water or in solid form. *Conversion by incineration converts solid matter into an inert solid form and into a gaseous form. Wastewater treatment has the function to convert into a separable solid form (mostly by sedimentation) or into a gaseous form. The essence in this context is that the solution is rarely a final one. It is just a conversion into something that can be transported acceptably on the next route.*

Contain: One of the ultimate fates is to contain the residues and leave them in place forever. *The deposit of radioactive waste in old salt mines is the best example. The material is deposited in the hope that it will never seep, creep, leach or migrate out to the environment. The difficulty of this notion is illustrated by the landfill, which supposedly was aimed at as an ultimate solution. Leakage from landfills has been identified as an essential problem that calls for treatment. Landfill is a treatment unit for many years until ultimately an inert residue has found its final resting place (until archaeologists become interested in the remains with the aim to learn about our behavior).*

Disperse: The only other ultimate fate is to disperse in the environment. *Gas from incineration has to be treated such that the residues in the flue-gas are acceptable from the environmental point of view. Wastewater treatment must create effluents with residues acceptable to the aquatic environment. The treatment of solid waste shall be such that the product can be either recycled, dispersed or contained.* There are no other options. (Except shooting into space).

It is thought provoking to imagine this in the context of the development over the last 100 years and the lack of reliable predictions of future technical development. If fusion should become an unlimited source of energy it may no longer be a matter of principle. However, as long as the prospects for energy supply based on fusion are technically remote, it would be unsafe to rely on such a potential: it would be the precautionary principle in reverse.

The Options

Society has very few options from a global point of view. The system has to be looked at in its entirety. Water can only be considered within the context of mass balances in societies. The flow of matter in society can no longer be looked at except in the context of resources, society, and environment. Cradle-to-grave, input-output, and mass-balance considerations have become indispensable tools. The instruments are available, though still under development and refinement.

Water Quantity

Sustainable usage of water is technically possible, where the circumstances are suitable. Unsustainable use is due to ignorance of or disregard for well known, fundamental hydrological principles.

The hydrologist enjoys the idea of the challenging issues and the problems associated with water and its use. Many things can be done better—a lot better. In this context misuse is defined as the ignorance of or the disregard for well-known, fundamental hydrological principles, such that the detrimental effects come as a surprise, including the statistics of extremes, whether droughts or floods.

Through history the misuse of water has led to some of the largest ecological and societal disasters. Mesopotamia developed an advanced society, as did Egypt, on the basis of water reclamation and vast irrigation systems. When society fell apart, so did the irrigation system and the semiarid land became desert.

An example is the disappearance of the Aral Sea. Upstream irrigation has diminished the flow to the Aral Sea to the extent that one of the largest lakes in the world is rapidly disappearing. These effects could have been foreseen on the basis of well known fundamental hydrological principles.

The greatest misuse of water resources is related to land use. Humans do not seem to have apprehended that changing upstream conditions may adversely affect the downstream water regime. Deforestation is the most striking example; which has pronounced effects in the temperate zone, but can have devastating effects in subtropical and tropical regions. Soil erosion, desertification and downstream flooding are the most visible detrimental effects. Urban areas generate increased runoff and, potentially, flooding downstream. The means for preventing these effects are available, but alleviation by reversal may be difficult.

However, it is feasible to achieve sustainable use of water because it is a renewable resource and availability is determined by flow, not quantity. It is the role of the hydrologist and the hydraulic engineer to know what constraints are involved and to communicate information about these to the decision makers.

Local constraints can be overcome by, e.g. long-range transport. The major constraints then become economics (and energy use).

Water Quality

While water flow can be made sustainable, it is frequently not sustainable when related to how we use other materials in society. In relation to water usage there are some basic principles that are wise to have in mind. Usage of water is intended to remove excess unwanted matter (or energy) at a given location. The problem is, however, the substances which are added, converted, transported and dispersed in the process.

It is frequently forgotten that water usage is deliberately polluting the water itself, with the intention to use the water as a means of transport away from the location of usage. It is equally forgotten that significant reductions in water usage may involve finding alternative options for the usage/transport of the matter which needs to be removed.

Due to lack of foresight, and looked at in hindsight, the solution was environmentally disastrous. Rivers became black, smelly and dead; the Rhine, the Thames and the Ohio rivers are three well-known examples. Lakes became green due to proliferate algae growth, with devastating effects to the lake ecology. It was not until recently that adequate measures were taken to reduce these effects. The solution is wastewater treatment where organic matter, nitrogen and phosphorus are removed from the water before discharge.

In urban water usage other substances are often added to the water, including some which are not suitable for removal at the treatment plant; e.g. metals and organic substances that are alien to the environment.

It is time to address the solutions to these problems in total and not individually. Some basic principles are worth having in mind: Any substance used in society will, ultimately, be detectable in wastewater. Water can be treated to any degree of purity, but at increasing cost. No matter how well treated, there will always be detectable residues in the water.

With the ever increasing improvements in chemical analysis we are able to, or will be able to, detect even the smallest traces of any substance used in society. That poses a problem, because the "zero standard" or the "nondetectable standard" is simply a desirable, but unattainable goal.

An example from Denmark will illustrate this point: The public demands water supplies containing no pesticides. That is impossible because today's analytical technique will ultimately lead to detection in groundwater, when pesticides are used in agriculture. Agriculture demands that permissible limits should be allowed. By contrast, agriculture has threatened not to receive sludge from wastewater treatment plants, due to the risks associated with residual organic chemicals in the sludge. They demand sludge with no chemical substances, which is equally impossible. Reason and consistency in views will ultimately prevail, but there will be turmoil on the political scene before that goal is reached. In all events, research into the effects of chemicals in the environment and the effects on man must be intensified.

In the 1970s, the US Congress decided to demand a "zero discharge policy", which is not feasible. It is common practice to demand BAT: "Best Available Technology". But, there will always be a technology that does better, probably at higher cost. It is a question of choosing a balanced solution in view of the totality of the system. To achieve such a balance another term has been introduced: "Best Available Technique, Not Entailing Excessive Cost": BATNEEC. I propose: "Best Available Technique, Not Entailing Excessive Cost, Nor Ecological Damage, Nor Anthropophagic Risk": BATNEECNEDNAR. This whole area is in need of renewed philosophical thinking. Hopefully, some fundamental principles, common sense and change of attitude will prevail.

Any substance in use in society will ultimately be detectable in the environment. We have to live with permissible limits. The only alternative is to ban the production of the chemical of concern, not just the usage.

Attitudes Versus Laws and Rules, and the Economic Dimension

People have three means of influencing the issues at stake:

i) influence as a consumer in the market place;
ii) influence on politics by voting and lobbyism;
iii) personal daily life behavior and attitudes.

In western societies it is a dominating belief that all aspects are determined by the market place. Economics rule the world. This is true, but only to an extent. It was in the 1950s that K. Galbraith in the book "The Affluent Society" demonstrated that even in the USA large companies did not strive to maximize profit, but managed to ensure economic survival. Beyond that a lot of other concerns dominated decision-making in the board-room. Prestige is an important human motivation. When you build your own house, do you build the cheapest or do you balance between cost and your personal value judgement?

The "Polluter Pays Principle" is the principle of nations who have the free market as their ideal. This includes nations where agriculture still is as heavily subsidized as during the crises of the 1930s. Even free enterprise nations have subsidized environmental measures as a means to facilitate environmental investments. Frequently, such subsidies have upset the rational approach to design, but it has kept the wheel turning for the rapid implementation of pollution abatement. Now—just to show how nations are consistent in their concept—a new approach is being advocated. The 3P-principle in reverse: *"Pay the Polluter Principle"*. Agriculture is being payed not to pollute with fertilizers and pesticides!

A few years ago, while industrial investments in pollution abatement were being minimized it was suggested that abatement was a waste of money from a production point of view. Changing attitudes within a decade demonstrate that change can be a matter of value judgements—not necessarily associated with economics nor by consumers and the market place—as long as there is a margin of survival. That is why pollution abatement and cleaner production is possible in the affluent world, but hard to apply in poor countries.

Figure 1. The development of the population of Egypt. A stable society for thousands of years. What will be the limiting factor for population growth and what will be the ethical, cultural and religious consequences of hitting a ceiling?

The public in a democratic society can influence politicians. In this regard, lobbyism has been very influential over the last three decades with respect to the environment. However, without controversial pressure groups the pace of development would have been much slower. Short-term, quick solutions have tended to be piecemeal, and the spectrum creates an array of laws and rules alien to the public. It is a good old saying that a law in conflict with the basic attitude of the public will not work.

The long-term solution cannot be reached without changing the attitudes of individuals and companies so that behavior is determined by accepted ethics and morale. The historic example is hygiene. With the development of understanding of contagious diseases the rationale for hygiene was successfully communicated to the public. It is a well established fact that preventive medicine has been more effective than all the hospitals together. When the proliferation of *Salmonella* in modern mass-produced food has little effect, in spite of the potential, it is because of the hygienic behavior, engraved during childhood by inherited rules from parents and the educational system.

Hygiene is a good example of knowledge, an attitude and behavior learned by the individual for the benefit of himself and his neighbors. Similarly,"Eco'ene" would be a good attitude to teach our children, for the benefit of all.

The future lies in learning from the experience of hygiene in the development of environmental behavior in individuals and companies, supplemented by laws, rules and economic incentives.

Traditional economic theory contains one inherent obstacle to environmental thinking: depreciation. The entrepreneurial vision does not reach beyond the depreciation period. There is no concern for the next generation. This is similar to the well-known shortcoming of the economic theory that values that cannot be expressed quantitatively tend to diminish in importance, in spite of the fact that they may be as important as the values that can be quantified. Economic theory needs to broaden its scope and deepen the axioms on which it is based.

Pressures On Water

Water demand is basically a local population issue. Urban water demand is a public health issue and rural demand is a food production issue.

Water is by nature a sustainable resource. It is recycled by nature. It is not a question of exhausting a limited resource, but a question of redirecting an available flow. The problem is that populations tend to grow at locations without adequate locally available water resources. The technical question becomes one of transport of water and/or food, which is relatively simple compared to limiting/relocating people.

Scientists and engineers are called upon to improve utilization of the available water flow. Unfortunately, mismanagement causes inadequate utilization, but improved management will simply extend the limit a bit in comparison with the increasing demand from an ever-increasing population.

Egypt is a case in point. A population in the order of 5 million people inhabited a confined area with a well-defined water resource, the Nile, for thousands of years in a stable society. We may not like the rule of the pharaohs, their treatment of their slaves nor the powers of the religious upper class: but for centuries the society maintained a culture, which we admire. Today, the benefits of industrialization are eaten up faster than they become available, by population growth out of control. No engineering measures nor scientific concepts can other than briefly postpone the disaster. It is time that the water professionals stop pretending that they have the solution. There is no solution in sight. Conflicts may be caused by shortage of water and food: but long before that, the conflict will relate to social frustrations and violence.

The conflict is now between the north and the south, the developed versus the developing world. Does the developed world have a moral obligation to developing nations or will self-interest prevail? Is it in the self-interest of the developed world to interfere or is that an encroachment on the rights of individual nations to determine their own lot—as is the view in China?

So far, the developed world "suffers" from a good conscience, due to billions provided as aid. The fact is that few nations provide more than 1 % of BNR, which is so little that it is of no significance at home, but the numbers look good. At the same time the developed world controls trade such that it is difficult to change the current pattern of developing nations providing raw materials to the industrialized world. The developing country needs to create its own basis for development—and that is happening in Southeast Asia right now.

It is hypocritical to shout "sustainability" when developing countries follow the course taken by the West just decades ago. It will undoubtedly, and understandably, be interpreted as another attempt to keep them at bay: by the nations who fought battles a 100 years ago for the right to export drugs. We can only

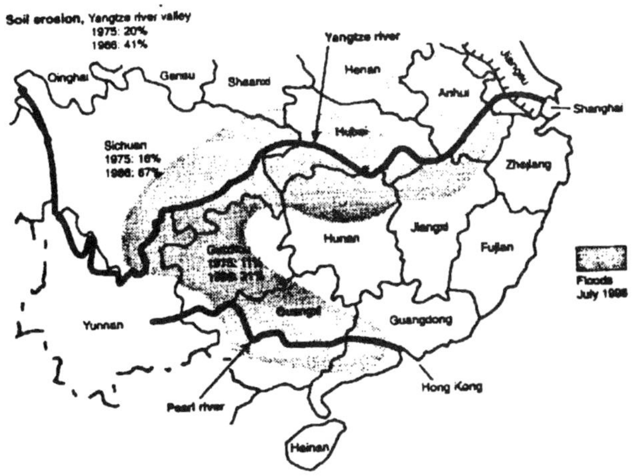

Figure 2. Floods in China, July 1996, and data on the development of soil erosion in the provinces of Sichuan and Guizhou and in the upper Yangtze-river catchment. Floods are not caused by heavy rain only, but also by improper land-use of the catchments upstream. A lesson not learned by man, in spite of numerous examples in the past and present. The consequences to food production and population migration are yet to be seen. (Data from Edwards: Patterns of China's Lost Harmony, 1994).

hope that they will take the time to save themselves and the rest of the world from the mistakes made in the west during industrialization.

Does Objectivity Exist?

It has been fashionable to claim that there is nothing objectively to hold on to in the political debate over environmental issues. It has been an opportune point of view for various pressure groups, but is not substantiated by experience.

In a public debate in the middle of a change of paradigm, it is understandable that one could get the impression that there is nothing factual to hold on to. However, this lack of facts is the result of missing knowledge, because past research did not rate such issues highly. However, the last 20 years have provided many examples of issues that were wide open to debate in the beginning with all kinds of claims being put forward. In a society, where attitudes towards environment have not yet changed there will always be an opposition ready to exploit any conceivable uncertainty. There will always be pressure groups ready to interpret first indications as indications of ultimate disaster. In this chaos of ideas, concepts and concerns, politicians must make decisions, and make mistakes. However, factual information becomes available with research and development, monitoring programs, and cause-effect studies. Compulsory environmental impact assessments play an important role in this regard. One might wish that politicians were more conscious of the need to provide the conditions for better information on which to make decisions as early as possible.

It is cheaper to invest in good information at an early stage than to make hasty, often costly decisions with a potential for waste.

The Role of Experts Versus Politicians and the Public

The environmental debate has been heated and controversial—often due to a lack of knowledge. In such a debate, the expert may play a significant role, as in similar debates involving ethical issues on top of scientific and engineering cause-effect relations. In this role, the expert may exploit the advantage and become the "politicizing expert", and outright dangerous.

By definition, the expert has knowledge in a specific field that others do not have. That position may be abused by the expert to get an influential position. The expert should have holistic understanding of the totality in order to stick to his expert analysis and "leave the politics to the politicians". One-sided, nearly religious conclusions can be drawn by experts to serve their own purposes. In which case, the expert is "politicizing" and dangerous.

The expert should be involved and aware of the totality of environmental issues and should share joint responsibility as a citizen. The expert should make sure he/she understands not only the expert matters, but also the much larger context from which he can ensure that his expert analysis takes all concerns into account. The representatives of the public, the politicians, set the basic paradigms and outline the scope to be encompassed by the expert analysis, and the politicians draw the conclusions facilitated by the expert analysis.

Defining the Precautionary Principle

The precautionary principle is based on the idea that any uncertainty should be interpreted as an urge to do more to safeguard the environment. It is similar to the court practice that any uncertainty of the evidence presented is judged to the benefit of the accused.

The principle is not new to the engineer, in fact it is common practice: Any uncertainty of material strength is accounted for on the "safe side" by safety factors. As long as the uncertainty is known and can be expressed in statistical terms the engineers have well-established tools with which to evaluate risk and advise against failure. These tools are in the process of being transferred to the environmental field. However, there are two problems associated with the precautionary principle:

i) One of the difficulties is the difference between "real risk" and "perceived risk". It is a fact that there is no consistency in public reaction to risk levels. It appears to be a utopian, expert wish that the public should consider risks equally. It seems to be difficult for statistics experts to accept that the public does not rate equally the risk of cancer from smoking, automobile accident, drinking beer, and the toxicity of an unknown chemical. As in most other human choices, there is a value judgement involved: some concept of a balance involving risk, cost, benefits, desires, traditions and a lot more;

ii) How does one handle challenges of which we are still ignorant? The risk cannot be expressed qualitatively nor quantitatively due to ignorance. Even ethical principles are difficult to apply, when we have to act on the basis of ignorance and suspicion. Decision-making becomes a question of intuition and beliefs. Again, the answer is to generate research and development at the earliest stage of the recognition of problems. It is a pity that the horizon of the politicians does not go beyond the next election.

The Commonwealth Versus the Individual

Western civilizations guarantee certain rights, which form the basis for regulations of relations between society and the individual. It is the perpetual conflict between a concern for the many versus concern for the individual. The expansion of the concern to include the environment and the long-range benefits beyond the horizon of the individual will increase the tension in the grey area of potential conflict.

The concern for the conflicts to be generated in confined areas, with population increase out of control, will demand tougher measures to control birth rates. By the basic thinking of most Europeans and Americans human rights issues may conflict with the one-child-policy in China. There is a tradition in China for a wider interpretation of the rights of the family, the community and Chinese society, relative to the individual. That has to be weighed against the rights of future generations not to suffer famine due to overpopulation, or the rights of neighboring nations not to be overrun by a population out of control. Europe is today facing such issues within its borders (Bosnia, Kosovo).

By any human standard, life must be protected. That is fundamental to medical ethics. Preventive medicine and antibiotics have increased survival rates. In the past, early childhood mortality was high. The social need for the parents to have descendants was paramount, especially boys, to care for parents in old age. How does concern for the next generation fit with efforts to rescue for a life in famine. By western standards there are no acceptable solutions, except birth control.

Ethical Issues: Towards A Stable Society

What is a sustainable society? Do we have knowledge about such societies from history, and is it relevant? When we say a sustainable society, we mean in a materialistic sense. The flow of matter and energy should be such that resources are not exhausted and the environment not polluted. There is a perpetual stable circulation of matter. In order to maintain that situation, a stable technology has to be envisaged. Will there be room for innovations? Does the stable society in a materialistic sense demand that centralized powers prevent innovations? Does materialistic stability also mean spiritual stability? Can innovations be concentrated on intellectual pursuits without spilling over into materialistic changes that defy stability?

Will a stable society be of the kind envisaged by Aldous L. Huxley in "Brave New World", just different, or does the intellect of man provide enough vitality and constraint at the same time to justify our present day concepts of an acceptable society? Most likely not, because we do not have the imagination to even get close to what it will be like; as our ancestors could not envisage a world without disease, with plenty of food, with heated houses, transport, radio, television and computers, etc. It will be difficult to secure the needs of a future which we can perceive only as an extrapolation of our own.

Professor Poul Harremoës
Department of Environmental Science and Engineering
Technical University of Denmark
Building 115
2800 Lyngby
Denmark

THE ECOLOGICAL LESSONS OF THE PAST:

An Anthropology of Environmental Decline

TIMOTHY C. WEISKEL

All civilisations depend ultimately on the ecological viability of their agricultural base, as the environmental archaeology of ancient civilisations makes clear. Expansionist Western industrial culture, dependent on resource-depleting petroleum-based agriculture, is only different in terms of its global scale. If the lessons of the past are not heeded its collapse will also be global.

It is often stated that our present ecological crises are totally unprecedented. We are told that 1988 saw "the hottest summer on record" in North America, that Boston harbour has never in its history been so polluted, and that the European seal virus epidemic is on a scale never before witnessed by man. By stressing this 'never before' aspect of events, it is sometimes argued that the experience of the past is largely irrelevant for policy planners. Since circumstances are so new, so the argument goes, past experience leaves us with little or no instruction as to how to formulate a practical public policy for the environment.

This is not altogether true. While particular types of industrial pollution may be new and the scale of ecological devastation may be greater now than previously, the modern world is *not* confronting completely unprecedented circumstances—numerous civilizations before our own have confronted environmental degradation and have paid the price. If we continue to tie our society's infrastructure and agricultural production to a declining resource base—as ancient civilizations did with such depressing regularity—we too will suffer the fate of unavoidable collapse.

The Ecological Decline of Ancient Civilizations

Many ecological catastrophes which have long been understood as 'acts of God' or 'natural disasters' were in fact largely generated or substantially aggravated by collective and cumulative human behaviour.[1] The repeated pattern of the rise and fall of ancient civilizations in the Mediterranean region is especially revealing in this respect. Recent archaeological research indicates that there was a substantial ecological component to the emergence and collapse of agricultural complexes in ancient Mesopotamia, Phoenicia, Palestine, Egypt, Greece and Rome.[2]

These civilizations had to solve the basic problem of producing food surpluses and collecting raw materials from rural areas to sustain large urban populations engaged in commerce, ritual, government and the arts. Over time the strategies that each society pursued to produce food and procure resources left their characteristic mark on the environment. Some of these strategies proved not to be sustainable and overtaxed the regional natural resource base resulting in the depletion of water, soil, or forest reserves. The general pattern was one of gradual emergence, brief flowering, and rapid collapse of civilizations, often taking the form in the final stages of devastating military struggles for the control of arable land or essential resources.

Techniques of agricultural intensification,—terracing, crop selection, animal husbandry, irrigation, and the like—were devised to meet repeated crises of production. Despite short-term improvements in output, however, the long-term consequences of these technologies were not foreseeable by early agricultural innovators. In subsequent decades or centuries, problems of over-grazing, watershed deforestation, soil erosion, siltation, water-logging, soil salinization and crop blight often left whole regions permanently destroyed for agricultural use.

Contemporary documents frequently explain these phenomena in terms of military, political or religious rivalry and conflict, perhaps most obviously because the élites that wrote such literature were part of military, political or religious institutions. These explanations inevitably over-simplify and distort a more fundamental understanding of the dynamic of agricultural civilizations.

Current archaeological research based upon scientific analysis of soil profiles, vegetation, and landscape evolution indicates that in the rise and fall of ancient civilizations there was at the base of nearly all sustained conflict a vital ecological component. Patterns of rivalry in the Mediterranean region could express either a momentary ecological crisis or a long-term depletion of some fundamental element of ecological capital. The ecological dislocations were frequently most visible in the 'peripheral' areas of the great Mediterranean empires, for it was here that the imperial powers established systems of commercial agriculture and proceeded to exact levels of agricultural production that exceeded the ecological capacity of the land.[3]

The Ecology of European Expansion

The European expansionism of the last five hundred years has overshadowed—indeed, nearly totally eclipsed—the lessons that we should have learned from the repeated decline and collapse of ancient agricultural civilizations. This has led to a potentially fatal cultural blind-spot as to the vulnerability of our current industrial system of agriculture. Because of its experiences between roughly 1450 and 1950—a period marked by seemingly unlimited expansion—the Western industrial world now finds itself conceptually ill-equipped to understand, and politically impotent to address, the problems of ecological adjustment that currently face all societies in our finite world.

Colonial discovery and settlement was to transform European agricultural techniques substantially in the New World. In the circumstances of comparative abundance, conservation of resources no longer seemed necessary. Whereas intensive techniques of resource husbandry had come to characterize the confined land-base of medieval European agriculture, the New World afforded rich new possibilities for agricultural expansion. In effect, since the discovery of the New World, predatory expansive agriculture and parasitic resource use has come to characterize European civilization, leading some emergent cultures to believe in a mythology of expanding 'frontiers'.

It is important to realize, however, that increases in agricultural output over most of this period were accounted for not so much by improvements in basic technology, as they were by the overall expansion of the land surface under cultivation. While total production rose dramatically, productivity per acre and productivity per unit of energy input often declined. Nevertheless, profits from total agricultural surpluses helped to finance the emergence of urban-based industrial systems. The dynamic of industrial growth served, in turn, to sustain the mythology of 'unlimited frontiers' and further transformed these formative frontier myths into a belief in perpetual economic growth. Having expanded upon the things of nature, the West came to believe that expansion was in the nature of things. Perpetual growth was considered both natural and good. The European experience of overseas expansion and the ensuing pattern of industrialization has engendered deep-seated habits of thought and images of cultural self-perception.[4] These images and mental metaphors leave the industrial world poorly equipped to construct a sustainable system of production. In effect, we are trying to sustain a 'frontier culture' in a post-frontier world.

Development and Ecological Degradation

In the present, just as in ancient times and in the age of colonial expansion, it is in the 'remote environments', distant from the centres of power, that the first indicators of environmental catastrophe become apparent. These regions are characterized by weak economies and highly vulnerable ecosystems, in our time just as they were in the past. The environmental condition of these regions constitutes an early warning mechanism for the ecological stability of the global ecosystem. If we begin to monitor this early warning system, we will recognize that the signs are not encouraging. Edward Goldsmith has summed up the overall situation:

"the last thirty years have been the most disastrous in the history of most, if not all, Third World countries. There has been massive deforestation, soil erosion and desertification. The incidence of floods and droughts has increased dramatically as has their destructiveness, population growth has surged, as has urbanisation, in particular the development of vast shanty-towns, in which human life has attained a degree of squalor probably unprecedented outside Hitler's concentration camps. With such developments, have come increased malnutrition and hunger; so much so, that today we are witnessing for the first time in human history, famine on a *continental scale*, with two-thirds of African countries to some degree affected."[5]

Several well-documented environmental and economic trends are of particular importance.[6] These include deforestation, the expansion of petro-chemical agriculture, the shift in weather patterns and perhaps climate in the semi-arid areas, continued population growth, and the penetration of local food markets with Western food surpluses through dumping or foreign aid. Deforestation is now becoming measurable from satellite in space. The scope of the transformation is massive (*see The Ecologist*, Vol. 17, No. 4/5). Yet despite the volumes of scientific studies that warn against the dangers, the pattern of deforestation has not been noticeably reversed by acts of policy in recent years. Third World countries involved in the process of forest loss are by now genuinely concerned about its impact, but they are frequently impotent to do anything more than monitor what has occurred and is occurring.

Is Development the Problem?

The issue of tropical deforestation may indicate whether institutions can adapt quickly enough to changing circumstance in the Third World. The World Bank intends to raise its funding for studies to preserve tropical forests from $138 million to $350 million by 1990. Moreover, it recognizes the need to fund projects that promote conservation, but it may well be that true conservation would require flat opposition to the 'development' programmes that have been launched and are likely to continue in the Third World. In any direct confrontation between the World Bank and the entrenched interests advocating conventional forms of trade, aid and development, it is not clear that the newly discovered environmental sensitivities of World Bank officials would prevail. Indeed, there are mounting signs that the policies of development that have caused the most rapid and irreversible forms of environmental degradation are likely to endure. In fact, these policies are likely to be applied in an accelerated manner in the coming years, despite the best intentions of the World Bank's officials. For these reasons environmentalists are now beginning to ask: Is there not a fundamental contradiction between environmental conservation and 'development' as it has traditionally been conceived? Or, as Goldsmith has put it: "Is development the solution or is it the problem?"[7]

The Underdevelopment Spiral

The destructive processes are so hard to stop because they are locked in what might be called: 'the underdevelopment spiral' (*see* Box). This is a syndrome of closely related social, economic and ecological phenomena which combine to cause a self-perpetuating cycle of environmental decline. As these phenomena interact they reinforce one another, accelerating a spiral of decline and making it increasingly difficult for any one party to intervene to arrest the process.

Additional elements may be present, and when they are, they tend to accelerate the pace of the spiralling interaction of the other elements. These added elements come into play when peasantries rightly seek to arrest the underdevelopment spiral, and take matters into their own hands. Typically this involves attempts to:

1) seek to resist direct government exactions;
2) seek better trade terms for items they produce;
3) seek control over arable land in order to pursue autonomous farming.

Elements of the 'Underdevelopment Spiral'

—the expansion of cash-crop agriculture in the Third World

—the secular decline of real prices of Third World commodities in world trade

—the growth of case exchange, collapse of local artisans and expanded merchandising of European manufactured goods.

—displacement of indigenous food crops by exogenous, hybrid, or petro-intensive cultivars

—the decline of small-holding agriculture and increase of 'landless peasants'

—the movement of Third World populations from rural to urban areas

—demographic expansion in Third World populations

—the industrialization of agriculture and overproduction in industrialized nations

—changes in local weather and micro-climate patterns

—local and regional food shortages and the growth of international food trade and aid

—destruction of agricultural or pastoral lands through overgrazing, loss of topsoil, salinization, and flooding, and waterlogging

—the fitful but secular increase in the price of petroleum and petroleum products

—the long-term increase in relative prices of manufactured goods

—the growth of Third World indebtedness

—the expansion of bureaucracies, the collapse of public services and the growth of corruption

Added Elements in Exceptional Cases

—the escalation of conflict in rural areas, diverting efforts from agricultural production

—the growth of an international arms trade, costing further drains on foreign exchange

—the outright destruction of crops, villages, and ground cover through ground combat, bombing, chemical defoliation or purposeful torching of biota to destroy an opponent's means of securing shelter, food or a livelihood.

Post-War Africa:
A Case-Study in Ecological Devastation

Throughout Africa and much of the rest of the Third World during the colonial period, rural regions became accustomed to imported manufactured goods that had become essential for their households or their agricultural pursuits.[8] Axes, machetes, hoes, pots, cotton cloth, and a whole range of petty manufactured goods, from matches to kerosene lanterns, penetrated into rural regions during the early years of colonial rule. At the same time, these regions had become accustomed to producing agricultural commodities like peanuts, coffee, cocoa and cotton to earn the money to purchase these products.

During the Second World War, however, the supply of European manufactured imports was cut off by scarce or non-existent shipping. This raised the price of imports considerably, and for some time many goods simply were not available at all. At the same time agricultural commodities produced for sale exceeded shipping capacity, and thus their price dropped in local markets. These combined phenomena created considerable pressure for economic and political change in the immediate post-war period. With the post-war increase in shipping capacity, there ensued a boom period of economic expansion. The colonial administrations could content themselves with building roads and public works while maintaining public order, and needed to do little positive planning to encourage economic growth. The cash-crop boom, although sustained for several years, remained nevertheless an inherently unsustainable phenomena, ironically because of its very success. So many peasants in the Ivory Coast, Ghana, Nigeria and the Cameroons turned to cocoa and coffee production that the mounting supply exceeded world demand, and the prices for these commodities began to fall in real terms during the 1950s. As Brazil and East African countries like Kenya entered the ranks of coffee and cocoa producers the purchase prices of these commodities declined even further.

Development, War and Ecology

In taking into account the environmental costs of various development strategies, scholars have tended to neglect the ecological impact of open conflict. Not all concerns are similarly myopic. Businesses involved in Third World agriculture projects regularly include considerations of 'security' as part of the operating costs they need to incur to protect their investments. If peasant resistance or sabotage to these projects becomes too expensive, the enterprise nearly always calls upon the state military apparatus to undertake the effort and expense of suppressing peasant opposition. The costs of repressing peasant revolts are readily calculable in terms of munitions and manpower, but the costs to the peasants' environment or to the world's ecosystem are usually overlooked. If we are to develop a reasonable means of assessing the environmental costs of development strategies, we should try to include this 'externality' in our calculations. The policy implications of undertaking these economic calculations would be significant, for whatever the economic virtues of pursuing conventional development strategies in regions like Central America, the ecological destruction involved in crushing peasant rebellions or launching counter insurgencies is massive and needs to be counted as a real cost in any cost-benefit analysis of proposed development strategies.

Urban Drift

Oversupply was at the root of the declining purchase price for these commodities and although the 'rational' economic response would have been for peasants to refrain from producing further cocoa or coffee or the like until the supply declined and the prices came up to a reasonable level, this was never really an option for most peasants as they had already made the infrastructural investment in the cocoa and coffee plantations. In the face of a depressed price, a peasant with fixed or escalating costs or other demands upon his income had basically two options open to him. Either to expand the scope and scale of his production to maintain or enlarge his income to meet his growing needs, or to leave cash-crop farming and go to the city in a hopeful search for non-agricultural work.

The option of engaging once again in food-stuff agriculture was largely precluded by the impact of the agricultural overproduction in the Western countries, particularly the United States. Through both aid and trade channels, the United States sought actively to export its agricultural surpluses.[9] Since most capital cities of Third World countries were constructed as ports or transshipment centres during the colonial period, it became very easy—and even appeared wise—to purchase American grain surpluses to feed growing urban populations in the Third World. When disasters like floods, earthquakes or typhoons damaged the remaining local agricultural systems, U.S. aid agencies provided relief supplies of food at cost.[9]

The self-perpetuating nature of the problem started to become apparent by the early 1960s. Peasants, trying to stay afloat economically, began to devote more and more of their arable land to cash-cropping which, in turn, provided less and less, in relative terms, the more they produced. From the 1960s onward, with relatively less land and labour devoted to foodstuff agriculture in Africa, not only its urban areas, but also major rural regions became dependent on substantial imports of foreign surpluses. According to Jennifer Whitaker, "food imports rose from 4 million to 24 million tons during the 1970s. By 1985, the continent was importing two-fifths of its food supply and about a third of its people depended wholly or partly on imported food"[10]. Meanwhile, those peasants or their children who had left the village began to swell the ranks of the urban unemployed or underemployed.[11]

The Debt Burden

Hungry urban populations prove to be politically volatile, and so urban-based political élites tended to continue to buy political tranquility in the short-term by purchasing food from the cheapest source—the surpluses of Western industrial countries. In order to gain the foreign exchange to purchase this foreign food, the state exhorted its cash-cropping peasants to produce ever greater quantities of export commodities. While exports expanded, commodity prices slumped with oversupply, and foreign exchange became scarce, despite expanded output, the states concerned therefore either had to seek food aid or incur foreign debt to purchase the food upon which they had come to depend. Indebtedness proved to be only a short-term solution, and many African countries soon found themselves devoting much of their foreign exchange earnings to servicing these debts. The overall debt burden for the continent rose from $14 billion in 1973 to an estimated $1255 billion in 1987.[12]

Faced with the departure of the young, able-bodied members of their households to the rapidly growing cities, families often resorted to having several children in the hope that some would remain to undertake the ever more demanding cash-crop work. Collectively this translated itself into a rapid spurt in population growth, particularly as this period also witnessed the arrival of rudimentary medical facilities and the equipment for clean water supplies in rural areas which contributed to the decline in infant mortality.

During the 1950s and 1960s, the area devoted to agricultural activity greatly expanded while the technologies applied to production changed little. In bush-fallow systems, fallow periods were shortened or eliminated altogether, and on the thin and nutrient poor soils this new pattern of usage rapidly exhausted arable lands. In addition, previously undisturbed forests began to be cleared as governments short of foreign exchange, extended rights to timber concessions for cutting and exporting tropical hardwoods and as peasants, often displaced from their own lands began to encroach upon the remaining areas of uncut forest.

Changing Weather Patterns

The removal of large portions of tropical forest cover and conversion of whole regions to cropland and grassland had the effect of changing the nature of the local hydrological regime. Water which previously was held in the canopy or locked in the root systems of vegetation in the forest, ran off at accelerated rates with little or no ground cover left to hold it. Dramatic flooding became a feature of the rainy season in many parts of Africa, and considerable topsoil was lost to sheet and rill erosion. Some regions of African rural areas have been so overgrazed, over-cropped and eroded that it is doubtful that agriculture can continue on these soils for much longer without considerable imported subsidies to rebuild soil structure and fertility. In areas where soil depletion and the change of ground cover has been in long-term decline, local weather patterns may be disrupted.[13]

Populations that have become dependent upon purchasing food in exchange for cash can find themselves short of food for reasons well beyond their control. Fluctuations in the price of oil affect both international shipping and internal transport costs of food as well as the total foreign exchange profile of individual countries. Thus, the fitful increases in oil prices have been translated in local terms in Africa into an increase in the price of imported food and occasionally a pattern of seasonal or chronic shortage. This is compounded as the world market in grains fluctuates with the purchasing habits of major industrial countries. When Soviet, Chinese or Indian harvests are poor and Western grain surpluses are bought up on the international market, prices rise beyond African countries' purchasing power. As we have seen in recent years, food shortages can become acute and famine widespread in these circumstances.

Green Revolution

'Modernized' agriculture based on 'green revolution' technology is offered by Western agricultural experts in the wake of these famines to meet the urgent and evident need for expanded food production. Generally these technological packages are based upon 'high-yield-varieties' (HYVs) of crops that have been selected to respond well to a combination of fertilizers and pesticides designed for their needs. In this regard, the HYVs are more appropriately labelled 'high response varieties' (HRVs). They respond well to the petro-chemical subsidies which they were engineered to use, but on their own their performance is often not even equal that of traditional varieties.[14] When one considers the probable rise in cost of the petro-chemical additives needed to produce a reasonable crop, along with the infrastructural investments required for irrigation systems, storage systems and mechanized equipment associated with the 'modernized' agriculture, it is clear that the choice of these technologies is not a wise one. Similar investments in roads, marketing facilities, or land directed towards traditional crops may well prove over the long run to be better spent than the money devoted to the alluring promise of 'miracle' crops. In the face of declining petroleum supplies, it is questionable whether development schemes based upon increasingly energy-intensive technologies can be sustained much longer.[15]

Problems of Perception

The problem of perception complicates the issue of achieving an adjustment between expanding demand and declining resources. In general, it is recognized that there are upper limits involved in the amount of information that an individual or a society can successfully absorb and act upon. After a certain threshold, as the environment becomes more complex, a society's ability to recognize or use information about its circumstance effectively declines. It is as though things become too complicated to know what is happening. This may well reflect our current position as regards the Third World. Ecological problems are global in scope, and we will need to develop a matching degree of perception. Tropical deforestation affects both local weather and world-wide climate patterns. Food from Iowa feeds both Boston and Burundi. Currently, it is in the Third World that global ecological crises become most pronounced, but it would be a major mistake of perception for Western leaders to assume that the problems in the Third World are merely the Third World's problem.

Cooperation for Stabilization

Broadly speaking, there seem to be two possible patterns that emerge. First, it is possible that a stable adjustment can be achieved. This would be possible if the entire population concerned perceives the problem of adjustment as a real one, and simultaneously cooperates to achieve stabilization. Such a resolution implies that the society can exercise a remarkable degree of self-imposed restraint.

Ecological Overshoot

An alternative resolution of current and future trends involves the phenomena of ecological 'overshoot' and subsequent collapse. This would most likely occur either if the society did not succeed in achieving cooperation in a transition to a stable adjustment. The time lag involved between the overshoot and subsequent collapse is not at all clear. Nor is it clear just where we are located at the current moment in either of these two possible scenarios. Some scientists feel that we may already have exceeded the earth's capacity to sustain present populations on a permanent basis.[16] We are perhaps beginning to witness the phenomena of collapse in indicative catastrophes ranging from repeated Bangladesh floods to recurrent African famine.

Whatever the time-scale, there are very definite signs of difficulty emerging. Briefly put, Western agriculture, and American agriculture in particular, has achieved enormous levels of output through a pattern of substituting human agricultural labour with fossil fuel-driven machines. The result is that while there has been a substantial decrease in the human labour used on American farms during this century, there has been an overall increase in the energy subsidy required to obtain one calorie of consumable food in the American food system.[17]

Moreover, the vast majority of the energy subsidy provided to the food system in the United States is not reflected in the amount of food energy consumed. The long range trend from the 1920s until the present suggests that increases in energy inputs into the system have been approaching a point of diminishing returns in terms of overall food production. If this trend persists, future increases in food output will require even higher rates of energy inputs. This is a disturbing circumstance, particularly in view of the decline in food self-sufficiency of large numbers of Third World countries.

Dead End

We cannot base our agriculture on fossil fuels and expect that agriculture to outlast the supply of this resource. Unless steps are taken in the near future to change to bio-sustainable forms of agricultural technology we can expect wide-scale dislocations including famine, disease and open-armed conflict on an ever-larger scale, as oil supplies decline and competition to control its use intensifies. We live in a highly industrialized, urban culture, but it is important to remember that there is no such thing as a 'post-agricultural' society. Policy decisions concerning agriculture, our environment and the future provision of public works (water projects, transport systems, land-use patterns, etc.) need to reflect this fundamental truth. Cultures that failed to understand this in the past have proved to be short-lived. We will be no exception to this pattern.

With our current patterns of resource-depleting, petroleum-based agriculture, it is as if we are travelling down a dimly lit, one-way street at 90 miles per hour, and we are just beginning to realize that the sign post we passed a while back said "Dead End". Action to avert future catastrophe in this circumstance must begin now on the part of all responsible political leadership. The remaining transition period cannot be very long. This is not a party-political issue. Nor is it simply a problem for the welfare of particular nation-states. It concerns our survival as a species.

This article is an edited version of testimony submitted to a hearing on the Environmental Protection Act of 1988 held by the United States Senate committee on Environment and Public Works.

References

1. A useful summary of ancient ecological problems can be found in J. Donald Hughes, *Ecology in Ancient Civilizations*, U. of New Mexico Press, Albuquerque, 1975. There has been some debate among archaeologists as to the relative importance of human agency or environmental change in explaining accelerated soil erosion in the circum-Mediterranean over historic time. While climate theories were previously favoured, theories stressing anthropogenic activity are gaining ascendency, for example, Kevin Greene, *The Archaeology of the Roman Economy*, U. of California Press, Berkeley, 1986, p. 85; S. Judson, 'Alluviation and

Erosion', in E. Fentess et al., *Excavations at Fosso della Crescenza 1962*, Rome, 1982, (Papers of the British School of Rome, No. 51), pp. 58–101; Claudio Vita-Finzi, *Archaeological Sites in Their Setting*, London, 1978, especially, pp. 8 and 11–13.

2. H. W. Lawton, and P. J. Wilke, 'Ancient Agricultural Systems in Dry Regions', *in* A. E. Hall, G. H. Cannell, and H. W. Lawton, (eds.), *Agriculture in Semi-Arid Environments*, Springer-Verlag, New York, 1979, pp. 1–44. Popular, if somewhat dated accounts include: Edward Hyams, *Soil and Civilization*, Harper and Row, New York, [1952] 1976; Vernon Gill Carter and Tom Dale, *Topsoil and Civilization*, University of Oklahoma Press, Norman, 1976.

3. See for example, W. Groenman van Waateringe, 'The Disastrous Effect of the Roman Occupation', *in* R. Brandt and J. Slofstra (eds), *Roman and Native in the Low Countries: Spheres of Interaction*, British Archaeological Reports s184, Oxford, 1983, pp. 147–157; Claudio Vita-Finzi, 'Roman Dams in Tripolitania', *Antiquity* 35, 1961, pp. 14–20; Z. Naveh and J. Dan, 'The Human Degradation of Mediterranean Landscapes', *in* F. Di Castri and H. A. Mooney (eds.), *Mediterranean Type Ecosystems*, Springer-Verlag, New York, 1973, pp. 373–390.

4. For a discussion of the impact of the frontier experience upon ecological consciousness in America see, Timothy C. Weiskel, 'Rubbish and Racism: Problems of Boundary in an Ecosystem', *The Yale Review*, Winter, 1983, pp. 225–244.

5. Edward Goldsmith, 'Is Development the Solution or the Problem?', *The Ecologist*, Vol. 15, Nos. 5/6, 1985, p. 210.

6. A number of global 'commissions' have been constituted to investigate the widening gap between the 'developed' and the 'underdeveloped' world, most notably the two chaired by Willie Brandt: Brandt Commission—Independent Commission on International Development Issues, *North-South: A Program for Survival*, M.I.T. Press, Cambridge, 1980; Brandt Commission, *Common Crisis: North-South Cooperation for World Recovery*, M.I.T. Press, Cambridge, 1983. For a critique of the Brandt Commission approach see, Teresa Hayter, *The Creation of World Poverty: An Alternative View to the Brandt Report*, Pluto, London, 1981. See also, United States Department of State and the Council on Environmental Quality, *Global 2000: A Report to the President*, Washington, 1980; World Commission on Environment and Development (Brundtland Commission), *Our Common Future*, Oxford University Press, 1987.

7. op. cit., supra 5.

8. For a more extended discussion of a case-study see, Timothy C. Weiskel, 'Toward an Archaeology of Colonialism: Elements in the Ecological Transformation of the Ivory Coast', *in* D. Worster (ed.), *The Ends of the Earth*, Cambridge, 1988, pp. 141–171.

9. For a summary of U.S. export food policy see, Alain Revel and Christophe Riboud, *American Green Power*, Johns Hopkins U. Press, Baltimore, 1986. For a critique of the impact of U.S. surplus exports see, Frances Moore Lappé, Rachel Schurman and Kevin Danaher, *Betraying the National Interest*, Institute for Food and Development Policy, San Francisco, 1987: For a discussion of the various impacts of foreign food dependence see, Timothy C. Weiskel, *Worldview* 23, 10, October, 1980: 'Food, Famine and the Frontier Mentality,' *Worldview*, 12, 12, December, 1981.

10. Jennifer Seymour Whitaker, *How Can Africa Survive?*, Harper & Row, New York, 1988, p. 18.

11. Urbanization rates are exceeding population growth in most of Africa. Mexico City is already the largest city in the world and growing at a rate of 1 million per year. Asia will have 23 cities with populations of 5 million or more by the year 2000. See, "Population Boom for Asian Pacific", *UPI News*, 7 June 1988, 17:47 pm.

12. op. cit., supra 11, p. 19.

13. Richard W. Franke, and Barbara H. Chasin, *Seeds of Famine: Ecological Destruction and the Development Dilemma in the West African Sahel*, Allanheld, Osmun, Montclair, 1980; Alan Grainger, *Desertification: How People Make Deserts, How People Can Stop and Why they Don't*,

International Institute for Environment and Development, Washington, 1985; Independent Commission on International Humanitarian Issues, *Famine: A Man-Made Disaster?*, Pan Books, London, 1985.

14. Kenneth A. Dahlberg, *Beyond the Green Revolution: The Ecology & Politics of Global Agricultural Development*, Plenum Publishing Corporation, New York, 1979; Tim Bayliss-Smith, and Sudhir Wanmali (eds.), *Understanding Green Revolutions: Agrarian Change & Development Planning in South Asia*, Cambridge University Press, 1984; Edward C. Wolf, 'Beyond the Green Revolution: New Approaches for Third World Agriculture', *Worldwatch Paper No. 73*, Worldwatch Institute, Washington, D.C., 1986.; Bernhard Glaeser, *The Green Revolution Revisited*, Allen & Unwin Incorporated, Winchester, MA, 1987.
15. Maurice B. Green, *Eating Oil: Energy Use in Food Production*, Westview Press, Boulder, CO., 1978; John Gever, et. al, *Beyond Oil: The Threat to Food and Fuel in the Coming Decades*, Ballinger, Cambridge, 1986; David Weir and Mark Schapiro, *Circle of Poison: Pesticides & People in a Hungry World*, Institute for Food & Development Policy, San Francisco, 1981; David Bull, *A Growing Problem: Pesticides and the Third World Poor*, OXFAM, Oxford, 1982; David Weir, *The Bhopal Syndrome: Pesticide Manufacturing and the Third World*, International Organization of Consumers Unions, Penang, Malaysia, 1986.
16. "Population bomb' still ticking,' *UPI News*, 3 Sept. 1988, 11:57 am.
17. John S. Steinhart and Carol E. Steinhart, 'Energy Use in the U.S. Food System', *Science*, 184, 19 April 1974, pp. 307–316.

ECONOMIC INTERDEPENDENCE AND THE WORLD FOOD SUMMIT*

AMARTYA SEN

The World Food Summit in November 1996 will be an important occasion for reexamining the task of eradicating chronic hunger and intermittent famines in the contemporary world. It is, of course, extremely difficult to calculate, even approximately, the number of people who are chronically undernourished. FAO's official announcement of a particular number (viz. that 'an estimated 800 million people are still chronically undernourished') is, inevitably, based on somewhat arbitrary assumptions. (FAO, 1996)[1] But the undeniable presence of large-scale hunger—affecting hundreds of millions of people—in a world of unprecedented overall prosperity is rightly regarded as an intolerable state of affairs. We have to change this situation radically. Few other tasks can compare in importance and urgency with this need.

The United Nations' Food and Agricultural Organization (FAO) deserves much praise for taking the initiative in arranging this summit. The issue now is how to proceed to make the deliberations of this summit match the nature of the task. The most important prerequisite of an appropriate response to hunger in the modern world is an adequately broad understanding of the economics of the food problem. Hunger is not just a matter of 'food and agriculture' and the focus of attention has to be rather wider than what has been traditionally interpreted as the FAO's own domain.

Hunger and economic interdependence

Undernourishment, starvation and famine are influenced by the working of the entire economy—not just food production and agricultural activities, and it is crucial to take adequate note of the interdependences that govern the incidence of hunger in the contemporary world. For example, a person may be reduced to starvation through unemployment, combined with the absence of a social security system of 'safety nets' (such as unemployment insurance). This can easily happen, and indeed even a serious famine can actually occur, despite a high and undiminished general level of food availability perhaps even a 'peak' level of food availability—in the economy as a whole.[2]

Just as one illustration (among many that can be given), we may consider the Bangladesh famine of 1974. (Sen, 1981, Chap. 9; Alamgir, 1980; Ravallion, 1987)[3] This occurred in a year of 'greater' food availability per head than in any other year between 1971 and 1976. The famine was initiated by regional unemployment caused by floods, which affected food output many months later, but which 'immediately' led to income-deprivation of rural labourers who would have been otherwise (that is, but for the floods) employed for wages in such activities as the transplanting of rice. The local starvation that followed from this was severely magnified by a steep rise in food prices as a result of exaggerated expectation of food shortage, leading to widespread hunger across the county.[4] Since the food shortage was overestimated and to some extent manipulated, the price rise was followed later on by a downward price correction, but by that time the famine had already taken its heavy toll.

Many other examples of such famines without a significant—or indeed any—decline in food availability have been studied. There are also numerous cases of chronic hunger without any corresponding chronic shortage of food supply.[5]

First step: food security and entitlement

If we see the commitment of eradicating hunger as one of providing food security (that is, of ensuring that everyone has the means of acquiring a standard amount of food), we must see the task as involving a great deal more than guaranteeing adequate food production and availability. The entitlements of those who produce food themselves does, of course, depend on what they 'themselves' manage to produce (though not directly on the 'national' output of food), but also on what they have to sell (for meeting other commitments such as clothing or medicine or creditors' demands) and what they can buy (sometimes people sell expensive foods such as animal products produced by them to buy cheaper calories from food grains, as pastoral people often do).

For those who do not themselves produce food (e.g., industrial workers or service providers), or do not own the food they produce (e.g., agricultural wage labourers), their ability to acquire food in the market depends on their earnings, food prices, and other necessary expenditures. The earnings depend on economic circumstance: employment and wage rates for wage labourers; production of other commodities and their prices and sales for craftsmen and service-providers; and so on.

What determines whether a person or a family goes hungry is their 'food entitlement' (the amount of food they can obtain, own and use), not just the total availability of food in the country or the region. Of course, the latter can influence the food entitlements of people (particularly through the effects of food supply on food prices), but it is one influence among many—there are other significant influences. Food security cannot be guaranteed simply by making sure that enough food is produced and supplied in the economy.

In fact, not only is the production of enough food 'not sufficient' to guarantee food security, it is 'not necessary' either. A country can purchase food from abroad if it has the means to do this (based, for example, on industrial production). If we compare food production per head in 1993–95 with that in 1979–81 in different countries in Asia and Africa, we find a 'decline' of 1.7 percent in South Korea, 12.4 percent in Japan, 33.5 percent in Botswana, and 58.0 percent in Singapore. We do not, however, observe any growing hunger in these economies, since they were, in addition to being richer anyway, experiencing fast expansion of real income per head through other means (such as industries or mining). The sharing of the increased income made the citizens of these countries more able to secure food than before, despite the falling food output. In contrast, even though there was little or no decline in food production per head in economies such as Angola (3.2 percent decline), Sudan (7.7 percent 'increase'), or Burkina Faso (29.4 percent 'increase'), they experienced considerable unfolding of hunger because of their general poverty and the vulnerable economic entitlements of many substantial groups. It is essential to focus on the actual processes through which a person or a family establishes command over food.

Shares and divisions

Even when the opportunities of international trade are absent or stifled, how the total food supply is shared between different groups within the country can also be crucially important. Families can be prevented by recreating lost incomes of the potential victims (for example, through the temporary creation of wage employment in specially devised public projects), giving them the ability to buy food in the market, making the available supply more equally shared. This makes the food supply serve the people more evenly than would have been the case in the absence of such public support and employment-creation; the support provided prevents leaving some people severely handicapped as purchasers of food, while others are quite unaffected by the economic hardship of the victims.[6]

To be effective, public support need not take the traditional form of herding people into 'relief camps' (with the accompanying problem of disrupting their economic activities and family life, and also, often, of unsettling the normal commerce and trade in the economy). Public policy can involve some division of labour with state support in creating in income and employment, combined with the operation of private markets for food and labour (with the government acting as one of the operators, but not the only one, nor necessary as the principal agent).

There is also the important issue of division of food and provisions within the family. The acquiring of food for the family as a whole still leaves unaddressed the question of sharing by different members of the family—women and men, girls and boys, children and adults, old and young. The arrangements for sharing within the family are, to a great extent, given by established conventions, but they are also influenced by such factors as the economic role and empowerment of women, and the changing value systems of the community at large.

Second step: beyond food security to nutritional adequacy

It is, thus, important to shift our focus from food production to the more crucial issue of food security. This is a very important move. But we have to move further. In fact, the task of eradicating hunger and undernourishment differs even from the broader issue of food security, since nutritional adequacy of a person depends also on other conditions, including epidemiology and health status.

For example, a person with a parasitic disease (common in the intestine of many people in tropical areas) may fail to be adequately nourished even if she gets to eat the amount of food that would have been more than sufficient for her had she been fully healthy. So epidemiological conditions and public health facilities may also be important for eliminating hunger in the modern world. (Osmani, 1993) Even in famines, most deaths actually occur from standard illnesses, and the extent of famine mortality can be radically reduced by preventing the spread of infectious diseases. On this see my *Poverty and Famines* (1981), Appendix D; Alex & de Waal, *Famines that Kill*, Oxford, Clarendon Press, (1989).

There is, thus, a strong case for the World Food Summit to extend its attention well beyond food and agriculture and to go outside the normal area of specialization of the FAO. The actions and policies that will be necessary to eradicate hunger in the modern world need integration with work in related areas, including those which fall within the domain of the ILO, the WHO, the UNCTAD, the WTO, and other international institutions. I recognize that there is an issue here of the division of the 'turf' between different international organizations. But there is a good case for the FAO to see itself as the 'prime mover' of this summit, rather than as a 'lone ranger'.

Food production and regional divisions

The FAO as an organization has often played an important part in emphasizing the role of food production in solving the problem of hunger. Even in the context of this World Food Summit, the FAO has forcefully presented the case for 'a new Green Revolution'. (FAO, 1995) In so far as food production is one of the determining influences on the entitlement and food security that people enjoy, this can be a positive contribution. But it has to be combined with activities on a much wider front, and this is exactly where the economic interdependences assume critical importance.

Before proceeding to identify some of the crucial economic interdependences, it is useful to comment briefly on a line of reasoning that has often been aired recently to argue that the world food production is presently under great danger of falling behind population growth, so that giving priority to the production-centered view may be (it is argued) the right focus at this time.

Table 1: Indices of food production per head by regions

Regions	1974–6	1979–81	1984–6	1989–91	1993–5
World	97.4	100.0	104.4	105.4	105.3
Africa	104.9	100.0	95.4	98.6	95.9
Asia	94.7	100.0	111.6	119.9	129.2
India	96.5	100.0	110.7	121.8	126.6
China	90.1	100.0	120.7	134.1	153.7
Europe	94.7	100.0	107.2	106.5	99.4
North and Central America	90.1	100.1	99.1	94.2	95.4
USA	89.8	100.0	99.3	95.0	97.6
South America	94.0	100.0	102.8	105.6	100.8

Is food production really losing the race with population growth? Table 1 presents the indices of food production per head (based on FAO statistics) for the world as a whole as well as for some of the major regions in terms of three-year averages (to avoid being misled by year-to-year fluctuations), with the average for 1979–81 serving as the base of the index (100); index values are given up to 1993–95. There is not only no real decline in world food production per head (quite the contrary), but also the largest increases have come in the more densely populated areas of the third world (in particular China, India and the rest of Asia). The African food output has, however, declined (on which more presently), and while that fall is no more than in North and Central America, the prevalence of poverty in Africa puts it in a very vulnerable situation.

Decline in food prices and economic incentives

It is also important to note that this rise in world food production has taken place despite a sharply declining trend in world food prices in real terms, as Table 2 indicates. This reflects the decline of economic incentives to produce more food in many areas of commercial food production in the world, including North America. Not surprisingly the biggest increases have come from regions (such as China and India) where the domestic food markets are relatively insulated from world markets. While it is true that over the last few years the world food prices have risen again, this rise has been far less than the precipitate declines that have been occurring on a long-term basis. Figure 1 shows that the rise in food prices between 1993 and 1995, which is often cited, is relatively small compared with the big fall since 1970. Indeed, food prices have gone up and down over the short run, but have had a strongly declining long-term trend, and there is nothing yet to indicate that the recent rise over a short period reflects that this long-run trend has been reversed.

In fact, in an impressive technical study of the International Food Policy Research Institute (IFPRI), it emerges that there might be very significant further decline in real world prices of food between 1990 and 2020. The study anticipates further 'declines' in food prices of about 15 percent for wheat, 22 percent for rice,

Table 2: Food prices in constant 1990 US$: 1970 to 1995

Food	1970	1995	%-age change
Wheat	219	154	–29.7
Rice	504	279	–44.6
Sorghum	207	103	–50.2
Maize	233	107	–54.1

23 percent for maize, and 25 percent for other coarse grains. (Rosengrant et al., 1995) Whether or not such further declines in real food prices take place (as anticipated by IFPRl), we can hardly overlook the disincentive effect that the lowering of world food prices have already had on food production. It is in this context particularly impressive to note that the world food output has continued to grow, nevertheless, well ahead of population growth. If more food had been produced (without curing the income shortage from which most of the hungry people in the world suffer), the selling of food would have been even more of a problem.

Like other economic activities, commercial production of food is also influenced by markets and prices. At this time, the world food production is being kept in check by the lack of demand and falling food prices, which in its turn reflects the poverty of some of the neediest people. Technical studies on the opportunity to produce more food (if and when the demand increases) outline very substantial opportunities of making the food production per head grow much faster in *per capita* terms. Indeed, yields per hectare have continued to rise in every region of the world, and for the world as a whole, it went up on average by about 42.6 kilograms per hectare per year during 1981–93. (Dyson, 1996, Table 4.6.) In terms of world food production, 94 percent of the rise of cereal production between 1970 and 1990 reflect increase in yield per unit of land, and only 6 percent that is due to area increase. (Dyson, 1996, Table 4.5) With greater demand for food, the intensification of cultivation can be expected to continue, especially since the differences in yield per hectare are still enormously large between the different regions in the world.

All this does not wipe out the need for slowing down the population growth (on which more presently). Indeed, the environmental challenge is not just that of food production—there are many other issues related to population growth. But it does indicate that there is little reason for any great pessimism that food output will soon start falling behind population growth.

Africa's political economy and systemic disruptions

The economic crisis in Africa is one of production and economic growth in general (not just of food production), and it looks like a food output problem only because Africa still concentrates so heavily on producing food. The roots of the African crisis relate to wars, political instability and economic uncertainties. Changes in these features are required on a really urgent basis.

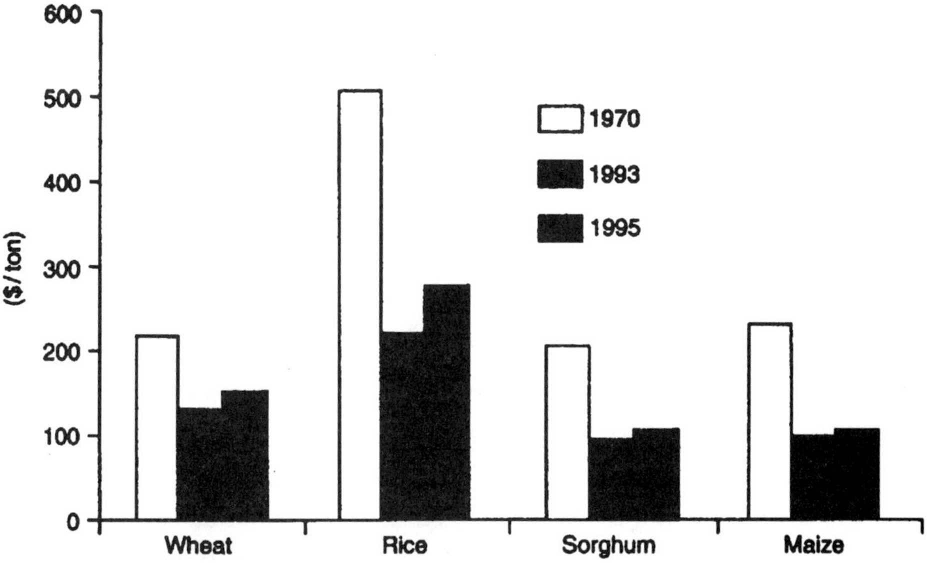

Figure 1. Food Prices in constant 1990 US $.

There is much opportunity for expanding food production in Africa, and they call for appropriate economic policies (including agricultural research, institutional reforms, and changes in relative prices). (Eicher, 1986; Mellor et al., 1987)[7] Africa must also consider the possibility of a more diversified pattern of output. Most countries in the world have achieved economic prosperity and eliminated economic insecurity through industrialization (and along with that, by putting agriculture on a higher technological basis), and this will no doubt have to happen in Africa as well. Given the climatic uncertainties, which is, if anything, more acute in Africa than elsewhere, there is also a classic issue here of not putting all your eggs in one vulnerable basket. Africa has already had many sad experiences of the effects of over-concentration on agriculture and of being periodically smacked by climatic disasters. Diversification is quite crucial for long-run economic stability and security of Africa.

The need for such economic and social change also includes the importance of overcoming military and civil strifes, expanding democratic governance, and also developing market institutions. The role of public policy must also cover the expansion of health care, family planning facilities, basic education (especially of women), and social security provisions. All these contribute—directly and indirectly—to nutritional security and economic strength and safety. What Africa needs is not a narrowly food-centered view of the economic task to be undertaken, but a broader appreciation of its problems and opportunities.

Varieties of economic interdependences

To solve the problem of hunger in the modern world, it is, in general, important to take an adequately broad view of the process of acquiring food and achieving nourishment. The processes involve the market mechanism as well as public services. They are influenced by such 'local' considerations as inequalities in the sharing of food and medical attention within the family as well as such 'global' matters as national and international trade and exchange. They involve the economics of buying and selling as well as the politics of inducing a government to take supportive action, rather than counterproductive restrictions.

I note some examples of important interdependences that we have to consider.

- *Interdependence between income and expenditure:* this is the most basic interdependence in understanding food deprivation. As already discussed, hunger is primarily a problem of general poverty and of deprivation of food entitlement and adequate health and social care—not just of the size of food production.

- *Interdependence between different sectors:* people can acquire food through trade and exchange. It is particularly important to pay attention to the exchange between labour power and commodities (given by employment opportunities and real wages) and between nonfood and food commodities (given by general economic opportunities of production and trade).

- *Interdependence between different countries:* food is bought and sold in world markets, and food self-sufficiency is not the only route—nor always the most economic route—to assure food security to all. While it is certainly the case that an economy under siege cannot rely on food imports, and it is also important not to ignore the imperfections of the world food markets, nevertheless the possibility of importing food has to be rationally considered along with other options. A 'siege mentality' can be psychologically distracting and economically foolish.

- *Interdependence between macro-stability and food security:* many past famines have been associated with high inflation, making the groups that fall behind (in the inflationary race) selected victims of starvation. (Sen, 1981, Chap. 6)[8] Prevention of severe and uneven inflation is, thus, an important part of achieving food security. This concern has to be supplemented by the need for the avoidance of substantial unemployment.

- *Interdependence between shared economic opportunities and distributional equity:* there is much evidence that inequalities in the distribution of food within the family, prevalent in many parts of the world, can be reduced by the economic empowerment of women and by reduction of gender inequality. In achieving this, an important role can be played by the possibility of female employment and by the sharing of property rights, and also by the spread of female education.[9] The relation between economic opportunities and the sharing of food also applies, obviously, between classes and regions.

- *Interdependence between women's empowerment and fertility decline:* the rate of growth of population is one of the factors that contribute to the intensification of the problem of food insecurity and hunger (even though the rate of growth of world population has definitely been slowing down). While some pressure groups concentrate too exclusively on the population issue ignoring other problems, and that bias must be avoided, there are good reasons to take note of the growth of population as one factor among many that deserve serious attention. There is considerable empirical evidence that the fertility rates can come down sharply as a result of greater power of young women within the family (often the most suppressed members of the household), since the lives that are most restrained and compromised by frequent child bearing and rearing are those of the younger women themselves. The crucially necessary empowerment of younger women includes the availability of family planning opportunities, but also the economic and social standing of women in general, governed by such factors as female employment and economic power and by women's education and social and political involvement.
- *Interdependence between political incentives and public policies:* it has been observed that famines do not occur in democratic countries with a relatively free press and active opposition parties (even when these countries are very poor). (Sen, 1984, Chap. 19; Drew and Sen, 1980; Rom and Sobhan, 1990)[10] Elections and the possibility of public criticism make the penalty of famines affect the rulers as well—not just the starving victims. The continuation of famines in sub-Saharan Africa has much to do with the disruption of democratic governance in that continent from the 1960s onwards, often connected with the mechanics of the cold war which produced client states of one super power or another in that battle-riddled continent. Even the provision of health care can be substantially influenced by the nature of the political process and deeply disrupted by military and sectarian governance.
- *Interdependence between wars and economic deprivations:* wars not only lead to the destruction of productive assets, but also to severe reduction of incentives to invest and undertake economic expansion. Sometimes war-related inflations also disrupt the entitlements of those who fail to keep up with the others. Furthermore, by increasing the power of military rulers and by encouraging their tendency to eliminate opposition and democracy, wars can also be indirectly disruptive of the efficient and smooth working of political incentives in the direction of protecting the vulnerable. Many famines in the world have been associated, in one way or another, with war and military activities.

A concluding remark

It is important to take an adequately wide view of the economic processes that lead to hunger and famines, which continue to kill millions of people and blight the lives of hundreds of millions. Given the deep-seated interdependences that influence economic and social deprivation and physical undernourishment, a narrowly food-centered view would defeat the purpose of finding practical ways of removing the persistence of hunger in the modern world. It is important not to lose sight of the diverse causal connections that powerfully sway the actual incidence of hunger and undernourishment.

The distinct influences identified here have been studied extensively over the last couple of decades. The findings that have emerged from recent research are extremely important for the deliberations in the World Food Summit. These economic interdependences have to be adequately seized for the ultimate elimination of the barbarity of continued hunger and starvation in the contemporary world.

Notes

* I am grateful for computational assistance to Arun Abraham, and for research support to the John D. and Catherine E. MacArthur Foundation.

1. On the difficult conceptual and empirical issues underlying such diagnoses, see the papers in S. R. Osmani (ed.) *Nutrition and Poverty*, Oxford, Clarendon Press, (1993), and Peter Svedberg, *Poverty and Undernutrition in Sub-Saharan Africa*, mimeographed, institute for International Economic Studies, Stockholm, 1996, to be published. See also Michael Lipton, *Poverty, Undernutrition and Hunger*, World Bank Staff Paper 597, (1983), and the papers of Sudhir Anand and Christopher Harris, of Partha Dasgupta and Debraj Ray, of S. R. Osmani, and of Peter Svedberg, in Jean Drèze and Amartya Sen (eds.)., *The Political Economy of Hunger*, Oxford, Clarendon Press, (1990).
2. Examples of such famines, and economic explanation of why they respectively occurred, are presented in my *Poverty and Famines*, Oxford, Clarendon Press, (1981), Chapters 6–10. See also Jean Drèze and Amartya Sen, *Hunger and Public Action*, Oxford, Clarendon Press, (1989), Drèze and Sen (eds.), *The Political Economy of Hunger*, Oxford, Clarendon Press, (1990), and its shortened version, Drèze, Sen and Athar Hussain, *The Political Economy of Hunger: Selected Essays*, Oxford, Clarendon Press, (1995).
3. See also the background material in Nurul Islam, *Development Planning in Bangladesh*, London, Hurst, (1977); A. R. Khan,"Poverty and Landlessness in Rural Bangladesh,"in Keith Griffin and A. R. Khan (eds.), *Poverty and Landlessness in Rural Asia*, Geneva, ILO, (1977); Michael Lipton, "Land Assets and Rural Poverty", World Bank Staff Working Paper 744, (1985).
4. The price movements and their consequences have been particularly well analyzed by Ravallion, *Markets and Famines* (1987); see also Alamgir, *Famine in South Asia* (1980).
5. On these general issues, see Drèze and Sen, *Hunger and Public Action* (1989) and *The Political Economy of Hunger* (1990), and S. R. Osmani (ed.), *Nutrition and Poverty*, Oxford, Clarendon Press, (1993), among other studies in this rapidly growing field of research.
6. On the mechanics of such procedures see Drèze and Sen, *Hunger and Public Action* (1989), Chapter 8, and the papers of Jean Drèze in Drèze and Sen, *The Political Economy of Hunger* (1990).

7. See also the papers of Jean-Philippe Platteau, Francis Idachaba, and Judith Heyer in Drèze and Sen, *The Political Economy of Hunger*, (1990).
8. The Bengal famine of 1943, in which between 2 and 3 million people died, was a "boom famine"—an example in fact of an inflation-driven decline of food entitlement (of those least able to participate in the economic boom). On this, see my *Poverty and Famines*, 1981, Chapter 6.
9. Some of the underlying issues are discussed in my "Gender and Cooperative Conflict" in Irene Tinker (ed.), *Persistent Inequalities*, New York, Oxford University Press, (1990). See also Megan Vaughan, *The Story of an African Famine: Hunger, Gender and Politics in Malawi*, Cambridge, Cambridge University Press, (1987); the papers of Barbara Harriss and Anne Whitehead in Drèze and Sen (eds.), *The Political Economy of Hunger* (1990); and also Bina Agarwal, *A Field of One's Own*, Cambridge, Cambridge University Press, (1995), among many other contributions in this rapidly expanding area of research.
10. See also Drèze and Sen (eds.), *The Political Economy of Hunger* (1990), especially the papers by N. Ram and Rehman Sobhan.

References

Alamgir, M., *Famine in South Asia*, Boston, Oelgeschlager, Gunn and Hain, (1980).

de Waal, A., *Famines that Kill*, Oxford, Clarendon Press, (1989).

Dyson, T. *Population and Food: Global Trends and Future Prospects*, London and New York, Routledge, (1996).

Eicher, C., *Transforming African Agriculture*, Hunger Project Paper 4, (1986).

FAO, *World Food Summit: Renewing Global Commitment to Fight Hunger*, Rome, (1996).

"Lessons from the Green Revolution—Towards a New Green Revolution", Food and Agriculture Organization of the United Nations (FAO), WFS 96/TECH/6, December 1995.

Mellor, J. W., Delgado, C. L. and Blackie, C. L. (eds.), *Accelerating Food Production in Sub-Saharan Africa*, Baltimore, Johns Hopkins University Press, (1987).

Osmani, S. R. (ed.), *Nutrition and Poverty*, Oxford, Clarendon Press, (1993).

Ravallion, M., *Markets and Famines*, Oxford, Clarendon Press, (1987).

Rosengrant, M.V., Agcaoili-Sombilla, M. and Pérez, N. D.,"Global Food Projections to 2020: Implications for Investment", Washington, DC, International Food Policy Research Institute, (1995).

Sen, A., *Poverty and Famines*, Oxford, Clarendon Press, (1981).

Sen, A., *Resources Values and Development*, Oxford, Blackwell, (1984).

CAN THE GROWING HUMAN POPULATION FEED ITSELF?

JOHN BONGAARTS

As human numbers surge toward 10 billion, some experts are alarmed, others optimistic. Who is right?

Demographers now project that the world's population will double during the next half century, from 5.3 billion people in 1990 to more than 10 billion by 2050. How will the environment and humanity respond to this unprecedented growth? Expert opinion divides into two camps. Environmentalists and ecologists, whose views have widely been disseminated by the electronic and print media, regard the situation as a catastrophe in the making. They argue that in order to feed the growing population farmers must intensify agricultural practices that already cause grave ecological damage. Our natural resources and the environment, now burdened by past population growth, will simply collapse under the weight of this future demand.

The optimists, on the other hand, comprising many economists as well as some agricultural scientists, assert that the earth can readily produce more than enough food for the expected population in 2050. They contend that technological innovation and the continued investment of human capital will deliver high standards of living to much of the globe, even if the population grows much larger than the projected 10 billion. Which point of view will hold sway? What shape might the future of our species and the environment actually take?

Many environmentalists fear that world food supply has reached a precarious state: "Human numbers are on a collision course with massive famines. . . . If humanity fails to act, nature will end the population explosion for us—in very unpleasant ways—well before 10 billion is reached," write Paul R. Ehrlich and Anne H. Ehrlich of Stanford University in their 1990 book *The Population Explosion*. In the long run, the Ehrlichs and like-minded experts consider substantial growth in food production to be absolutely impossible. "We are feeding ourselves at the expense of our children. By definition farmers can overplow and overpump only in the short run. For many farmers the short run is drawing to a close," states Lester R. Brown, president of the Worldwatch Institute, in a 1988 paper.

Over the past three decades, these authors point out, enormous efforts and resources have been pooled to amplify agricultural output. Indeed, the total quantity of harvested crops increased dramatically during this time. In the developing world, food production rose by an average of 117 percent in the quarter of a century between 1965 and 1990. Asia performed far better than other regions, which saw increases below average.

Because population has expanded rapidly as well, per capita food production has generally shown only modest change; in Africa it actually declined. As a consequence, the number of under-nourished people is still rising in most parts of the developing world, although that number did fall from 844 million to 786 million during the 1980s. But this decline reflects improved nutritional conditions in Asia alone. During the same period, the number of people having energy-deficient diets in Latin America, the Near East and Africa climbed.

Many social factors can bring about conditions of hunger, but the pessimists emphasize that population pressure on fragile ecosystems plays a significant role. One specific concern is that we seem to be running short on land suitable for cultivation. If so, current efforts to bolster per capita food production by clearing more fertile land will find fewer options. Between 1850 and 1950 the amount of arable land grew quickly to accommodate both larger populations and greater demand for better diets. This expansion then slowed and by the late 1980s ceased altogether. In the developed world, as well as in some developing countries (especially China), the amount of land under cultivation started to decline during the 1980s. This drop is largely because spreading urban centers have engulfed fertile land or, once the land is depleted, farmers have abandoned it. Farmers have also fled from irrigated land that has become unproductive because of salt accumulation.

Moreover, environmentalists insist that soil erosion is destroying much of the land that is left. The extent of the damage is the subject of controversy. A recent global assessment, sponsored by the United Nations Environment Program and reported by the World Resources Institute and others, offers some perspective. The study concludes that 17 percent of the land supporting plant life worldwide has lost value over the past 45 years. The estimate includes erosion caused by water and wind, as well as chemical and physical deterioration, and ranks the degree of soil degradation from light to severe. This degradation is least prevalent in North America (5.3 percent) and most widespread in Central America (25 percent), Europe (23 percent), Africa (22 percent) and Asia (20 percent). In most of these regions, the average farmer could not gather the resources necessary to restore moderate and severely affected soil regions to full productivity. Therefore, prospects for reversing the effects of soil erosion are not good, and it is likely that this problem will worsen.

Despite the loss and degradation of fertile land, the "green revolution" has promoted per capita food production by increasing the yield per hectare. The

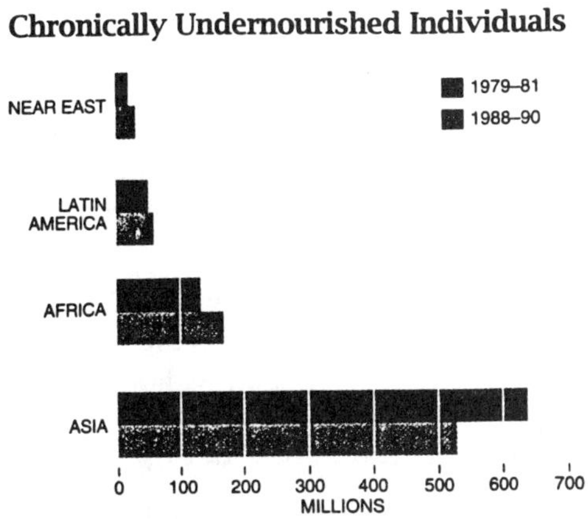

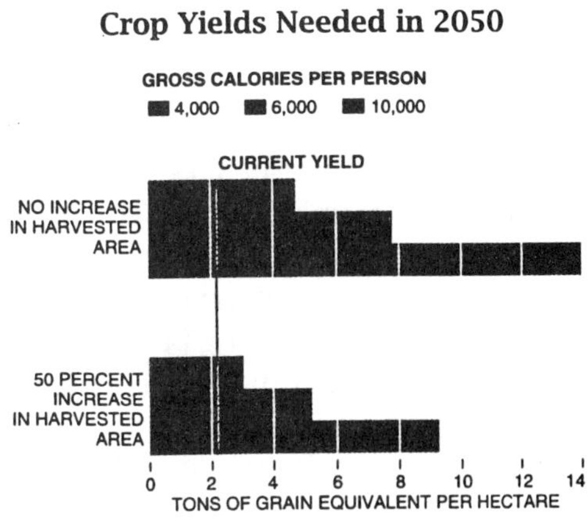

INCIDENCE OF CHRONIC UNDERNUTRITION fell in the developing world from an estimated 844 million sufferers in 1979 to 786 million in 1990, showing evidence of dramatic nutritional improvements in Asia *(left)*. Agricultural productivity must improve to continue this trend *(right)*. Even if more land is harvested in 2050, the average yield must rise sharply as well to offer the projected Third World population of 8.7 billion the current diet of 4,000 gross calories per day.

new, high-yielding strains of grains such as wheat and rice have proliferated since their introduction in the 1960s, especially in Asia. To reap full advantage from these new crop varieties, however, farmers must apply abundant quantities of fertilizer and water.

Environmentalists question whether further conversion to such crops can be achieved at reasonable cost, especially in the developing world, where the gain in production is most needed. At the moment, farmers in Asia, Latin America and Africa use fertilizer sparingly, if at all, because it is too expensive or unavailable. Fertilizer use in the developed world has recently waned. The reasons for the decline are complex and may be temporary, but clearly farmers in North America and Europe have decided that increasing their already heavy application of fertilizer will not further enhance crop yields.

Unfortunately, irrigation systems, which would enable many developing countries to join in the green revolution, are often too expensive to build. In most areas, irrigation is essential for generating higher yields. It also can make arid land cultivable and protect farmers from the vulnerability inherent in natural variations in the weather. Land brought into cultivation this way could be used for growing multiple crop varieties, thereby helping food production to increase.

Such advantages have been realized since the beginning of agriculture: the earliest irrigation systems are thousands of years old. Yet only a fraction of productive land in the developing world is now irrigated, and its expansion has been slower than population growth. Consequently, the amount of irrigated land per capita has been dwindling during recent decades. The trend, pessimists argue, will be hard to stop. Irrigation systems have been built in the most affordable sites, and the hope for extending them is curtailed by rising costs. Moreover, the accretion of silt in dams and reservoirs and of salt in already irrigated soil is increasingly costly to avoid or reverse.

Environmentalists Ehrlich and Ehrlich note that modern agriculture is by nature at risk wherever it is practiced. The genetic uniformity of single, high-yielding crop strains planted over large areas makes them highly productive but also renders them particularly vulnerable to insects and disease. Current preventive tactics, such as spraying pesticides and rotating crops, are only partial solutions. Rapidly evolving pathogens pose a continuous challenge. Plant breeders must maintain a broad genetic arsenal of crops by collecting and storing natural varieties and by breeding new ones in the laboratory.

The optimists do not deny that many problems exist within the food supply system. But many of these authorities, including D. Gale Johnson, the late Herman Kahn, Walter R. Brown, L. Martel, the late Roger Revelle, Vaclav Smil and Julian L. Simon, believe the world's food supply can dramatically be expanded. Ironically, they draw their enthusiasm from extrapolation of the very trends that so alarm those experts who expect doom. In fact, statistics show that the average daily caloric intake per capita climbed by 21 percent (from 2,063 calories to 2,495 calories) between 1965 and 1990 in the developing countries. These higher calories have generally delivered greater amounts of protein. On average, the per capita consumption of protein rose from 52 grams per day to 61 grams per day between 1965 and 1990.

According to the optimists, not only has the world food situation improved significantly in recent decades, but further growth can be brought about in various ways. A detailed assessment of climate and soil conditions in 93 developing countries (excluding China) shows that nearly three times as much land as is currently farmed, or an additional 2.1 billion hectares, could be cultivated.

The Potential Impact of Global Warming on Agriculture

The scientific evidence on the greenhouse effect indicates that slow but significant global warming is likely to occur if the emission of greenhouse gases, such as carbon dioxide, methane, nitrogen oxide and chlorofluorocarbons, continues to grow. Agriculture is directly or, at least in some cases, indirectly responsible for releasing a substantial proportion of these gases. Policy responses to the potentially adverse consequences of global climatic change now focus primarily on hindering emissions rather than on halting them. But considering the present need to improve living standards and produce more food for vast numbers of people, experts doubt that even a reduction in global emissions could occur in the near future.

In a 1990 study the Intergovernmental Panel on Climate Change estimated that over the next century the average global temperature will rise by three degrees Celsius. The study assumes that agriculture will expand considerably. This forecast of temperature change is uncertain, but there is now broad agreement that some global warming will take place. All the same, the effect that temperature rise will have on human society remains an open question.

Global warming could either enhance or impede agriculture, suggest Cynthia Rosenzweig of Columbia University and Martin L. Parry of the University of Oxford. Given sufficient water and light, increased ambient carbon dioxide concentrations absorbed during photosynthesis could act as a fertilizer and facilitate growth in certain plants. In addition, by extending the time between the last frost in the spring and the first frost in the fall, global warming will benefit agriculture in cold regions where the growing season is short, such as in Canada and northern areas of Europe and the former Soviet Union. Moreover, warmer air holds more water vapor, and so global warming will bring about more evaporation and precipitation. Areas where crop production is limited by arid conditions would benefit from a wetter climate.

If increased evaporation from soil and plants does not coincide with more rainfall in a region, however, more frequent dry spells and droughts would occur. And a further rise in temperature will reduce crop yields in tropical and subtropical areas, where certain crops are already grown near their limit of heat tolerance. Furthermore, some cereal crops need low winter temperatures to initiate flowering. Warmer winters in temperate regions could therefore stall growing periods and lead to reduced harvests. Finally, global warming will precipitate a thermal swelling of the oceans and melt polar ice. Higher sea levels may claim low-lying farmland and cause higher salt concentrations in the coastal groundwater.

Techniques used to model the climate are not sufficiently advanced to predict the balance of these effects in specific areas. The most recent analysis on the impact of climatic change on the world food supply, by Rosenzweig and Parry in 1992, concludes that average global food production will decline 5 percent by 2060. And they anticipate a somewhat larger drop in the developing world, thus exacerbating the problems expected to arise in attempts to feed growing populations. In contrast, their report predicts a slight rise in agricultural output in developed countries situated at middle and high latitudes.

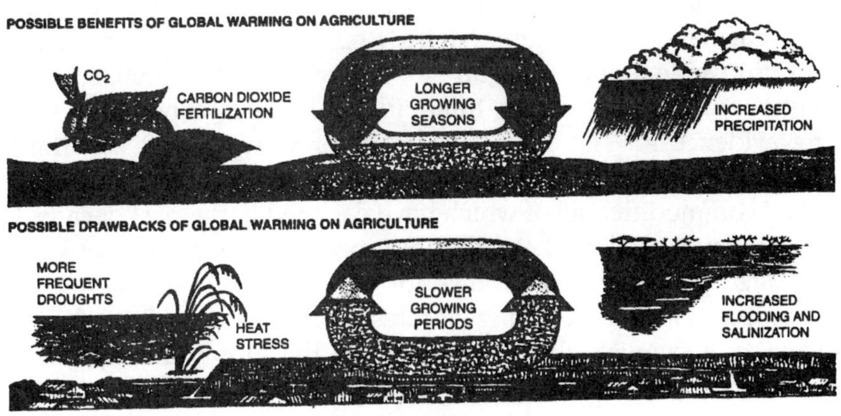

Regional soil estimates indicate that sub-Saharan Africa and Latin America can exploit many more stretches of unused land than can Asia, the Near East and North Africa.

Even in regions where the amount of potentially arable land is limited, crops could be grown more times every year than is currently the case. This scenario is particularly true in the tropics and subtropics where conditions are such—relatively even temperature throughout the year and a consistent distribution of daylight hours—that more than one crop would thrive. Nearly twice as many crops are harvested every year in Asia than in Africa at present, but further increases are possible in all regions.

In addition to multicropping, higher yields per crop are attainable, especially in Africa and the Near East. Many more crops are currently harvested per hectare in the First World than elsewhere: cereal yields in North America and Europe averaged 4.2 tons per hectare, compared with 2.9 in the Far East (4.2 in China), 2.1 in Latin America, 1.7 in the Near East and only 1.0 in Africa.

Such yield improvements, the enthusiasts note, can be achieved by expanding the still limited use of high-yield crop varieties, fertilizer and irrigation. In *World Agriculture: Toward 2000,* Nikos Alexandratos of the Food and Agriculture Organization (FAO) of the United Nations reports that only 34 percent of all seeds planted during the mid-1980s were high-yielding varieties. Statistics from the FAO show that at present only about one in five hectares of arable land is irrigated, and very little fertilizer is used. Pesticides are sparsely applied. Food output could drastically be increased simply by more widespread implementation of such technologies.

Aside from producing more food, many economists and agriculturalists point out, consumption levels in the developing world could be boosted by wasting fewer crops, as well as by cutting storage and distribution losses. How much of an increase would these measures yield? Robert W. Kates, director of the Alan Shawn Feinstein World Hunger Program at Brown University, writes in *The Hunger Report: 1988* that humans consume only 60 percent of all harvested crops, and some 25 to 30 percent is lost before reaching individual homes. The FAO, on the other hand, estimates lower distribution losses: 6 percent for cereals, 11 percent for roots and 5 percent for pulses. All the same, there is no doubt that improved storage and distribution systems would leave more food available for human nutrition, independent of future food production capabilities.

For optimists, the long-range trend in food prices constitutes the most convincing evidence for the correctness of their view. In 1992–93 the World Resources Institute reported that food prices dropped further than the price of most nonfuel commodities, all of which have declined in the past decade. Cereal prices in the international market fell by approximately one third between 1980 and 1989. Huge government subsidies for agriculture in North America and western Europe, and the resulting surpluses of agricultural products, have depressed prices. Obviously, the optimists assert, the supply already exceeds the demand of a global population that has doubled since 1950.

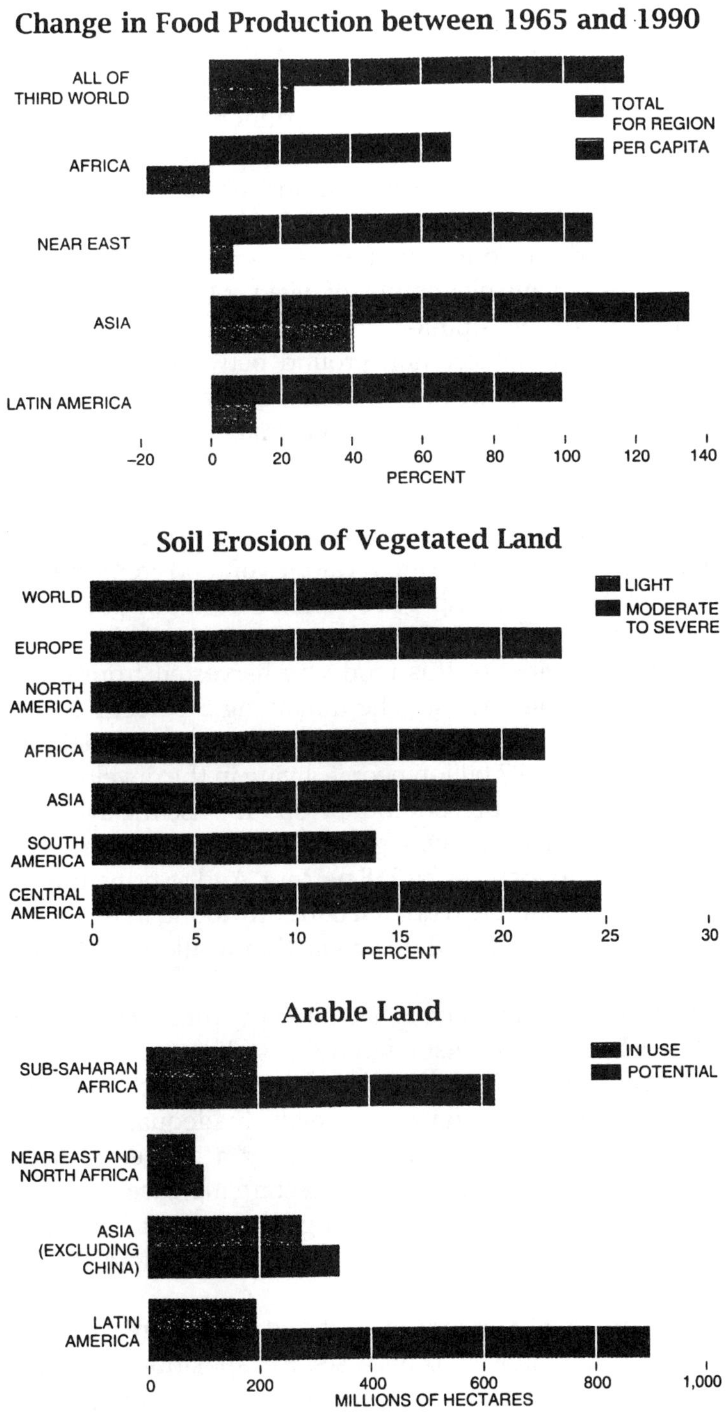

TOTAL FOOD PRODUCTION rose nearly 120 percent between 1965 and 1990 in the developing world. Per capita food production showed little change in regions outside Asia (*top*). Soil erosion has debased much of the land worldwide on which that food was produced (*middle*). Buy many Third World nations have vast holdings that could be farmed successfully if given more water and fertilizer (*bottom*).

Taken together, this evidence leads many experts to see no significant obstacles to raising levels of nutrition for world populations exceeding 10 billion people. The potential for an enormous expansion of food production exists, but its realization depends of course on sensible governmental policies, increased domestic and international trade and large investments in infrastructure and agricultural extension. Such improvements can be achieved, the optimists believe, without incurring irreparable damage to global ecosystems.

Proponents of either of these conflicting perspectives have difficulty accepting the existence of other plausible points of view. Moreover, the polarity between the two sides of expert opinion shows that neither group can be completely correct. Finding some common ground between these seemingly irreconcilable positions is not as difficult as it at first appears if empirical issues are emphasized and important differences in value systems and political beliefs are ignored.

Both sides agree that the demand for food will swell rapidly over the next several decades. In 1990 a person living in the developing world ate on average 2,500 calories each day, taken from 4,000 gross calories of food crops made available within a household. The remaining 1,500 calories from this gross total not used to meet nutritional requirements were either lost, inedible or used as animal feed and plant seed. Most of this food was harvested from 0.7 billion hectares of land in the developing world. The remaining 5 percent of the total food supply came from imports. To sustain this 4,000-gross-calorie diet for more than twice as many residents, or 8.7 billion people, living in the developing world by 2050, agriculture must offer 112 percent more crops.To raise the average Third World diet to 6,000 gross calories per day, slightly above the 1990 world average, food production would need to increase by 218 percent. And to bring the average Third World diet to a level comparable with that currently found in the developed world, or 10,000 gross calories per day, food production would have to surge by 430 percent.

A more generous food supply will be achieved in the future through boosting crop yields, as it has been accomplished in the past. If the harvested area in the developing world remains at 0.7 billion hectares, then each hectare must more than double its yield to maintain an already inadequate diet for the future population of the developing world. Providing a diet equivalent to a First World diet in 1990 would require that each hectare increase its yield more than six times. Such an event in the developing world must be considered virtually impossible, barring a major breakthrough in the biotechnology of food production.

Instead farmers will no doubt plant more acres and grow more crops per year on the same land to help augment crop harvests. Extrapolation of past trends suggests that the total harvested area will increase by about 50 percent by the year 2050. Each hectare will then have to provide nearly 50 percent more tons of grain or its equivalent to keep up with current dietary levels. Improved diets could result only from much larger yields.

The technological optimists are correct in stating that overall world food production can substantially be increased over the next few decades. Current crop yields are well below their theoretical maxima, and only about 11 percent of the world's farmable land is now under cultivation. Moreover, the experience gained recently in a number of developing countries, such as China, holds important lessons on how to tap this potential elsewhere. Agricultural productivity responds to well-designed policies that assist farmers by supplying needed fertilizer and other inputs, building sound infrastructure and providing market access. Further investments in agricultural research will spawn new technologies that will fortify agriculture in the future. The vital question then is not how to grow more food but rather how to implement agricultural methods that may make possible a boost in food production.

A more troublesome problem is how to achieve this technological enhancement at acceptable environmental costs. It is here that the arguments of those experts who forecast a catastrophe carry considerable weight. There can be no doubt that the land now used for growing food crops is generally of better quality than unused, potentially cultivable land. Similarly, existing irrigation systems have been built on the most favorable sites. Consequently, each new measure applied to increase yields is becoming more expensive to implement, especially in the developed world and parts of the developing world such as China, where productivity is already high. In short, such constraints are raising the marginal cost of each additional ton of grain or its equivalent. This tax is even higher if one takes into account negative externalities—primarily environmental costs not reflected in the price of agricultural products.

The environmental price of what in the Ehrlichs' view amounts to "turning the earth into a giant human feedlot" could be severe. A large inflation of agriculture to provide growing populations with improved diets is likely to lead to widespread deforestation, loss of species, soil erosion and pollution from pesticides, and runoff of fertilizer as farming intensifies and new land is brought into production. Reducing or minimizing this environmental impact is possible but costly.

Given so many uncertainties, the course of future food prices is difficult to chart. At the very least, the rising marginal cost of food production will engender steeper prices on the international market than would be the case if there were no environmental constraints. Whether these higher costs can offset the historical decline in food prices remains to be seen. An upward trend in the price of food sometime in the near future is a distinct possibility. Such a hike will be mitigated by the continued development and application of new technology and by the likely recovery of agricultural production and exports in the former Soviet Union, eastern Europe and Latin America. Also, any future price increases could be lessened by taking advantage of the underutilized agricultural resources in North America, notes Per Pinstrup-Andersen of Cornell University in his 1992 paper "Global Perspectives for Food Production and Consumption." Rising prices will have little effect on high-income countries or on households possessing reasonable purchasing power, but the poor will suffer.

In reality, the future of global food production is neither as grim as the pessimists believe nor as rosy as the optimists claim. The most plausible outcome is that dietary intake will creep higher in most regions. Significant annual fluctuations in food availability and prices are, of course, likely; a variety of factors, including the weather, trade interruptions and the vulnerability of monocropping to pests, can alter food supply anywhere. The expansion of agriculture will be achieved by boosting crop yields and by using existing farmland more intensively, as well as by bringing arable land into cultivation where such action proves economical. Such events will transpire more slowly than in the past, however, because of environmental constraints. In addition, the demand for food in the developed world is approaching saturation levels. In the U.S., mounting concerns about health have caused the per capita consumption of calories from animal products to drop.

Still, progress will be far from uniform. Numerous countries will struggle to overcome unsatisfactory nutrition levels. These countries fall into three main categories. Some low-income countries have little or no reserves of fertile land or water. The absence of agricultural resources is in itself not an insurmountable problem, as is demonstrated by regions, such as Hong Kong and Kuwait, that can purchase their food on the international market. But many poor countries, such as Bangladesh, cannot afford to buy food from abroad and thereby compensate for insufficient natural resources. These countries will probably rely more on food aid in the future.

Low nutrition levels are also found in many countries, such as Zaire, that do possess large reserves of potentially cultivable land and water. Government neglect of agriculture and policy failures have typically caused poor diets in such countries. A recent World Bank report describes the damaging effects of direct and indirect taxation of agriculture, controls placed on prices and market access, and overvalued currencies, which discourage exports and encourage imports. Where agricultural production has suffered from misguided government intervention (as is particularly the case in Africa), the solution—policy reform—is clear.

Food aid will be needed as well in areas rife with political instability and civil strife. The most devastating famines of the past decade, known to television viewers around the world, have occurred in regions fighting prolonged civil wars, such as Ethiopia, Somalia and the Sudan. In many of these cases, drought was instrumental in stirring social and political disruption. The addition of violent conflict prevented the recuperation of agriculture and the distribution of food, thus turning bad but remediable situations into disasters. International military intervention, as in Somalia, provides only a short-term remedy. In the absence of sweeping political compromise, hunger and malnutrition will remain endemic in these war-torn regions.

Feeding a growing world population a diet that improves over time in quality and quantity is technologically feasible. But the economic and environmental costs incurred through bolstering food production may well prove too great for many poor countries. The course of events will depend crucially on their governments' ability to design and enforce effective policies that address the challenges posed by mounting human numbers, rising poverty and environmental degradation. Whatever the outcome, the task ahead will be made more difficult if population growth rates cannot be reduced.

Further Reading

Poverty and Hunger: Issues and Options For Food Security in Developing Countries. World Bank, 1986.

Energy, Food, Environment: Realities, Myths, Options. Vaclav Smil. Clarendon Press, 1987.

World Agriculture: Toward 2000. Nikos Alexandratos, New York University Press, 1988.

World Resources 1992–93. World Resources Institute. Oxford University Press, 1992.

WHERE THE POOR LIVE:

Are the Assumptions Correct?

ROBERT W. KATES
AND VIOLA HAARMANN

Can those concerned with the fate of the hungry and poor and with the fate of the Earth reconcile their differing interests? In the 20 years since the United Nations Conference on the Human Environment was held in Stockholm, three views have been widespread. Early on, environmental concerns were viewed as an imposition of the developed "North" on the undeveloped "South," asking sacrifices of developing countries that industrialized countries never made in the course of their own economic growth. Moreover, according to this view, if the fate of the Earth is in doubt, it is threatened far more by the insatiable resource depletion and pollution of the industrialized countries than by the modest per-capita consumption of developing countries. A second view conceded the long-term interest of poor people in maintaining their environmental resources but noted the pressures on poor people to exploit their meager resources excessively and thereby degrade them for future use. Still another, rapidly spreading view asserts that poor people and threatened environments are inextricably linked and that a common strategy to address their fates is both desired and needed. The 1992 United Nations Conference on Environment and Development (UNCED) in Rio de Janeiro this June is intended to provide such a common strategy.

This evolution from perceiving conflict between two great social concerns to recognizing their complementarity[1] mirrors a long-term trend to link global problems. This trend arises both from a genuine appreciation of the interaction and multiple causation that underlie great world problems and from the widespread desire to find common ground among all good causes. Thus, there has been a series of reports on issues of global concern published by commissions convened by eminent persons, from the Palme Report by the Independent Commission on Disarmament and Security Issues on the threat of war, to the Brandt Report by the Independent Commission on International Development Issues that linked peace and development, to the Brundtland Report by the World Commission on Environment and Development on conjoining peace, development, and environmental efforts.[2] From "spaceship Earth" to "our common future," the search for the links between great global problems is a tenet of globally oriented and socially concerned groups and individuals.

But if the people at the center of the political spectrum seek to link the fate of the hungry and poor with that of the environment, those on the left of the spectrum view conflict rather than complementarity as the central theme of analysis. The view that the poor and the hungry are relegated by the rich and the powerful to marginal lands that are degraded and cannot sustain intensive use fits easily with theories of center-periphery conflict, rural exploitation, and underdevelopment. So, it is not surprising that a recent spate of reports, books, and collections of papers, spanning a broad spectrum of ideological perspectives and academic traditions, has found common cause between the fate of the Earth and that of the hungry and poor.[3] The inclination to link these issues is not wrong, for in the rush to address the broad concerns of development and the environment, the fate of the hungry of the world could easily be ignored or forgotten.

At the same time, a review of the many pertinent reports and papers reveals how limited and selective is the literature that carefully documents the causal relationships between poverty and environmental degradation, while an implicit assumption of a strong relationship between the two is widely apparent. Thus, an assessment of what global overviews, country comparisons, and local and regional case studies exist that link poor people to threatened environments should provide insights into the validity of this assumption.[4]

Matching Poverty and Environment

How many poor and hungry people are there in the world? Which environments are most threatened? Does the geographical distribution of hungry and poor people match that of threatened environments? To answer such questions, analysts use "poverty lines" to separate the poor from the less needy. But poverty lines are difficult to create, and there is a large amount of literature to guide their creation.[5] To draw these lines, some use *absolute* measures of poverty, such as insufficient resources and income to provide minimal household necessities—mainly food in developing countries. Others use *relative* measures, which designate the poor as being at the lower end (for example, the lowest quintile) of a national or regional distribution of resources and income. Thus, it is not surprising that recent estimates of global poverty and hunger differ by as many as 750 million people (see Table 1. Poverty estimates range from 630 million "extremely poor" people, through 780 million "poorest of the poor," to 1,225 million people "living in absolute poverty." Similarly, by one hunger measure, almost one-half billion people live without enough energy to maintain minimal activity and

Table 1: Estimates of Poverty and Hunger

	Poverty						Hunger			
	Extremely poor[a]		Poorest of the poor[b]		Living absolute poverty[c]		Energy deficient for maintenance[d]		Energy deficient for work[e]	
Region	Number of people (millions)	Percentage of total population	Number of people (millions)	Percentage of total population	Number of people (millions)	Percentage of total population	Number of people (millions)	Percentage of total population	Number of people (millions)	Percentage of total population
Asia					675	25				
South Asia	300	29	390	37			197	17	572	50
China	80	68	84	8			36	3	72	6
East Asia	40	4	33	7			97	9	70	14
Sub-Saharan Africa	120	30	156	30	325	62	141	26	239	44
Middle East and North Africa	40	21	39	15	75	28	12	5	31	10
Latin America	50	12	78	17	150	35	42	10	58	13
Total	630	14[f]	780	17[f]	1,225	23[f]	477[g]	9[f]	1,053[g]	20[f]

[a] The extremely poor are people who fall below a poverty line of $275 (purchasing power parity) per capita per year.

[b] The poorest of the poor are defined as the poorest 20 percent of the total population of all developing countries.

[c] The people estimated to be living in absolute poverty are based on the Worldwatch Institute's country estimates of absolute poverty and other social and economic indicators. Estimates should be viewed as midpoints in a range of plus or minus 10 percent.

[d] Updated to 1990 by the World Hunger Program, which used earlier food-energy estimates by the UN Food and Agriculture Organization that use a 1.2 times the basal metabolic rate for adults and adolescents (minimal activity) and a more generous requirement for children to determine food income or consumption adequacy. East Asian data include data for Southeast Asia.

[e] Updated to 1990 by the World Hunger Program, which used earlier World Bank food-energy estimates that use 90 percent of the estimated dietary requirement for active work as the level of food-income adequacy. East Asian data include data for Southeast Asia.

[f] Percentage of world population for various years, including Europe, North America, and Australia.

[g] These numbers are global totals rather than column totals.

SOURCES: World Bank, World Development Report 1990: Poverty—World Development Indicators (New York: Oxford University Press, 1990), 29; H. J. Leonard, "Environment and the Poor: Development Strategies for a Common Agenda," in H. J. Leonard, ed., Environment and the Poor: Development Strategies for a Common Agenda (Washington, D.C.: Overseas Development Council, 1989), 18; A. B. Durning, Poverty and the Environment: Reversing the Downward Spiral (Washington, D.C.: Worldwatch Institute, 1989), 20.

growth, while by another measure, more than a billion hungry and poor people have insufficient energy for work.

There also is little consistency in the way environments are classified. Climate, vegetative cover, resource use, and land forms, for example, are all freely mixed in discussions of global environmental problems and modified by terms such as fragile, vulnerable, and marginal.[6] Here, however, threatened environments are defined as those areas currently undergoing extensive degradation or transformation that elicits widespread concerns; degraded environments are those areas in which the essential features of the environment are maintained but at very low levels of biological or resource productivity; and marginal environments are those areas that are unsuitable for some intended resource use. The literature indicates that the three major environments of global concern are the highlands, tropical and subtropical drylands, and rain forests. (Wetlands, especially coastal wetlands, in the developing world are also threatened, but they have been much less studied.[7]) Taken together, these three threatened environments occupy more than half of the world's land area and are home to one-quarter of its people.

Highlands occupy about 23 percent of the Earth's land area and contain about 10 percent of the world's population.[8] They fringe the main mountain axes and the lower slopes of volcanic mountains. For the latter, in particular, which have very rich soils, population densities can be quite high. But in general, human populations are sparse in highland environments. These are "high-energy" environments because gravity can move large amounts of material downhill. Semi-arid and arid productive lands constitute about 22 percent of the Earth's land area and contain about 14 percent of its population. By one estimate, more than 60 percent of such land is desertified, or has lost at least 25 percent of its natural productivity.[9] Other estimates of desertification are much lower, however.[10] Tropical moist forests occupy about 8 percent of the world's land.[11] Half of these forests are in Latin America, and the remainder are shared about equally between Africa, Asia, and the Pacific region. By one estimate, about 0.5 percent of this forest is being cleared annually, though some estimates of deforestation are up to three times larger.[12] Perhaps as many as 200 million people live in or near the tropical moist forests.[13]

There are no global data bases that allow one to test the widespread notion that impoverished people are concentrated in threatened environments. Ideally, one would want to compare the distribution of poor people against the distribution of threatened environments or of those parts of threatened environments that are already degraded. Estimates of the numbers of poor people are available only by country—and even then not for many countries—and not for regions within countries.[14] Although maps and regional estimates of threatened environments exist, estimates of their areas differ greatly.[15] Thus, attempts to test the hypothesis that impoverished people are concentrated in threatened environments must rely on very rough estimates, often countrywide aggregates.

A noteworthy attempt to compare the distribution of poor people and "areas of high ecological vulnerability" was recently made by H. Jeffrey Leonard.[16] As a surrogate for ecological vulnerability in rural areas, he used the International Food Policy Research Institute's (IFPRI) estimates of land of low agricultural potential[17] and assumed that such land was arid, unfertile, or steep. As a surrogate for urban areas, Leonard used peri-urban squatter settlements, for which the classic cases (particularly in Latin America) are precariously perched on hillsides subject to landslides, flooding, and pollution. Combining those with the World Bank's estimates of poverty,[18] he concluded that 57 percent of the rural poor live on lands of low agricultural potential and 76 percent of the urban poor live in squatter settlements.

Leonard's attempt to map the poor onto threatened environments should be viewed as a restatement of a hypothesis rather than as an answer to the question of whether one can correlate poverty with threatened environments. Even if one accepts the crude and preliminary estimates of IFPRI, 43 percent of the rural poor live in areas of high agricultural potential. Moreover, it is not appropriate simply to equate land of low agricultural potential and squatter settlements with areas of high ecological vulnerability. In many parts of the world, land of low agricultural potential is little used, is appropriately used for pastoralism, or is forested. Although subject to erosion, desertification, and deforestation, these areas are not necessarily more vulnerable than are intensively used lands of high agricultural potential that are also subject to erosion, flooding, and, in the case of valuable irrigated lands, salinization and waterlogging. Similarly, although squatter settlements are crowded and have few or no urban services and amenities, their ecological setting may not be very different from that of the urban settlement they surround.

If it is difficult to match poor people against threatened environments, it is possible to match poor countries against them. Poor countries may be considered to be those included in the United Nations' aggregation of the 42 least developed nations and the World Bank's classification of 41 low-income economies whose annual per-capita income is less than $500.[19] There are many available statistics of land to determine the distribution of threatened environments.[20] For drylands, for example, one may use as an indicator the estimated areas of savannas and grasslands; for highlands, there are estimates of areas in steep slopes; and for rain forests, there are estimates of areas in tropical moist forests. From these sets of data (see the map in Figure 1), one can determine whether poor countries contain more than their share of threatened environments.

According to the data, the incidence of poverty and dryness correlate closely in developing countries, with 78 percent of the savanna grasslands found in low-income countries that contain only about one-half of the total area of developing countries (see Table 2). And these drylands are even more concentrated in the poorest, least developed countries; although these latter account for only 20 percent of developing country area, they contain 63 percent of the drylands. Highland areas with steep slopes, however, appear to be about equally distributed among poorer and wealthier developing countries, whereas tropical moist forests are somewhat concentrated in low-income countries (71 percent) but not in the least developed countries (16 percent).

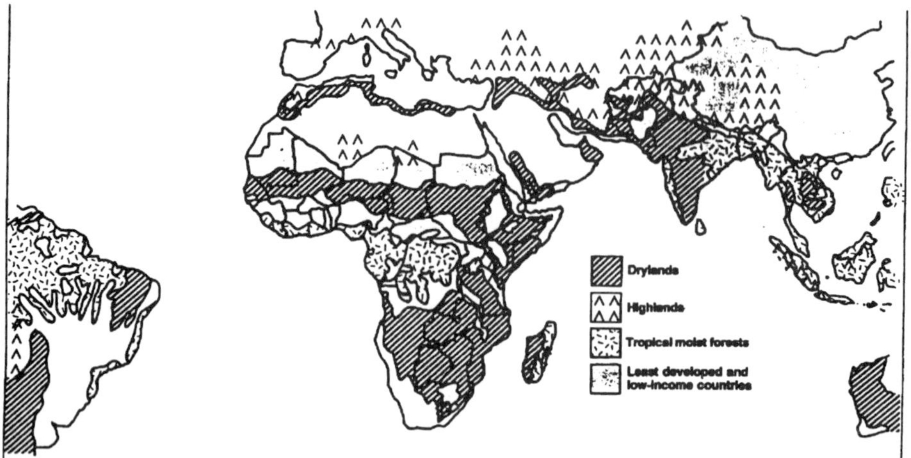

SOURCES: Environmental data are taken from *Alexander Weltatlas*, 3rd ed. (Stuttgart: Ernst Klett, 1979), 85. Country data are taken from United Nations Conference on Trade and Development, *The Least Developed Countries: 1990 Report* (New York: United Nations, 1990); and World Bank, *World Development Report 1991: The Challenge of Development—World Development Indicators* (New York: Oxford University Press, 1991).

Figure 1. The distribution of least developed and low-income countries and threatened environments.

Thus, low-income countries have more than their share of drylands and tropical moist forests, but not of highlands. Overall, the poorest countries are primarily dry. This is a modest conclusion at best because country-level comparisons, though suggestive, are too nonspecific to draw conclusions about the poverty, hunger, and environment nexus.

Local and Regional Studies

To help one examine the links between poverty, hunger, and the environment, there are some case studies in which the connections between the impoverishment of people and the degradation of the environment have been specified and placed in the context of actual livelihoods and particular places, but they are few

Table 2: Drylands, Highlands, and Tropical Moist Forests in the Least Developed and Low-Income Countries

Area	(A) All developing countries (millions of square kilometers)	(B) Low-income countries (millions of square kilometers)	B/A (percent)	(C) Least developed countries (millions of square kilometers)	C/A (percent)
Total Area	76.5	39.9	52	15.5	20
Drylands	2.7	2.1	78	1.7	63
Highlands	11.1	4.5	41	2.2	20
Tropical moist forests	3.1	2.2	71	0.5	16

SOURCE: World Resource Institute, *World Resources 1990–91* (New York: Oxford University Press, 1990), 286–87, 306–07; United Nations Conference on Trade and Development, *The Least Developed Countries: 1990 Report* (New York: United Nations, 1990); and World Bank, *World Development Report 1991: The Challenge of Development—World Development Indicators* (New York: Oxford University Press, 1991).

and far between.[21] A search of the literature, including the last 10 years of issues of 40 journals dealing with aspects of environment and development, yielded only 30 studies that described specific links, and these were primarily the work of anthropologists and geographers. All of these studies reported positive findings linking poverty to environmental degradation. The absence of negative findings can be viewed as evidence for a strong connection, for a bias in researchers' case study selection to those for which positive links are likely, or for a bias against publishing reports of negative findings.

More than half of the studies were from drylands, one-third were from tropical moist forests, and only four were from highland regions.[22] The studies included areas within 25 countries, with one-third from Asia, one-fifth from Latin America and the Caribbean, and the remainder from Africa. Yet, with remarkable agreement irrespective of location, the case studies tell repeatedly how poor people lose their entitlement to environmental resources, how those resources are further degraded, and how such loss and degradation lead to further impoverishment.

Entitlement may be defined as the access enjoyed by a household to needed environmental resources by virtue of the household's socially recognized rights.[23] Among the many socially recognized rights are ownership, tenancies of various forms, customary allocations, and various common property rights. The environmental resources are primarily land, water, natural vegetation, and wildlife. Stocks and flows of such resources are often complex, subtle, and seasonal and are not easily recognized by outside observers or land and water project managers. Poor people lose their entitlement to such resources by displacement, by division, and by degradation.

The poor are displaced by activities that, in the name of development or commercialization, deprive the hungry of their traditional entitlement to the common property resources that are essential to their survival. Large-scale agriculture, hydroelectric development, export forestry, tourism, and wildlife preservation dispossess poor people of their resource access directly by expropriation and by destroying or limiting access to small but crucial seasonal resources. The poor are also displaced by wealthier claimants to land who use both legal and illegal means. Finally, the poor are displaced by each other because limited land and employment opportunities often force the young from poor families to migrate in search of land or employment.

The entitlements of the poor are divided and reduced because of their need to share resources with their children or to sell off portions of their resources to offset extreme losses, such as crop failure, illness, or death; to fulfill social requirements, such as marriage dowries; or to provide simple subsistence. Throughout the developing world, the number of landless and land-poor households is increasing.

The resources of the poor are degraded by excessive or inappropriate use, by failure to restore or to maintain protective works, and by the loss of productive capacity because of natural hazards.[24] Those who are poor in resources often press what little resources they do have to levels of production or reproduction—

for example, by reducing fallow, overgrazing, or excessive fuelwood removal—that cannot be sustained. The degradation of many common property resources, however, occurs because of excessive use, not by the poorest users, but by the wealthier ones, who often have large herds of livestock, require large diversions of water, or use the forest commercially.

The poor also may lack the ability to restore or maintain protective works, such as terraces or drainage canals; the means to hire people with specialized skills or to make needed inputs; or the access to public programs of resource improvement and renewal. Of course, the poor are also responsible for some of the inappropriate uses that degrade the marginal lands that are the migration destinations for the displaced, dispossessed, and disinherited. Finally, natural hazards, such as diseases, droughts, floods, landslides, and pests, also degrade the environment, causing it to lose the capacity to provide particular resources.

Many studies report that these processes occur at the same time or in serial fashion, producing a spiral of impoverishment and degradation.[25] Resources are reduced by processes of displacement, are increasingly shared by division, and are then subject to degradation through the excessive use of the remainder. Such a spiral is suggested by the multiple pathways of Figure 2. At the top of Figure 2 are four major driving forces in the destructive sequences of the poverty-environmental degradation spiral. Two are driven by forces external to the case-study locales: natural hazards and development and commercialization. Two are internal to the communities studied: population growth and existing poverty. These latter two are linked by the strong correlation between household poverty and birth rates. To some extent, all the driving forces are implicated in the various directions that impoverishment-degradation spirals assume,[26] but three major spiral sequences (numbers 1, 2, and 3 in Figure 2) emerge from the case studies. In each spiral, two of the driving forces dominate.

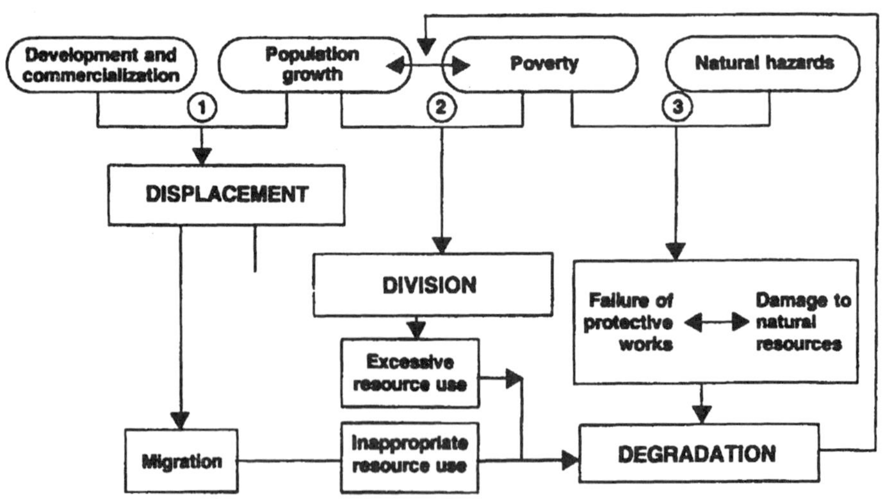

Figure 2. Impoverishment and degradation spirals.

In the first sequence, poor people driven by development activities or commercialization and by population growth are displaced from their resources by richer claimants or by the competition for existing land or employment. For the displaced, either division of the remaining resources or forced migration to other, usually more marginal areas generally ensues. In the second sequence, meager resources are further divided because of population growth and existing poverty. Resources are then degraded by excessive use of divided lands or by inappropriate use of environments unable to sustain the requisite resource use. And in the third sequence, even in the absence of further displacement and division, poverty creates poor households unable to maintain needed protective works or to restore resources, while natural hazards, such as disease, drought, flood, soil erosion, landslides, and pests, further degrade these natural resources.

The case-study environments and localities differ in their particular combinations of displacement, division, and degradation; in the driving forces that set them in motion; in the ameliorative measures undertaken; and in their ecological and social contexts. Grouping these case studies according to the three categories of major threatened environments (drylands, highlands, and rain forests) helps to address some of these important differences. In each environment, two or three characteristic sequences of displacement, division, and degradation, creating spirals of impoverishment and degradation, are initiated by the driving forces of development, commercialization, population growth, poverty, and natural hazards.

The Dryland Cases

People's access to land and the availability of water determine the productive use of dryland. Rainfall patterns, infiltration rates, run-off, surface- and groundwater reserves, and their significant variability from year to year—all structure settlement, cultivation, and pastoralism. It follows that the severity of entitlement losses and displacement is directly linked to the loss of access to favorably watered land and to resource reserves that serve to moderate the frequent ecological crises generated by natural events, particularly droughts. In this sense, dryland case studies tell of three major, typical sequences of displacement, division, and degradation:

- the sequence of people's displacement and subsequent disruption caused in long-standing livelihood structures through the introduction or encouragement of "development" activities in prime areas of water availability with subsequent loss of needed common property resources and environmental degradation in less suitable areas;[27]
- the sequence of people's displacement—driven by population pressure or resource expropriation—from regions of higher agricultural potential because of limited opportunities for land or employment, leading to encroachment and expansion into areas of lower agricultural potential with subsequent excessive or inappropriate resource use;[28] and

- the sequence of degradation and displacement by extreme natural events, most often drought, leading to a dramatic reduction in resources from direct losses and from the disposal and sale of family holdings to compensate for such losses, and to subsequent environmental degradation from excessive use and inability to restore or maintain the degraded resources.[29]

Water supply is the key factor for the use of drylands, and the development of large-scale water projects often sets in motion the first sequence, especially when many people are affected by large dams of national significance on major rivers. Construction of high dams has long been a tempting undertaking for developing countries. These projects are envisaged as the mechanism to promote industrialization (with hydroelectric power) and agriculture (with irrigation) and ostensibly to create employment opportunities, increase food self-sufficiency, and strengthen foreign-exchange earnings. The result of such large-scale development projects is often that the original rural population is largely pushed away from the prime locations near the river, which are now organized into irrigation projects, and into more marginal, far less productive areas further inland. The limited capacity of these lower productivity habitats to sustain increased populations results in increased environmental pressure, declines in primary productivity of natural resources, and increasing poverty for the populations forced to rely on them for survival.

Other case studies tell of livelihood disruption through development and the alienation of water as a common property resource. Examples from India and Kenya show that attempts to secure alternative income sources can lead to serious water access problems. Thus, the development of a small-scale leather cottage industry in parts of dryland India has caused an alarming decline in water quality and crop yields that is endangering food security.[30] In dry parts of eastern Kenya, riverbed sand mining for urban construction by Nairobi entrepreneurs has reduced the land's water retention capacity, lowered the water table, and blocked access to sufficient dry-season water supplies.[31]

In all marginal environments, common property resources (that is, nonprivatized access to vegetation and water resources under a well-established system of local administrative rules and regulations) constitute a significant proportion of the total land resources on which the rural poor, in particular, depend and reduce the inequality in private property resources. Common property resources allow land- and resource-poor populations to supplement their livelihood with extractive activities, such as water collection, grazing, collecting natural products, and tree harvesting, and thus create stabilized livelihood opportunities. Development in these areas, however well intentioned, often works to the disadvantage of those who depend most on these resources.

In the second sequence, the main story is one of competition among different livelihood systems in transitional zones between greater and lesser environmental advantage. Typical examples come from Kenya, with its gradually declining land and water resource potential from the primarily agricultural highlands to the primarily pastoral lowlands.[32] In Kenya, the large populations in the limited areas of high agricultural potential are forcing more and more people to move onto the drier lowlands, where they settle near dependable water sources (perennial streams and swamps) and attempt to farm land that is often best suited for pastoralism. Thus, the displaced, in turn, partly displace the local pastoralists from access to the water sources and dry-season pastures. This increasing agropastoral competition for land and water has caused resource overuse, degradation, and economic decline, particularly among the pastoralists, and marginal subsistence among the agriculturalists.

In the third sequence, the spiral of displacement and degradation is turned around. In the arid reaches of the Sahel beyond the agronomic dry boundary or in the arid north of Kenya, drought, desertification, and the loss of crops and animals force cultivators and pastoralists to migrate south or cluster around water sources, thus increasing population pressure, displacement, division, and degradation in less marginal areas.[35]

The Highland Cases

The salient environmental feature of the productive use of highlands is the vertical, complementary organization in both the physical processes and the livelihood systems operating in the different altitudinal zones. Threatened areas within highland environments are those with reasonable agricultural potential where high population densities result in tough competition over the limited amounts of arable land. Consequently, farmers' landholdings become increasingly smaller, and there is a trend to use more marginal fringes (steeper slopes) of the environment that are poorly suited for sustained use. This trend exacerbates the problems of deforestation and soil erosion on a broad front. As in the drylands, case studies of highlands reveal three major sequences of displacement, division, and degradation:

- the sequence of population growth in regions with limited arable land leading to division of land into smaller and smaller units, extension of grazing and agriculture onto the steep slopes and forested uplands, and an inability to hire or provide the labor needed to maintain and preserve a productive resource use;[34]
- the sequence of expropriation of desirable highlands for commercial development with subsequent concentration of populations on smaller areas or displacement into less productive, marginal areas and excessive or inappropriate use;[33] and
- the sequence of degradation and displacement by natural hazards—such as soil erosion, landslides, and flooding—which, because of a lack of restorative efforts, lead to further degradation and excessive use.[36]

In Nepal, Peru, and Kenya,[37] small-holder hill farmers have to engage in a multiplicity of activities, away from as well as on the farm, to maintain the minimum levels of income necessary to meet short-term needs and to minimize the risk from failure of any one activity. They lack both the funds and the labor necessary to undertake intensive slope, soil, and vegetation conservation practices and to secure alternative sources of food, fodder, fertilizer, and fuel in response to environmental signs of degradation. Thus, deforestation causes not only erosion and soil loss but also the decreased availability of composted forest products and manure.

Soil loss and erosion result in reduced availability of plant nutrients and a subsequent decline in crop yields. The ability to "transfer fertility" from the forest to arable land is lost, and the potentially complementary functions of these altitudinal zones are disrupted. Instead, a downward spiral of increasing decline is set in motion. Nepal is a particularly striking example in this respect.

The Rain Forest Cases

Significantly, almost all of the world's tropical moist forests are in the developing countries . . . [and these] tropical moist forests are typically prone to certain management problems, owing to their unique ecology. They are usually situated on agriculturally marginal soils, [and] they have more limited potential for recovery following disturbance than most other forest types.[38]

At the same time, however, tropical moist forests have become prime targets for vast development projects, both commercial and on the level of small-holder farming resettlements. Two major sequences of displacement, division, and degradation stand out in this environment:

- the sequence of displacement of indigenous hunter-gatherers or poor farmers or collectors by commercial activities, leading to destruction of the forest and subsequent degradation and inability to sustain the requisite resource use;[39] and
- the sequence of displacement by spontaneous or planned immigrants because of limited opportunities for land or employment, driven by population pressure or resource expropriation, leading in turn to displacement of indigenous hunter-gatherers or poor farmers and resulting in destruction of the forest and subsequent degradation.[40]

In tropical moist forests, the huge expansion of such commercial land uses as logging, ranching, and mechanized cultivation projects and industrial competition for the very resources that sustain traditional extractive agroecosystems cause the problem. In such forests, where the vegetation locks in soil nutrients—the major production potential—access to particular vegetative formations, through a multitude of extractable products such as fruits, fibers, and wood, provides a sizeable supplement to the inhabitants' subsistence. Thus, in Brazil, traditional populations' access to successional palm forests is threatened by large commercial enterprises.

State-sponsored development, like the Tucurui Dam and Reservoir in Amazonia, also displaces people.[41] At the same time, the magnitude of these projects has enormous environmental effects, through the decomposition of thousands of square miles of flooded rain forest or riverine galleria forest or through large earth movements and land leveling for irrigation projects, which severely damage vulnerable soil structures and result in erosion and loss of natural fertility.

Linking the Fates

How justified is the link between poor people and threatened environments? On one hand, no available data base exists to match poor people with threatened environments. The data permit matching only poor countries (not numbers of people) with environments and yield, at best, some broad generalizations. Clearly, poverty is not evenly distributed in the world. The familiar North-South distinction is really a latitudinal distinction between temperate lands and the tropics and semitropics, where most of the world's poor reside. The environments of the poorest countries are excessively dry, and poorer countries are both dry and wet. Hills, however, are seemingly shared among both poor and wealthy nations.

If the global overviews are at best suggestive, the case studies are more informative, albeit selective. The universality of the driving forces of development, commercialization, population growth, poverty, and natural hazards and the repetitive sequences of displacement, division, and degradation, which are all found in widely varying places and environments, are convincing. Some large, even if unknown, fraction of the world's hungry and poor is trapped in a spiral of impoverishment and environmental degradation.

For the poorest fifth of humankind, maintaining access to the natural resource base and the inputs needed for agriculture, herding, or fishing is becoming increasingly difficult because of growing population, increased competition for land, and development. Food-poor households are more often having to cope with the deterioration of their resources, the loss of crucial access to common resources, and restriction to the most ecologically marginal lands. Because of growing populations in land-poor countries, the children of the hungry become landless laborers or move onto marginal lands and degrade them in the process. Even in the land-rich countries of Africa, this process is well under way in the regions of highest land productivity. And in small ways, everywhere, the desperate search for fuelwood for energy, for pasture in times of drought, and for additional land taken from poorly protected reserves makes life more difficult, degrades the commonly shared resources, and results in extensive deforestation and desertification.

There are a variety of remedies espoused for the environmental plight of the hungry.[42] At the local level, agroforestry and other agricultural techniques have demonstrated their ability to sustain productivity, provide fuelwood, limit soil erosion, and increase food and income. Everywhere, there is a heartening increase in the organization of the hungry and dispossessed on their own behalf and an emergence of local advocacy groups to strengthen their voice. But, although promising methods exist for limiting the environmental damage of human activities, these methods cannot replace a basic commitment to alleviate the root causes of unsustainable development.

Latitude and the Wealth of Nations

Differences between the industrialized world and the less industrialized countries, or between the richer and poorer countries, are often referred to by the seemingly more neutral designation of "North-South" differences. As Figure 1 below demonstrates, however, this designation is incorrect. In a plot of each nation's per-capita gross national product against the latitude of its capital city, a line drawn thorugh the median point within five degree columns does not simply decline from north to south. Rather it suggests a complex but overall "U"-shaped relationship. This relationship between income and latitude appears to be more than an artifact because it is reproduced to a similar extent in both the Northern and Southern Hemispheres, though it is distorted somewhat in the northern subtropics by the wealth of the oil-producing countries. If there is any meaning to this relationship, latitude appears as a surrogate for the great climatic belts. Thus, the great differences in national incomes are not really North-South differences but are more aptly described as temperate-tropical differences.

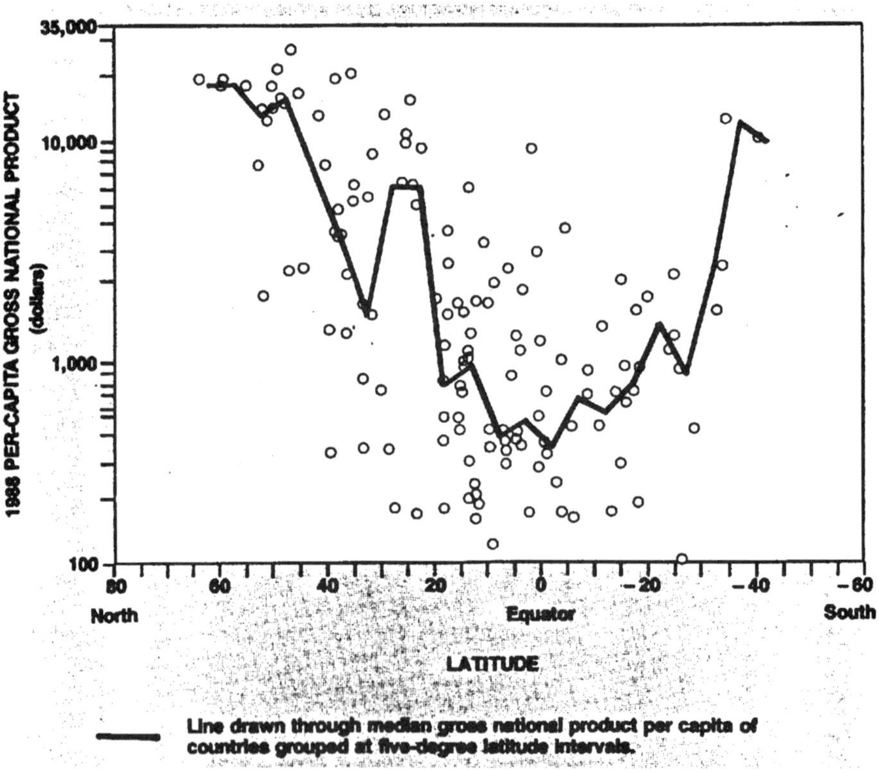

Note: The points of latitude are for the capital cities of 141 countries and territories.
SOURCE: Compiled and plotted by Tracy Zafian from data in World Bank, World Development Report 1989 (New York: Oxford University Press, 1989), 165–65.

Eliminating the most obvious cause of unsustainable development, land shortage, requires the redistribution of land that is unused, little used, or in excessively large holdings to small holders of land and migration and resettlement to sparsely occupied but appropriate lands. Unfortunately, land reform has not succeeded in many places. Progress in slowing population growth needs to be maintained and broadened. Much ongoing environmental degradation in the developing world results from agricultural activities and other land uses driven by short-term needs for survival, resettlement, and foreign-exchange earnings. Specifically, it will be necessary for national and international agencies to recognize that foreign debt should not be reduced simply by destroying natural resources and that expansion of populations into ecologically marginal land is a waste of both natural and human capital.

This June, the United Nations Conference on Environment and Development will meet in Rio de Janeiro, Brazil. This intergovernmental conference is designed to highlight the common future of both developing and industrialized countries. The delegates will seek to ratify some global bargains and to identify others needed. The discussions will probably equate poor people with poor countries or poor countries with all developing countries. But this global overview has shown that considerable differences and stratification exist among developing countries. The case studies reveal that the interests of poor people are not always the same as the interests of poor countries, for in the interest of "development," the poor may grow poorer. Over the long term, an international trust fund should be developed specifically to assist the poorest peoples to cope with the inevitable changes under way.

Yet it is not the absence of remedial measures nor the confusion of interests that some people fear most, but rather that the hungry will be out of sight and out of mind during the immense struggle to direct the fate of the Earth. The fate of the Earth directly concerns the well-being of the wealthy and articulate, while the fate of the poor and the hungry seems to make claims only on their altruism and solidarity. Thus, the ultimate question is not whether poor people or poor countries match up with threatened environments but whether the world can jointly sustain efforts to overcome poverty and save the Earth.

Notes

1. S. Davies, M. Leach, and R. David, "Food Security and the Environment: Conflict or Complementarity?" (Discussion paper for the Institute of Development Studies, University of Sussex, Brighton, U.K., 1991).
2. Independent Commission on Disarmament and Security Issues, *Common Security: A Blueprint for Survival* (New York: Simon and Schuster, 1982); Independent Commission on International Development Issues, *North-South: A Program of Survival* (Cambridge, Mass.: MIT Press, 1980); and World Commission on Environment and Development, *Our Common Future* (Oxford, England, and New York: Oxford University Press, 1987).
3. P. Blaikie, *The Political Economy of Soil Erosion in Developing Countries* (London: Longman, 1985); P. Blaikie and H. Brookfield, eds., *Land Degradation and Society* (London: Methuen, 1987); A. B. Durning, *Poverty and the Environment: Reversing the Downward Spiral* (Washington, D.C.: Worldwatch Institute, 1989); G. C. Gallopin, P. Gutman, and H. Maletta, "Global

Impoverishment, Sustainable Development and the Environment: A Conceptual Approach," *International Social Science Journal* 41, no. 3 (1989):375–97; H. J. Leonard, ed., *Divesting Natures's Capital* (New York: Holmes and Meier, 1985); H. J. Leonard, ed., *Environment and the Poor: Development Strategies for a Common Agenda* (Washington, D.C.: Overseas Development Council, 1989); P. D. Little and M. M. Horowitz, eds., *Lands at Risk in the Third World: Local-Level Perspectives* (Boulder, Colo.: Westview Press, 1987); and M. Redclift, *Sustainable Development: Exploring the Contradictions* (London: Methuen, 1987).

4. This paper is based on the 1991 World Hunger Program Research Report (RR–91–2): R. W. Kates and V. Haarmann, *Poor People and Threatened Environments: Global Overviews, Country Comparisons, and Local Studies* (Providence, R.I.: Alan Shawn Feinstein World Hunger Program, Brown University, 1991). An earlier version of this report entitled *Hunger, Poverty and the Human Environment* appeared in the 1991 CASID (Center for Advanced Study of International Development, Michigan State University) Distinguished Speaker Series. Thanks are due to Barbara DeMaio for helping to edit and format this article.

5. A. Sen, *Poverty and Famines: An Essay on Entitlement and Deprivation* (Oxford, England: Clarendon Press, 1982); and M. Lipton, *Poverty, Undernutrition, and Hunger* (Washington, D.C.: World Bank, 1989).

6. For an excellent discussion of fragility and fragile lands, see B. L. Turner II and P. A. Benjamin, "Fragile Lands: Identification and Use for Agriculture" (Paper prepared for the Bellagio Conference on Institutional Innovations and Sustainable Agricultural Development: Into the Twenty-First Century, at the Graduate School of Geography and George Perkins Marsh Institute, Clark University, Worcester, Mass., 14–18 October 1991).

7. See, for example, World Resources Institute and International Institute for Environment and Development, *World Resources 1986* (New York: Basic Books, 1986), 146–52; and M. Williams, Ed., *Wetlands: A Threatened Landscape* (Oxford, England: Basil Blackwell, 1990). For high-potential agricultural land and irrigated areas in particular, see M. Yudelman, "Sustainable and Equitable Development in Irrigated Environments," in H. J. Leonard, ed., *Environment and the Poor: Development Strategies for a Common Agenda*, pages 61–85, note 3 above. Urban environments often combine environmental problems of both developing and industrialized countries. See, for example, T. Campbell, "Urban Development in the Third World: Environmental Dilemmas and the Urban Poor," in H. J. Leonard, ed., idem, 165–87.

8. E. P. Eckholm, *Losing Ground: Environmental Stress and World Food Prospects* (New York: W. W. Norton, 1976), 74.

9. J. A. Mabbutt, "A New Global Assessment of the Status and Trends of Desertification," *Environmental Conservation* 11, no. 2 (1984):103–13.

10. In one study made 15 years ago, semi-arid grasslands were estimated to contain about 10 percent of the world's population, and about 50 million people were thought to be affected in "lands threatened by desertification." R. W. Kates, D. L. Johnson, and K. Johnson Haring, "Population, Society and Desertification," in Secretariat of the UN Conference on Desertification, ed., *Desertification: Its Causes and Consequences* (Oxford, England: Pergamon Press, 1977), 261–318.

11. G. Ledec, "The Political Economy of Tropical Deforestation," in H. J. Leonard, ed., *Divesting Nature's Capital*, pages 179–226, note 3 above.

12. World Resources Institute and International Institute for Environment and Development, *World Resources 1988–89* (New York: Basic Books, 1988), 71; World Resources Institute, *World Resources 1990–91* (New York: Oxford University Press, 1990), 101–06; and N. Meyers, "Tropical Forests: Present Status and Future Outlook," *Climatic Change* 19 (1991):3–32.

13. World Resources Institute and International Institute for Environment and Development, page 71, note 7 above.

14. World Bank, *World Development Report 1990: Poverty—World Development Indicators* (New York: Oxford University Press, 1990).

15. World Resources Institute, page 123, table 8.1, note 12 above.
16. H. J. Leonard,"Environment and the Poor: Development Strategies for a Common Agenda," in H. J. Leonard, ed., *Environment and the Poor: Development Strategies for a Common Agenda,* pages 3–45, note 3 above.
17. J. W. Mellor, *Agricultural Development Opportunities for the 1990s* (Washington, D.C.: International Food Policy Research Institute, 1989).
18. World Bank, *Social Indicators of Development 1988* (Baltimore, Md.: Johns Hopkins University Press, 1988).
19. United Nations Conference on Trade and Development, *The Least Developed Countries: 1990 Report* (New York: United Nations, 1990); and World Bank, *World Development Report 1991: The Challenge of Development—World Development Indicators* (New York: Oxford University Press, 1991).
20. World Resources Institute, pages 286–87 and 306–07, note 12 above.
21. The authors are indebted to B. Bowonder, H. Brookfield, D. Campbell, T. Downing, G. Gallopin, J. Gritzner, M. Norowitz, N. Jodha, J. Mathews, D. Pimentel, P. Porter, and A. Rapp for taking time to direct us to the relevant literature and to C. Radel for helping the authors to find and digest it.
22. For case studies of drylands, see B. Beckman,"Bakolori: The Menace of a Dam,"*Development* 3 (1985):24–32; B. Bowonder and K. V. Ramana, *Environmental Degradation and Economic Development: A Case Study of a Marginally Productive Area* (Bella Vista, Hyderabad, India: Centre for Energy, Environment and Technology, Administrative Staff College of India, 1986); D. J. Campbell,"Land-Use Competition at the Margins of the Rangelands: An Issue in Development Strategies for Semiarid Areas,"in G. Norcliffe and T. Pinfold, eds., *Planning African Development* (Boulder, Colo.: Westview Press, 1981), 39–61; A. B. Durning, *Apartheid's Environmental Toll* (Washington, D.C.: Worldwatch Institute, 1990); U. J. Herren, "'Droughts Have Different Tails': Responses to Crises in Mukogodo Division, North Central Kenya, 1950s–1980s," *Disasters* 15, no. 2 (1991):93–107; M. M. Horowitz and M. Salem-Murdock,"The Political Economy of Desertification in White Nile Province, Sudan,"in P. D. Little and M. M. Horowitz, eds., *Lands at Risk in the Third World: Local-Level Perspectives* (Boulder, Colo.: Westview Press, 1987), 95–114; F. N. Ibrahim, "Ecology and Land Use Changes in the Semiarid Zone of the Sudan,"in Little and Horowitz, pages 213–29, note 3 above; N. S. Jodha,"Common Property Resources and Rural Poor in Dry Regions of India," *Economic Political Weekly* 21, no. 27 (1986):1169–81; N. S. Jodha,"A Case Study of the Degradation of Common Property Resources in India,"in Blaikie and Brookfield, pages 196–207, note 3 above; R. W. Kates, "Drought in the Sahel: Competing Views as to What Really Happened in 1910–14 and 1968–74,"*Mazingira* 5, no. 2 (1981):72–83; P. D. Little,"Land Use Conflicts in the Agricultural/Pastoral Borderlands: The Case of Kenya," in Little and Horowitz, pages 195–212, note 3 above; M. O'Leary, "Ecological Villains or Economic Victims: The Case of the Rendille of Northern Kenya,"*Desertification Control Bulletin* (United Nations Environment Programme), no 11 (1984):17–21; M. U. A. Tennakoon,"Desertification in the Dry Zone of Sri Lanka,"in R. L. Heathcote, ed., *Perception of Desertification* (Tokyo: United Nations University, 1980), 4–33; B. P. Thomas-Slayter and R. Ford, "Water, Soils, Food, and Rural Development: Examining Institutional Frameworks in Katheka Sublocation,"*Canadian Journal of African Studies* 23, no. 2 (1989):250–71; J. T. Thomson,"The Politics of Desertification in Marginal Environments: The Sahelian Case," in H. J. Leonard, ed., *Divesting Nature's Capital,* pages 227–62, note 3 above; M. Watts, "Drought, Environment and Food Security: Some Reflections on Peasants, Pastoralists and Commoditization in Dryland West Africa," in M. H. Glantz, ed., *Drought and Hunger in Africa* (New York: Cambridge University Press, 1987), 171–211; and B. Wisner,"Man-Made Famine in Eastern Kenya: The Interrelationship of Environment and Development," in P. O'Keefe and B. Wisner, eds., *Land Use and Development* (London: International African Institute, 1977),

194–215. Summaries of these case studies can be found in Kates and Haarmann, pages 41–57, note 4 above.

For case studies of tropical moist forests, see J. N. Anderson, "Lands at Risk, People at Risk: Perspectives on Tropical Forest Transformations in the Philippines," in Little and Horowitz, pages 249–67, note 3 above; W. H. Durham, *Scarcity and Survival in Central America: Ecological Origins of the Soccer War* (Stanford, Calif.: Stanford University Press, 1979), 62–101, 127–58; S. B. Hecht, A. B. Anderson, and P. May, "The Subsidy from Nature: Shifting Cultivation, Successional Palm Forests, and Rural Development," *Human Organization* 47, no. 1 (1988):25–35; Ledec, note 11 above; M. E. Lopez, "The Politics of Lands at Risk in a Philippine Frontier," in Little and Horowitz, pages 230–48; note 3 above; Monosowski, "Brazil's Tucurui Dam: Development at Environmental Cost," in E. Goldsmith and N. Hildyard, eds., *The Social and Environmental Effects of Large Dams*, vol. 2, *Case Studies* (Camelford: U.K.: Wadebridge Ecological Centre, 1986), 191–98; M. Painter, "Unequal Exchange: The Dynamics of Settler Impoverishment and Environmental Destruction in Lowland Bolivia," in Little and Horowitz, pages 164–91, note 3 above; L. Potter, "Degradation, Innovation, and Social Welfare in the Riam Kiwa Valley, Kalimantan, Indonesia," in Blaikie and Brookfield, pages 164–76, note 3 above; and S. C. Stonich, "The Dynamics of Social Processes and Environmental Destruction: A Central American Case Study," *Population and Development Review* 15, no. 2 (1989):269–96. Summaries of these case studies can be found in Kates and Haarmann, pages 62–70, note 4 above.

For case studies of highland regions, see P. Blaikie and H. Brookfield, "The Erosion Problem in Crowded Nepal: Crisis of Environment or Crisis of Explanation?" in Blaikie and Brookfield, pages 37–48, note 3 above; J. L. Collins, "Labor Scarcity and Ecological Change," in Little and Horowitz, pages 19–37, note 3 above; P. O'Keefe, B. Wisner, and A. Baird, "Kenyan Underdevelopment: A Case Study of Proletarianization," in P. O'Keefe and B. Wisner, Eds., *Land Use and Development* (London: International African Institute, 1977), 216–28; and D. Seddon, "Population Pressure and Environmental Deterioration," in D. Seddon, ed., *Nepal: A State of Poverty* (New Delhi: Vikas Publishing House, 1987), 86–117. Summaries of these case studies can be found in Kates and Haarmann, pages 58–61, note 4 above.

23. Sen, note 5 above.

24. See Blaikie and Brookfield's (note 3 above) useful definition of degradation "as a reduction in the capability of land to satisfy a particular use." This definition is, of course, anthropocentric because it uses human welfare as a measure. It is also highly variable because reductions in this biophysical capability of land will differ greatly according to the use of the resource. Blaikie and Brookfield suggest that degrading processes are both natural and human-induced and are subject to both natural restoration and human repair. They offer a measure of net degradation defined as (natural degrading processes + human interference) – (natural reproduction + restorative management).

25. See, for example, Durning, note 3 above; and O'Keefe, Wisner, and Baird, note 22 above.

26. These driving forces all appear in the various "boxes and arrows" of schematic models drawn by various authors, such as Leonard, note 16 above, and Blaikie and Brookfield, note 3 above. Authors of case studies tend to differ in their choice of which driving forces to emphasize and where in the spiral of poverty and degradation they choose to begin their analyses. Four emphases in particular stand out: the role of external, even very distant forces; the inequality of growth and modernization; the pressure of population; and the destruction of local and indigenous livelihood and knowledge systems. These, in turn, are linked to four traditions: neoliberal economics focused more on growth than equity; neo-Marxist analysis that minimizes the stress of population growth and emphasizes the effects of distant capitalists; a biocentric ecology that favors plants and animals rather than people; and a cultural ecology that tends to study small societies and to focus on indigenous knowledge.

27. See Beckman; Bowonder and Ramana; Horowitz and Salem-Murdock; Jodha; and Thomas-Slayter and Ford, all in note 22 above.

28. See Campbell; Durning; Herren; Little; Tennakoon; Wisner; Beckman; and Horowitz and Salem Murdock, all in note 22 above.

29. See Ibrahim; Kates; O'Leary; Thomson; Watts; and Herren, all in note 22 above.

30. See Bowonder and Ramana, note 22 above.

31. See Thomas-Slayter and Ford, note 22 above.

32. See Campbell; Little; and Wisner, note 22 above.

33. See Ibrahim; and O'Leary, note 22 above.

34. See Blaikie and Brookfield, note 3 above; Collins; O'Keefe, Wisner and Baird: and Seddon, note 22 above.

35. See O'Keefe, Wisner, and Baird, note 22 above.

36. See Blaikie and Brookfield, note 3 above; Collins; and Seddon, note 22 above.

37. See Blaikie and Brookfield, note 3 above; Seddon; Collins; and O'Keefe, Wisner, and Baird, note 22 above.

38. Ledec, page 182, note 11 above.

39. See Anderson; Hecht, Anderson, and May; Monosowski; and Stonich, note 22 above; and Ledec, note 11 above.

40. See Durham; Lopez; Painter; and Potter, note 22 above; and Ledec, note 11 above.

41. See Monosowski, note 22 above.

42. See, for example, J. A Gritzner, *The West African Sahel: Human Agency and Environmental Change* (Chicago: Committee on Geographical Studies, University of Chicago, 1988); A. Kiriro and C. Juma, eds., *Gaining Ground: Institutional Innovations in Land-Use Management in Kenya* Nairobi: ACTS Press, 1989); and World Resources Institute and International Institute for Environment and Development, pages 215–33, note 12 above.

PART 2

Looking at Land Use Patterns: Agriculture, Forestry, and Pastoralism

All energy on which life on earth is based derives from the sun. Plants are able to convert the solar energy through photosynthesis into storable form, and that stored energy is the primary link in the food chain. All animals are ultimately dependent on plants for food. Carnivores, lions for example, may feed entirely on the flesh of other animals, but those animals, zebras for example, are herbivores, feeding on grasses. Land and land use are central for humans, since the ultimate dependence for most human populations is on the plant produce of the soil. The study of plant food production in modern countries, such as the United States, has typically been done by agronomists, horticulturalists, foresters, etc., and not by anthropologists, who in turn have long recognized the central importance of plant cultivation and of pastoralism, the herding of livestock, by people in the lesser developed and non-Western areas of the world. As the world becomes more global and interconnected, the distinctions among who studies what becomes more blurred, and the issues confronting food production and use of forest products become more similar. The articles in this section address issues common to agriculture, forestry, and pastoralism in the United States, other areas of the developed world, and areas of the developing world.

The lead article in the first subsection, Issues and Overviews, is by Evan Eisenberg and is about a leading environmentalist in the United States in regard to agriculture, Wes Jackson. The article documents his ecological perspective on farming in the central plains of the United States. Jackson has been a pioneer in the argument for a naturalistic view of the prairie and its plant production capacity. He has demonstrated that in that environment what is crucial is the root system of the plants that both hold the topsoil in place and have fundamental importance in the maintenance of soil fertility. Yet the agricultural practices in the area by Euro-American settlers have typically taken, as the first step, the uprooting of the plants and disregard for the importance of native plants. Through his understanding of the natural principles of plant-soil interaction in the region, Jackson, however, has shown how a much more sustainable agriculture is possible.

The second article, by Diane Curtin, takes a critical stance toward the Green Revolution, which has been exported to many areas of the developing world following World War II. The Green Revolution, based on high-yield strains of food crops, was hailed as a major cure for the world's ills in relation to food production. For a variety of ecological and social reasons, the Green Revolution has fallen far short of its predictions, and in addition, it has increasingly been seen by some as a tool of exploitation of developing countries by those more developed. Curtin develops the argument that the Green Revolution is, in fact, a form of "war" on poorer populations of the world, especially on women who do most of the agricultural work in societies found in those areas.

Biotechnology has increasingly been hailed as a successor to the Green Revolution, taking the manipulation of genetics to greater lengths toward increased yields of crops for food. Just as the Green Revolution has become controversial, biotechnology has now come under scrutiny and criticism. The third article in this section, by Jack Kloppenburg and Beth Burrows, assesses biotechnology in terms of its place in agriculture, producing a substantial argument that it is wholly incompatible with sustainable agriculture. That article is followed by one looking at similar issues to the Green Revolution and biotechnology but in relation to tropical rain forests and deforestation. In that piece, Alastair Gunn looks principally at the ethical issue of who has the right to criticize deforestation and the use of forest products. The observation made by some developing countries is that the nations of the northern hemisphere, i.e., the wealthier ones, exploited their forests to degraded levels but now have the affrontry to tell nations of the southern hemisphere that they should not exploit but preserve their forests. The point of the essay is principally to raise environmental awareness at "home," in the northern hemisphere both for what has happened to forest resources there and what is not happening in developing areas of the world.

The second sub-section of this part of the Reader contains four case studies that probe in detail issues concerning land and land use patterns in particular societies. The first of those, by Francisco Pichón, looks at land use strategies in the Amazon Basin in the South American country of Ecuador. Based on a survey of small-scale farms in that area, he found that the limitations of the environment, i.e., soil conditions, climate, etc., were not as constraining as were the socioeconomic resources available to the farmers. In other words, the social and cultural components of the farming enterprise were not nearly as marginal as some scenarios would have it. The second article, by Dee Mack Williams, looks at the grasslands in Inner Mongolia and at the factors that lead to land degradation there. What he found was privatization of land, that is, conversion of land to individual ownership, leads to grassland degradation to a greater extent than cases in which community standards and functions take precedence. Secure land tenure and equitable access to community resources were found to be particularly important.

The two final articles are on land use patterns in relation to agroforestry and deforestation, both in East Africa. The article by Sara Scherr documents changes in land use patterns in an area near Lake Victoria in Kenya in which conservation ideas have taken some hold. Tree planting has led to highly desirable outcomes, such as increased provision of household food and enhanced food security. She shows that improvements in agroforestry practices derive from a variety of sources, including community standards and knowledge. The article by Solon Barraclough and Krishna Ghimire describes deforestation in Tanzania, showing that the relationship between deforestation and population level and associated poverty is not as direct as it might always seem. Local-level case studies in the country of Tanzania show that the relationships are more dynamic and are closely tied to social and cultural factors. In general, the more local-level the use of forest products, the more conservation is practiced, and conversely, the more world market and state policies come into play, the more exploitative and degrading agroforestry practices tend to be.

BACK TO EDEN

EVAN EISENBERG

The Atlantic Monthly, *November 1989. Agriculture as we know it, Wes Jackson argues, is destroying our topsoil and polluting our environment. Human survival, he maintains, may depend on our learning to use the secrets of the prairie.*

Although agriculture has given us cities, wealth, the arts and sciences—what we call civilization—on the whole it has been a mistake. That is Wes Jackson's view. He takes a long view, stretching back to the moment some 400 million years ago when a bony planet began to clothe itself in soil. That process, which allowed intricate forms of life to wriggle up out of the oceans and plant themselves on land, began to reverse itself about ten thousand years ago, with the arrival of a form of life called the farmer. It is reversing now at a roller-coaster, sick-in-the-pit-of-the-stomach rate, as if a film we had been watching all day were run backward in seconds. If most of us do not feel sick, perhaps that is because we are not watching.

From a hillock on the Kansas prairie Wes Jackson *is* watching, and what he sees, and says, and does, makes him the most radical of America's agricultural prophets. While most of the others save their thunderbolts for agribusiness, having mainly kind words for traditional farming. Jackson attacks all agriculture as we know it. This business, he says, of scratching the earth and sticking annual seeds in it is a recent notion, and on balance a bad one. Except under the most favorable conditions of climate, topography, and virtue, it has involved a rapid squandering of the earth's hardest-won resource: soil. As Jackson puts it—one prophet taking a swipe at another—"the plowshare may well have destroyed more options for future generations than the sword."

Jackson is not a madman. He is not a flatland druid who would sacrifice human beings to tall grass; he does not ask that ninety-nine out of a hundred of us volunteer to starve while the others revert to hunting and gathering. What he is, is harder to say. I am tempted to use the language of the discipline he was trained in—genetics. If you crossed Isaiah with a bison, you might get something like Wes Jackson. If you crossed Charles Darwin with William Jennings Bryan, splicing in a few choice genes from Will Rogers, you might come even closer. If you crossed a U.S. Department of Agriculture experiment station with Gandhi's Tolstoy Farm, you might get something resembling the Land Institute, which is the outfit in Salina, Kansas, that Jackson and his wife, Dana, direct.

Now, suppose you crossed a wheat field with a prairie, and a product flourished: you would have done what Wes Jackson and the Land Institute have been trying to do for twelve years and will keep trying to do for another hundred, if necessary.

The job may take that long. Although wheat field and prairie are both, botanically, mainly grass, they are otherwise as dissimilar as any two things under the sun. The typical wheat field is tilled several times a year; by definition, the native prairie has never been tilled at all. While the wheat field wastes topsoil, the prairie creates it. Heat, frost, and drought often catch the wheat field flat-footed; the prairie, which has seen all these things before, takes them in stride. Left to its own devices, the wheat field is next to defenseless against pests and pathogens that are naturally kept in check on the prairie. Like the wheat field, the prairie is a sprawling factory for turning sunlight into fiber, starch, fat, and protein; but while the prairie relies on today's sunlight, the wheat field lives on mummified sunlight in the form of fertilizers and pesticides derived from oil and gas. The prairie lives on income, the wheat field on capital. What is more, the prairie saves.

From the point of view of the ecologist, the prairie is doing everything right. From the point of view of the hungry ecologist, it has a serious flaw. What Jackson has in mind is a prairie that can feed the world.

Jackson likes to tell the story of the old Sioux Indian who watched a pioneer plowing up the prairie sod, examined the furrow, and said, "Wrong side up." Traditionally, the joke is on the Indian.

A century after the Indian's alleged remark, a third of our farmland topsoil, accumulated over millennia, is gone. For every bushel of corn that Iowa grows, it sheds at least two bushels of soil. Farm runoff is causing algal bloom in our lakes and is poisoning our drinking water. Nitrogen fertilizers may be helping to thin the ozone layer. In 1948, at the dawn of the chemical age, American farmers used 15 million pounds of insecticides and lost seven percent of their crops to insects; today they use 125 million pounds and lose 13 percent. Because most agricultural chemicals are made from fossil fuels, we expend three calories of energy to produce each calorie of food we eat. (If processing and transportation are included, ten calories are expended for each one we consume.) In contrast, Tsembaga farmers in highland New Guinea, using Neolithic slash-and-burn methods, expend less than a tenth of a calorie for each calorie they eat. For each American man, woman, and child, our agriculture ingests 160 pounds of nitrogen, phosphate, and potash fertilizer each year, and 325 gallons of water each *day*. The Ogallala aquifer of the Great Plains, which starts just west of Salina, held enough water forty years ago to fill Lake Huron; in another twenty years it may be too low to pump. The great dams of the West are silting up. Sloppy irrigation and poor drainage (worsened by the compaction that heavy machinery causes) have caused saline or sodic conditions on 10 percent of our crop and pasture land. Should we escape the fate of the ancient empires that were done in by erosion, we may instead face that of the Sumerians, whose civilization crumbled as their soil turned salty.

Many people have noticed these problems and offered piecemeal solutions. Jackson thinks the only real solution is to turn agriculture on its head. The field of waving grain he sees in his mind's eye would be plowed maybe once in five years. It would be a mixture, or polyculture, of three or four crops nourishing to human beings or livestock. Like many existing crops, these would yield at least 1,800 pounds of edible seed per acre. Unlike existing crops but like most prairie plants, they would be perennials. Their roots, undisturbed for years at a stretch, would hold the soil in place, and would become adept at finding moisture and sustenance. The only chemicals applied would be those nature applies to the prairie. Agronomists would call this field a herbaceous seed-bearing perennial polyculture. With Jackson's permission, the rest of us would call it a domestic prairie. Presumably the old Sioux Indian would call it "right side up."

Visiting the Land Institute

Jackson got the idea shortly after his classroom burned down. The Land Institute, dedicated to finding "sustainable alternatives in agriculture, energy, waste management, and shelter," had enrolled its first class of students in September of 1976. The fire came in October. Instead of going home, the students decided that fate had handed them their first assignment in sustainable shelter. Elbow to elbow with the Jacksons, they built a new, solar-heated classroom from scratch. The institute's modest library having been destroyed, they pooled their beer money and bought a dictionary. Like prairie scoured by fire, the school regenerated from its roots. By Christmas of 1977 it had completed a structure sound enough to meet in, along with an office sound enough for Jackson to sit and think in.

The fire had put the fear of God in him. In a world where you have to run as fast as you can just to stay in place, the fire said, you had best know where you are going. He jotted down a list of the world's infirmities. Of those he thought he might be able to do something about, soil erosion came out on top. But what unique contribution could the Land Institute make to a problem that had plagued farming since farming began?

That fall Jackson had taken his students to an 8,616-acre classroom called the Konza Prairie, near Manhattan, Kansas. Now, chewing over in his mind's maw the tough, springy, soil-mongering vegetative structure of the tall-grass prairie, he began wondering where it fit in the grand scheme of the plant kingdom as colonized by man. He sketched out a chart dividing that kingdom according to four dichotomies: monoculture or polyculture, woody or herbaceous, annual or perennial, grown for vegetative matter (leaves, stems, wood) or grown for fruit and seed. Four of the sixteen possible combinations described woody annuals, the botanical counterparts of griffins and manticores. That left twelve combinations—of which eleven had some embodiment in farming or forestry.

Like a Count Basie piano solo, the chart was most intriguing where it held back. Just above the line describing the prairie and other mixed pastures ("polyculture, herbaceous, perennial, vegetative") was another line ("polyculture, herbaceous, perennial, fruit/seed") that had a thundering blank where its embodiment in farming should have been. The blank was intriguing because the change in a single parameter—from "vegetative" to "fruit/seed"—made a great difference.

I said before that from the point of view of the hungry ecologist, the prairie has a serious flaw. Human beings have been seed-eaters for many thousands of years, perhaps since our long morning on the African savanna. Seeds, meant by nature to provision an infant plant, are concentrated pellets of energy and protein. Meant to stay intact until the signals are right for germination, they are readily traded, transported, and stored against winters and lean years. The basis of human proliferation is not our own seed but the seed of grasses. Corn, wheat, and rice produce plenty of seed, big bluestem and switch-grass precious little. Quadrupeds can live on the stems and leaves of prairie grasses, and bipeds can live on the quadrupeds, but that arrangement leaves bipeds perched on a high, narrow ledge of the trophic pyramid. We are now too plentiful to fit. We need bread, brought forth directly, as the Hebrew blessing has it, "from the earth."

Could a new kind of prairie bring forth that bread? Jackson had every reason to think the answer was no. A plant has only so much energy to spend. Unlike a national economy, it can't run a deficit; when it has earned what it can from the sun, it has no place left from which to borrow (in nature, that is: with man as broker, a plant can get a loan from long-dead plants, in the form of fertilizer). In view of this hard fact, two main schools of economic thought have arisen among flowering plants. One advises throwing most of one's energy into seeds; the other gives precedence to roots. Plants adopting the seed strategy won't live long—most will be annuals—but they may, with luck, have a lot of progeny. Plants with strong root systems will generally be perennials, able to survive drought, wind, frost, even fire; often they will reproduce, slowly but steadily, by budding from creeping rhizomes (underground stems) or roots. This is good policy on the prairie, where conditions can be extreme and competition tough.

On a level evolutionary playing field, untampered with by humans beings, seed strategists don't do well under prairie conditions. But seed strategists are opportunists, always on the lookout for disturbed soil and open patches. Such were the weeds that dogged the heels of early human beings, and some of them were domesticated as grains. By paying human beings to husband their seeds and plant them each year in soil disturbed by the plow—in effect, giving us a commission—they have displaced root strategists from many prairie zones, including the Great Plains. Unluckily, the soil disturbance these seeds thrive on has led to loss—at times cataclysmic—of the soil they and we depend on.

The great virtues of the prairie are the virtues of its roots. Breed the prairie grasses for seed, and you lose those good roots. What is gained at one end is lost at the other, and your high-yielding perennial turns out to be an annual.

Things might turn out that way, Jackson allowed. But biology is not a zero-sum game. Sometimes an organism or a collocation of organisms does manage, by dint of greater efficiency or a nimbler juggling of resources and needs, to beat the system. With the stakes so high, it was worth finding out if the system could be beaten in this case.

I first came across Wes Jackson in the pages of *Manas*, a journal published weekly, anonymously, and more or less single-handedly from 1948 until the end of 1988 by a man in Los Angeles named Henry Geiger. (He died two months after the last issue appeared.) In a style like white marble Geiger wrote about his heroes. A handful of names kept recurring: Socrates, Gautama Buddha, Thoreau, Gandhi, Abraham Maslow, Wendell Berry, Wes Jackson. I liked most of these people well enough to want to investigate the others.

In June of 1987 my wife and I attended the Ninth Annual Prairie Festival at the Land Institute. We walked the institute's native prairie—long, loping swells, frothy with wildflowers—with a botanist from the University of Kansas, a former Land student, who knew the names of all the plants and what the Indians had done with them. We toured test plots of Illinois bundleflower. Maximilian sunflower, eastern gama grass, levmus, milo, and Johnson grass in assorted crosses and polycultures, each plot interpreted with humor and assurance by the intern in charge of it. We sat through symposia and stumbled through square dances in a big graying barn made from salvaged railroad ties. The fiddler would play the tune slowly until everyone had learned the steps, and then floor the pedal. At the last symposium, the grand finale, a panel that included two eminent ecological historians, Jackson's friends Donald Worster and Angus Wright, talked about how the Protestant ethic of work and production was exhausting the earth. The audience joined in their lament. After a while a big man who had been slumped like a giant marionette on a ledge off to the side of the barn jumped up, came over to the folding table the panelists were sitting at, and said,"Well, I think this nonsense has gone on long enough."He gave a gap-toothed grin."When the oil runs out, we're going to need the work ethic more than ever."

For a moment only the barn swallows up in the rafters responded. Then came bursts of startled, illuminated laughter.

The big gap-toothed man was Wes Jackson.

If you cut out the contiguous United States and balanced it on a finger, you would be pointing to the Land Institute. The country's center of gravity is here, and you can feel it. There is a sense of poise and deliberateness, something in the air that counsels against rashness.

We came back the next year at a bad time, well into July. Trouble, in the shape of heat and drought, had already arrived. Each day the digital thermometer across the parking lot from our motel exercised all three digits. Salina is a town of great decorum, with limestone churches on every corner and trees artfully pruned to disguise tornado damage. At Salina's edge Interstate 135 runs south across an asphalt bottomland, shopping malls clinging to it like cottonwoods to a creek. We turned off it onto a gravel road that cut through fields of winter wheat harvested the month before and fields of sorghum just coming up. A tractor disking in stubble cast a shadow of dust on the sky.

A sign carved in wood marked an archipelago of small wooden buildings in a sea of shrubs, vines, vegetables, and flowers. Shipshape and stained a self-effacing siena, the buildings looked homemade. To the west paths led through thickets of burr oak to a bluff overlooking the Smokey Hill River. The field to the east, which the previous spring had been a fairground dense with tents, was empty, but the field beyond it was dense with oats. On the other side of the road, to the north, the prairie that a month ago must have been spangled with purple cone-flower, plains larkspur, penstemon, and wild indigo was monochrome, save for the occasional yellow silphium dawdling like the last string of balloons after a party. The Land Institute sits on the western fringe of the tallgrass-prairie zone. The midgrass starts here, and the shortgrass of the High Plains starts a hundred miles to the west. The tallgrass prairie alone once mantled 400,000 square miles; now it is patches and tatters. The institute has some native prairie and some reclaimed from the plow, which it manages by spring burning every few years. At the time of my second visit so little rain had fallen that even the big bluestem was small. The sod was shut like a fist. It seemed to be husbanding its strength, making sure the soil did not slip through its fingers.

We found Dana in her office next to the classroom, and Dana tracked down Wes. If Wes looks like the giant troll who knocks over trees when he walks, Dana looks like the elf who sets up housekeeping in the stumps. This is a function not of her size, which is average, but of a deftness in the turn of her nose and the cut of her chin, an amused light in her eyes. There is nothing deft about the cut of Wes's features. They do not look cut; they do not even look hewn. They look found. You can read the history of Kansas in his features, which are as rough as cottonwood limestone, the blocky stuff the settlers built their houses from when they got tired of living in sod dugouts.

Jackson took us on a tour of the year's experiments. Since the institute's 205 acres had just been augmented by a seventy-two-acre field about two miles down the road, he decided to drive us around. For all his steeping in the small-is-beautiful ethic, he is still farm boy enough to be proud of having a spread too big to walk. He does not salaam to technology, however; this can be seen in the way he treats his car. The landscape of north-central Kansas is flat, but not that flat. He stampeded his old Subaru over rutted prairie, bouncing from experimental plot to plot; because it was lunchtime, no intern was in danger of being plowed under. Braking beside a stand of *Tripsacum dactyloides*, or eastern gama grass, a tall prairie bunch-grass that looks like a kind of oversexed wheat but is actually a cousin of corn, Jackson jumped out and rummaged among the seed heads. Shortly he displayed in one large hand a normal seed head, in the other a sport: the pistillate or gynomonoecious form, a recently discovered mutant that has four times as many seeds as garden-variety eastern gama grass, most of them in soft, loose glumes instead of hard, tight hulls. The pistillate form may be the best break that Jackson's team has had in twelve years of work (though, as Jackson is quick to point out, Robert Dayton, then the head of the USDA Plant Materials Center at Manhattan, Kansas, is the man who found it). It has even furnished Jackson's daughter Laura with the germ of a doctoral thesis at Cornell.

In another plot shorter, weedy plants that branched like poplars were growing in buried pots. They were the grandchildren of a cross between milo (*Sorghum bicolor*), a common grain sorghum that is not winter-hardy outside the tropics and is grown as an annual, and Johnson grass (*Sorghum halepense*), a weed that is perennial with a vengeance. Johnson grass has crab-claw rhizomes that spread it like a nasty rumor. It is tolerated by ranchers (it makes decent pasture so long as it isn't grazed in hot summers or after the first frost, when the prussic acid in it can kill livestock) and abominated by farmers. On occasion farms overrun by it have been abandoned. Here its toughness was being enlisted in the effort to breed a perennial grain sorghum. Because these plants were still on the weedy side—throwing out too many rhizomes and dispersing their seed too readily—they were being backcrossed with milo, the cultivated parent. The pots were good agronomic practice and would prevent a lynching by local farmers.

In one big plot three species could be distinguished. One was eastern gama grass. Another, even taller, was giant wild rye (*Leymus racemosus*), a bunchgrass with huge flowering spikes. It was a long way from home. A native of Russia—its seeds were eaten in drought years, when the wheat failed—it was among 4,300 accessions from the USDA Plant Introduction Station in Pullman, Washington, from which candidates were chosen for the institute's breeding program. The third plant was not a grass but a legume—a relative of beans, alfalfa, and locust trees—with frilly leaves like a locust or a mimosa. Like many legumes, Illinois bundle-flower (*Desmanthus illinoensis*) is high in protein; it is a favorite of livestock. Also like many legumes, it has rhizobium bacteria on its roots which "fix" nitrogen from the air, changing it into a form that plants can use. Its first-year yields in monoculture had approached the standard of 1,800 pounds per acre set by middling Kansas wheat.

So far the trio had cohabited happily enough, though of the two grasses, only the gama seemed to be getting any extra nitrogen from the deal. Because each crop ripens in a different month, farmers would be able to harvest the three separately by machine. But even hand harvesting would be impracticable unless the plants could be bred, like other cultivated grains, to be shatter-resistant—that is, to keep their seed heads intact until harvest, instead of letting the heads burst and expel their seed, as wild plants do.

What I saw on that drive, and on my walks later, was interesting. So was what I did not see. When Jackson first imagined his polyculture, he wanted to model it on the prairie's own triumvirate of grass, legume, and composite. Leymus, Illinois bundleflower, and Maximilian sunflower seemed to be a combination that was worth checking out. The grass and the legume would provide starch and protein, the composite oil. The legume would fix nitrogen, while allelopathic chemicals released (some scientists believed) by the sunflower would keep weeds at bay.

Like most ideas in plant breeding, this did not work. The leymus never quite got a foothold. A monoculture of Maximilian sunflower was not encouraging either; by the third year it was yielding hardly enough seeds to satisfy a mourning dove. As things stood, the entire family of Compositae was in disgrace. The favored triculture now was leymus, eastern gama grass, and bundleflower. which could be said to mimic the prairie's mix of cool-season grasses, warm-season grasses, and legumes. But Jackson was not ready to give up on the sunflower ingredient of his original mix. "We didn't know how much variation was in it." he told me later, referring to genetic variation within the species, which would be revealed by collecting and growing more strains. "We didn't have the horsepower."

The horses at the Land Institute wear shorts and T-shirts. Some wear feed caps. Others wear wide-brimmed, flat-topped straw hats with black ribbons made by Gohn Brothers, Amish and Plain Clothing, Middlebury, Indiana. They are frisky but pull like Belgians. Paid small stipends, they take rooms in Salina and carpool or cycle to the Land. At noon they forage in the garden (not in the research plots) and cook up communal ratatouilles. They eat at a big table in the classroom, which is also the library: current issues of *BioScience, Bulletin of the Atomic Scientists, Kansas Business News, The Furrow, Agrichemical Age, Conservation Biology, High Country News,* and *Small Farmer's Journal*, among other publications, are handy for browsing. But mostly they talk. They start classes in February, working in the greenhouse only in the afternoons, but by summer they have traded their books for spading forks. Each has taken on a project from an agenda drawn up by Jackson and his staff. One helps Jon Piper, the Land's ecologist, monitor the species mix of the native prairie, in hopes of uncovering its polycultural secrets. Another works with Peter Kulakow, the chief plant breeder, on the sorghum-breeding program. A third takes charge of the triculture. A fourth, issued a butterfly net and specimen pins, studies insect populations on the prairie and in the plots. A fifth works with Mary Handley, the Land's plant pathologist and Kulakow's wife, to culture and identify pathogens. Eight to ten interns work at the Land Institute in a typical year. In the fall they will take turns planting their data in a desktop computer. Before they leave, in December, sturdy articles will be harvested.

Since the first edition of Jackson's book *New Roots for Agriculture* appeared, in 1980, a few people not at the Land have begun working toward the domestic prairie, among them Peggy Wagoner, at the Rodale Research Center, and a number of Land alumni at the Kerr Center for Sustainable Agriculture, in Poteau, Oklahoma. But most breeders with an interest in prairie plants are working on forage and feed crops. It is a lot easier to breed perennials for livestock than for human beings, because livestock can live on the leaves and other fibrous parts, while people need the seeds. Also, people are pickier. It is a measure of Jackson's ambition that he wants to change the human diet, and at its very kernel. He is going straight for the staff of life. While acknowledging that wheat and other annuals will always find a niche on the most level, least erodible land, he wants to replace them with perennials everywhere else.

Thom Leonard, an ex-baker who runs the Grain Exchange—a cabal, based at the Land, to save and spread the seeds of rare varieties of wheat, corn, barley, and so on—thinks perennial grains might sneak into the supermarket as secondary ingredients in breads and breakfast cereals, as amaranth (a high-protein annual) has done. "You do into the grocery stores here in Salina, you can buy a loaf of amaranth bread that has five or ten percent amaranth in a wheat stock," he told me. "I think that's probably a realistic way of beginning to incorporate it. The bread grains we know are pretty unique, and I don't think in terms of protein structure either eastern gama grass or leymus would make the kind of bread we're used to having."

Nevertheless, a couple of years ago some of the interns substituted eastern gama grass for its tamer cousin in a cornbread recipe. How was it? "It was great," Leonard said, "I could eat it every day." He smiled. He paused. "It was pretty good."

But prairie rolls and prairie flakes are as yet fantasy. For that matter, so is the domestic prairie. Aware that Americans don't like to wait decades for results, Jackson admits he'd like to have something to show in the meantime. Eastern gama grass, planted as a perennial monoculture, would make a good forage crop and a good first step. This is a goal nonvisionary enough to have attracted public funds. At the USDA Southern Plains Range Research Station, in woodward, Oklahoma, breeders are working (as Jackson is) to improve seed harvest and germination, which is naturally slack in a prairie grass that propagates by means of rhizomes rather than seeds. The discovery of a seed-rich mutant was a big step forward. Another promising forage crop for perennial monoculture is winter-hardy grain sorghum, and again public funds have been enlisted; people at the Land are collaborating with people at Kansas State. The good news here is how lustily milo miscegenates with Johnson grass. "But if we are content to stop there," Jackson told me, "we will have missed the point, see. We've got to keep in mind the domestic-prairie notion."

The Education of a Maverick

When Wes Jackson was in college, he thought of grass, when he thought of it at all, as something for the fullback to sit down on when robbed of the support of his legs. He played defensive and offensive guard for Kansas Wesleyan, starting on championship teams in his junior and senior years. "I've just been a long time growing up," he told me. He was grown up enough by his senior year to marry Dana Percival, an English major from Abilene whose grandfather had driven cattle over the Chisholm Trail and whose father was born in a dugout built of prairie sod. (Their prairie pedigrees were well matched: Wes's great-grandfather, the second white man in Douglas County, rode with John Brown at Black Jack Creek.) After a few years coaching and teaching, first at a Kansas high school, then at Kansas Wesleyan. Jackson found his attention beginning to stray from scrimmage lines to breeding lines. Despite a childhood steeped in Scripture, he had long had an interest in evolution. "I tell you what I believed firmly: I hung on tight to 'Ye shall know the truth and the truth shall set you free.'" In 1964 he went on to North Carolina State for a doctorate in genetics.

Some interesting things were going on at N.C. State. Researchers were looking at a quantitative gene in corn that might offer perpetual high yields from open-pollinated seed. Instead of buying hybrid seed each spring, farmers would be able to do as their fathers had done: save seed corn from each harvest to plant the following year. Strange to relate, this marvel has yet to reach the farmers' hands. Richard Lewontin, a population geneticist now at Harvard, formerly at N.C. State, has suggested that the research has languished because seed companies—for obvious reasons—are not interested in funding it. Jackson tends to agree. In any case, he learned while in grad school that in the realm of science, truth alone is not enough to make you free. You also need funding.

Jackson learned some other things, most of them from his major professor, Ben Smith. Smith would wander into Jackson's corner of the lab late at night, sweetening the air with pipe smoke and tales of N. I. Vavilov. Vavilov, the titan of Soviet genetics, was perhaps the first scientist to see cultivated plants as products of natural as well as human selection. Before running afoul of T. D. Lysenko, in the thirties, and dying in a prison on the Volga, he set up research stations where eye-poppingly wide crosses were made between wild plants and domestic, domestic and domestic, wild and wild. Long after the smell of tobacco had left his clothes, Jackson remembered those stories. He remembered, too, an off-the-cuff comment of Smith's that went something like this: "Wilderness is the standard against which agriculture should be judged."

Jackson went back to teaching and coaching at Kansas Wesleyan. There he and Dana got caught up in the swim of the environmental movement. Having been raised, as Dana says, "Kansan, Republican, Protestant, a real category," they could hardly swallow the sixties whole: "Wes marched against the Vietnam War, but we were never long-haired hippies." Much in sixties thinking, however, jibed with their Kansan, Republican (circa 1858), Protestant idealism and Wes's biological perspective. His textbook, *Man and the Environment*, an anthology that drew on historical, literary, and scientific sources, sold smartly. He was invited to found an environmental-studies program at California State University in Sacramento. At the age of thirty-seven he became a full professor: a perfect specimen of *Homo academicus*, preserved in tenure.

As the gel began to harden in the Central Valley sun, the Jacksons got fidgety. Did they really want to see their three children growing up and themselves growing old on the freeways, amid the alien tomatoes? Could Wes remain happy when half his view of the world was blocked by a blackboard? He took an unpaid leave, and they packed up the wagon and went back to Kansas.

As soon as they arrived, their toes started rooting. Shortly before the move to California they had bought three acres on the Smoky Hill River, just south of Salina, and built a house "out of junk," burrowed into a slope like the dugout Dana's father grew up in. Here they now settled down. Dana planted a garden: "We had three kids and no money, so the garden was serious business for me." The kept chickens and a cow. While the life they lived—self-sufficient, close to the land, sparing of resources and of energy not their own—was dictated mainly by thrift, it clicked with the spirit of the age. The young people whose minds had been exercised by Wes's lectures had bodies as well. Why not start a school that would exercise both, giving the students a chance to practice what was preached to them? But here no one would preach, as this would be less a school than a place of joint exploring for "sustainable alternatives in agriculture, energy, waste management, and shelter."

In September of 1976, having raised $16,000 with the help of a Salina lawyer and built a classroom (also of junk), the Land Institute enrolled its first class. In October the classroom turned to ashes, and a year or so later Jackson started thinking hard about dust.

The War Against Erosion

In the days of the dust bowl, soil erosion was hard to miss. A midsize grass, wheat, had been planted on shortgrass prairie as if this were the most natural thing in the world. The Great War had cut off wheat exports from Russia to a devastated Europe, and America had tried to meet the demand by plowing up as much of its own steppe as possible. When drought came, as sooner or later it would, nature's recoil was spectacular. The midgrass was rejected like alien tissue, and the soil went with it. As the sky grew dark over the Capitol dome, even manicured senators had to take notice. The Soil Conservation Service arose and declared a "war without guns" against erosion. Led by the eminent scientist Hugh Bennett, the SCS may have been the most spirited and enlightened agency in the history of the republic. It had the support of an extraordinary private group called the Friends of the Land in which such men as Aldo Leopold, Paul Sears, Gifford Pinchot, E. B. White, and Liberty Hyde Bailey had a hand. Writers, publicists, and filmmakers were enlisted to get the message out. Walking the land with big farmers and small farmers—five million of them, by the time Bennett retired—SCS agents helped them to find the best ways to safeguard their soil. Windrows of trees, contour plowing, and strip cropping did much to lay the ghost of the Dust Bowl. By the 1950s yields were so high that erosion was assumed to be under control. The Friends of the Land disbanded and the SCS relaxed. By the early 1970s, with the world grain market bullish, President Richard Nixon's Secretary of Agriculture, Earl Butz, was exhorting farmers to "produce fully" on "every harvestable acre."

Today the soil is going faster than it was in Dust Bowl days (about 25 percent faster, Jackson estimates, though other experts say methods of assessment have changed so much that comparison is pointless). But it is going more quietly. No avenging clouds swirl over Washington (at least not yet: the current drought has produced some of the worst dust storms in decades). As has usually been the case, most of the emigrating soil is taking the water route. Ironically, Jackson told me, water erosion is actually worse. Wind, in most instances, just moves the soil around; rain takes it and loses it in rivers (where it becomes sediment and the nutrients in it become pollution) or buries it at sea. The government is not ignoring soil erosion. Among its favorite remedies is "conservation tillage," or "low till," which means cultivating less and spraying more: the weeds that escape the tractor get blasted with Lasso or 2,4-D. To Jackson, the Department of Agriculture's logic seems worthy of the Pentagon: "We had to poison the soil to save it."

Closer to the mark is the three-year-old Conservation Reserve Program, which pays farmers to keep vulnerable slopes in grass or trees, ungrazed and uncut. (If the seed-production problem can be solved, Jackson thinks eastern gama grass would be an excellent grass to keep them in.) At a cost of billions, the program aims to take 40 to 45 million acres out of production: "A drop in the bucket, when you consider that we have 400 million acres of till land."

It is, admittedly, an important drop. One of the crucial findings of the first systematic survey of soil erosion in the United States, the National Resource Inventory of 1977, was that half of all sheet, rill, and wind erosion on nonirrigated cropland was taking place on roughly one tenth of that land. Today, with more than 28 million acres signed up, the Conservation Reserve Program is estimated by the USDA to be reducing cropland erosion by more than 574 million tons a year, from a previous annual total of some 2.6 billion tons.

The soil being saved is not, Jackson notes, the nation's best. Some of it is just subsoil, the topsoil having absconded long ago. But Jackson's real objection is crankier. "It's taking out the most highly erodible land. Now, intuitively that sounds right. But consider this as a thought experiment. Let's say that we were going to take out of production the *best* land. And this may be wrong, but I'd like to hold it as an item of discussion: Good land makes poor farmers. Poor land makes good farmers. In other words, in the areas of the world where the agricultural base, the resources, are marginal and difficult, like the Andes, or hard won, such as in Holland, there people take care of the land. Illinois is relatively flat, and has good soils and so on, and that's where they're drenching the place with chemicals and you know they just treat it like dirt. Treat the soil like dirt.

"So let's say we were to take Illinois and Iowa out of production and turn the farmers there into park rangers. Again, this is a thought experiment. Now we have some of the more marginal places, where we would be forced to take care of it better, and we would be forced into rotation, maybe a little more contour farming. Let's say that we were going to take a hundred million acres of the top-quality land out of production, leaving three hundred million acres. Now, that would be the statement of the nation that we are no longer interested in a cheap-food policy." There are better ways of making sure hungry people get fed, Jackson thinks. "Why should the people be allowed to buy as many VCRs and new cars as they do at the expense of the American landscape? I just don't believe in a cheap-food policy. It hurts landscapes and farmers and the culture at large."

Jackson is only half in jest; he can propose this kind of thought experiment because, he says, the system has "a lot of slack" in it. For the same reason, he can propose transforming much of America's cropland into domestic prairie without imagining that his prairie will be as prolific as a corn field. "We have a lot of good land in this country. A few years ago eighty-eight percent of all agricultural vegetable matter was going into the cattle and pig welfare program." In any case,

Jackson says, conventional field crops will produce high yields only as long as the soil under them and the oil pumped into them hold out. In a century or two the figures could look different. Jackson's tortoise could be chugging along while the agribusiness hare lies gasping.

Some think the hare is about to get a whole new lease on life. A great deal of money, enough to float hundreds of Land Institutes, is being invested in a discipline promising technical fixes for all the problems of agriculture, from gullies to weevils. On the subject of genetic engineering (or any other subject, but this one in particular) Jackson is not reticent. It is for him the Great Satan, the homuncular incarnation of everything that is wrong with science, agriculture, and the modern world. Like most environmentalists, he worries about an experiment gone wrong, some recombinant golem from whose forehead the aleph can not be erased. "Ninety-nine of these experiments will be harmless," he says, "and then one will be in the category of ozone destruction." What worries him even more, though, is what might happen if everything goes according to plan.

For one thing, the imperious infant is sucking up all the funding, so that not only mavericks like Jackson, who would naturally be underfunded anyway ("That's how I think you know you're in the right camp: if you're underfunded," Jackson says), but also mainstream plant ecologists and old-fashioned plant breeders are left high and dry. More than self-interest or sentiment moves Jackson: this is no mere turf fight. The genetic engineers, Jackson writes in his most recent book, *Altars of Unhewn Stone*, "are offering the 'specific problem-specific solution' approach as the infallible recipe. This approach assumes that everything outside the specific problem for which they intend to splice in a solution can be held still, that nothing else will wobble; or if it does, that they can splice in a correction for that, too."

Not only interactions between the plant and its environment but even interactions between genes get left on the cutting-room floor. In the real world, however, most traits are the joint handiwork of several genes, including regulator genes (which tell the others when and whether to kick in, and suppress vestigial genes—so that, for example, hens don't grow teeth) and episodal genes (which kick in at a strategic juncture to produce, say, a jolt of a certain growth hormone). After a splice has been sewn in, Jackson writes, "an untold amount of breeding work remains before the genetic background is shaken down enough to accommodate the newly introduced gene and its regulators." And this is simply to make the gene fit the plant. How the altered plant will fit its environment is another matter.

No doubt biotechnology will solve some problems. Given the cut-and-paste approach and the interests that are paying for the scissors, they are likely to be the wrong problems, solved in a Pyrrhic or at best Procrustean way. If a crop is threatened by weeds, the easiest thing to splice in is not the ability to repel them allelopathically but resistance to a herbicide that kills the weeds; that way farmers can spray at will. Such resistance is also the profitable thing to splice in if, as often happens, the company funding the research also makes the herbicide.

"All of this," Jackson writes, "is high tech research and I suspect that any outfit that gives you a crop with a spliced-in gene is going to demand a patent and some kind of royalty payment. It is doubtful that their primary concern will be the high energy cost of American agriculture. One also doubts that they will care greatly about the national and global soil loss problem."

What they are likely to care about is bigger yields—purchased, in the grand manner of the Green Revolution, at great cost to fuel, soil, water, economic equity, and genetic diversity. The cost will be greatest in the Third World.

"When you take the agricultural technology out of the developed world," Jackson told me one afternoon in his office, "it's coming out of a fossil-fuel-intensive infrastructure. You parachute it into the Third World, and if it's to be successful, you've got to build that fossil-fuel-intensive infrastructure there, too. On the other hand, if you have an ecosystem approach, one thing we know is that whether we're talking about some plant growing near Point Barrow, Alaska, or a plant growing in the Amazon Basin, they all follow the same laws of evolution and natural selection. The principles of agro-ecology, I think, should hold in Chad, in the Soviet Union, and on farms in north-eastern Ohio."

Jackson is not dead set against biotechnology now and forever. Plainly it offers some tempting shortcuts to the grain-bearing prairie that he sees in the distance. If a genetic engineer drove up and handed him a fine specimen of eastern gama grass with a spliced-in gene for shatter resistance, he would not slam the door. He would need some time to think about it. In fact, he believes the nation needs some time to think about the whole business.

In a paper that is as yet unpublished Jackson lists nine categories of ecological trouble that biotechnology might unleash. For each category, he told me, "here is the claim on the part of the technologist that says, 'This isn't gonna hurt,' and then there is what I call the citizen's response. I gave that paper up in Minnesota. I was very careful to lay out all of this, and then at the end I said, 'Now, look, you people worked very hard to follow me, but you couldn't stay with me, could you?'"

The question spun in the air between us once, twice, and landed like a cat. "'That's my point.'" he said quietly. "'That it's going to require more of us as citizens than we can handle.'" He crouched at his desk and rapped out the dactyls with a fist: "'*This is a threat to democracy!*'

"So I see nothing wrong with calling for a moratorium. Good grief, in a basketball game they call time out—you know, 'Gotta think about this, the clock's running out and we're behind.' But it's unthinkable in this culture to call time out and say, 'Let's think about it.'"

Seeding a Domestic Prairie

For all their talk about a hundred-year research program, the Jacksons do not imagine that the Land Institute will single-handedly bring the domestic prairie to fruition. One can say that the first priority of the Land is to produce plausible moles who will enter the land-grant system of state agricultural colleges and bore from within. Young people drawn to a mom-and-pop institute, however, may not be drawn to large institutions. Drawn to them by duty, such moles are less likely to bore than to be bored. Of the interns who have worked at the Land, only about one in five has sought an advanced degree in an agricultural subject. More have done a hitch with the Peace Corps or joined the staff of an advocacy group. A few years ago a decision was made to seek out interns with a calling in teaching and research, and this seems to be working. Several recent interns even came primed with master's degrees in agronomy.

"Anybody that's in education," Dana told me, "is just living by a thread of hope that education in the long run makes a difference. You can't prove it." But you can prove mathematically that the Jacksons and others speaking for sustainable agriculture *have* made a difference, even in the land-grant schools. Twelve years ago perhaps two or three of these schools offered credit courses in sustainable agriculture. Today at least seventeen have such courses, and several offer full-scale programs. The Land Institute is collaborating with Kansas State on a number of projects—notably the effort to breed a winter-hardy sorghum, begun by the late Dan Rodgers and carried on by Paula Bramel-Cox. Miguel Altieri, at Berkeley, Stephen Gliessman, at Santa Cruz, and Charles Francis, at the University of Nebraska, are working on (mostly annual) polyculture; though their inspiration comes from peasant agriculture in Latin America, they are big fans of Jackson's.

In fact, hardly anyone in the land-grant schools will say anything nasty about Jackson, at least for the record. Even Clenton Owensby, a rangeland scientist in the agronomy department at Kansas State, who once expressed skepticism about Jackson's ideas in *The Wall Street Journal*, says he thinks perennial polyculture is "an admirable goal, with some modification." The basic modification he has in mind is to scrap the perennial part. "Most of the prairie-type systems with which I'm very familiar are perennial by virtue of asexual reproduction. These plants are relatively nonflowering"—in other words, nonseeding. "Now, there are some good ideas. The idea of integrating species with different rooting zones into a profile is a very reasonable approach; in fact, that's what the native prairie does. But I think that strategy would be much better applied to annuals than to perennials if you're planning to harvest the seeds."

The official line on the Land Institute is perhaps best stated by George Ham, the head of Kansas State's agronomy department. "I think their work is very good," he told me. "They're looking at something that's going to have a payoff in the long term, much longer than most of our projects here at Kansas State." Pressed to enlarge on the difference, he said. "We feel that the system of crop production that we've been suggesting to farmers is a sustainable system, and I think Wes Jackson may not agree."

The sudden popularity of sustainable agriculture has some of its stalwarts worried. Dana says, "The land grants are being forced by the public to acknowledge the importance of sustainable agriculture; then they're defining it in as broad terms as they can." They are defining it to mean reducing the inputs of fuel, water, and chemicals to wheat fields, corn fields, and soybean fields. According to

Jackson, nine tenths of what huddles under the umbrella of sustainable agriculture is "low input," while maybe one tenth takes nature as its model.

Nature is an elusive model, Jackson admits. The prairie is a young ecosystem, and some scientists have even suggested that it is man-made—though most now think that lightning, without the Indians' help, set enough fires to keep the tallgrass from becoming forest. When the historian William Irwin Thompson visited the Land last year, he declared that there is no such thing as nature: "nature is the cultural horizon." Jackson's response was that nature, although a slippery standard, is still "the best standard we've got."

The idea of a domestic prairie is tailored to the prairie states. In time, specific mixes of perennials might be bred for specific regions, and for soil types and microclimates within regions. But getting too specific would be a pain in the neck for breeders and farmers alike. A farmer can't plant, and a seed house is not about to market, a different mix for every square foot of his land. Yet this is what nature does. Our image of native prairie is of endless sameness, but in truth the flora shifts kaleidoscopically, inch by inch, like some of the better minimalist music. Although chance (where a seed falls) plays a role in the arrangement, adaptation (whether a seed survives, whether a rhizome thrives) has the last word. Such diversity is not something agriculture suffers gladly. "The agriculturist seeks to homogenize the environment." Jackson says. "*Homo* is a good name for us—homogenizer of landscapes. When you plow, you homogenize, because you've destroyed the plant-soil interactions, plant-water interactions. And when you move to a chemical approach to agriculture, that's further homogenization, because you have the herbicides that knock out all the weeds." The domestic prairie would buck that trend, but after a certain point farming is farming. Good yields of specific, marketable crops are what you need to harvest, and you need to harvest them mechanically. So, Jackson says, while "there will be much more diversity represented than we find in an agricultural landscape, you will find less heterogeneity than there is in a prairie. A domestic prairie will be somewhere in between."

A natural prairie adapts not only in space but also in time. It adapts to changes in rainfall and temperature, to invasions of insects and microbes. It can do so because each population of plants has a deep pool of genes to draw on. The gene pool of a human-selected population is almost always shallower. Jackson has described the problem this way: "Successful plant breeding tends to narrow the genetic base of any crop roughly in proportion to its success. The better adapted to the human purpose a population is, the narrower its genetic base and, therefore, the fewer options it has available to meet future threats." Eighty percent of all hybrid corn grown in the United States in 1970 came from closely related lines, which is what laid it open to the calamitous blight of that year. Yet peasant agriculture in many parts of the world has, until lately, kept alive a good deal of diversity. Before the Green Revolution bulldozed them under. 30,000 varieties of rice had been grown in India, and ethnobotanists have reported more than fifty kinds of potato in a single village in the Andes.

In one of our conversations I asked Jackson if, while breeding his optimal mix for high yield, high protein, easy harvesting, and so on, he would leave in enough diversity to provide some of the adaptive properties of a natural ecosystem. He said yes, but that was the "long-range ideal"; compromise would be necessary in the short run. "The question becomes, how much of a purist do you want to be?" Some of his customers, I suggested, would want something fairly uniform, something they recognized as farming; others would want to go back to nature. "And it'll be all over the place," Jackson said. "I would love to have humans arguing over things like that." He laughed. "It's this kind of diversity that keeps the discussion going. It's pluralism. And I believe in pluralism." He lowered his voice. "With certain restrictions."

A few more years as dry as the last two have been on the plains might turn a lot of hardheaded farmers into purists. Comparing their gouged yields and blown-away fields with the unfazed prairie, they might decide they wanted to get as close to nature as possible. I asked Jackson if the domestic prairie would stand up better to cyclical drought—or permanent drought caused by the greenhouse effect, if it comes to that—than a conventional field. Clearly the soil would hold up better, but would the crops themselves?

He said, "That's my bet. Better than a corn field. Better than a wheat field. More ecological genes are present in the perennials recently domesticated than would be in plants that have been domesticated for a hundred centuries."

He went on: "One of my worries is that the response to the greenhouse effect is going to be an exercise of more heroic cleverness, rather than a stepping-back to say, 'What has nature been doing here?' The philosophy that got us into the trouble is not the philosophy that is going to get us out of the trouble."

Jackson can be found somewhere in the middle of a broken line of people walking toward Eden. Each has his own vision of the place. All hope to see man and nature reconciled, but they disagree on the terms. What should be the proportion of human cleverness to nature's wisdom? As Jackson says, that depends on how much of a purist you want to be. At the extreme of purism is Masanobu Fukuoka, who almost half a century ago gave up a career in microbiology to go home and work his father's farm on the island of Shikoku. A prodigal's return to tradition was not what he had in mind. He thought farmers worked too hard. He thought they fussed and interfered too much with the workings of nature, instead of letting nature work for them. To test his theory, he abstained from pruning his father's orchard. The trees died. He continued to experiment, and learned that if fruit trees grew up without ever being pruned, with weeds and wild or feral vegetables growing freely among them, they would flourish and fruit prolifically without the aid of pesticides. He learned that he could broadcast his winter barley seed in the standing rice just before it was harvested, and his rice seed in the standing barley, and never have to plow at all. He learned that he could get yields as high as any in the prefecture with half the work that traditional Japanese farmers put in, and none of the chemicals that industrial farmers use. Young people come to his mountain, where they dwell in huts, wear loincloths, and learn how to be self-sufficient "quarter-acre farmers."

At the opposite extreme, admitting the machine into the garden, is the New Alchemy Institute, in Falmouth, Massachusetts. These people are at once more traditional than Fukuoka and more modern. Inspired by the elegant juggling of nature that is found in the best traditional technology, they have enlisted computers, advanced polymers, and geodesic domes in the quest for ever more elegant legerdemain. John Todd and Nancy Jack Todd, two of the institute's founders (they have since gone on to found Ocean Arks International), studied Indonesian farming at first hand under the tutelage of Margaret Mead; its deft interweaving of earth, water, trees, grains, fish, and livestock, which has kept the soil fertile for a thousand years, still inspires them. The New Alchemists' ingenuity lies in matching old wisdom to new science, to create such things as argon-insulated bioshelters, small-scale aqua-culture tanks, and vacuum-bagged constant-camber wood-epoxy fishing boats, all of which were designed to be made cheaply and simply anywhere in the world.

Jackson is in touch with both extremes. He talked with Fukuoka at a conference of permaculture designers (disciples of the Australian Bill Mollison) in the Pacific Northwest and "felt friendly toward him," though he wondered how many of Fukuoka's ideas would survive translation to American climates and mindsets. The Jacksons are friendly with the Todds, both couples being part of the hard core of mom-and-pop institutes working toward a "sustainable future." Which extreme is Jackson closer to? In his willingness to take nature as his model, he is very close to Fukuoka. In some ways he traces nature's lines even more faithfully (because less romantically) than the Japanese sage does. On the other hand, he is no Luddite and he is surely no Taoist. The scientific methods in use at the Land Institute are rigorous. Jackson is comfortable with machines. He is confident that machines will be invented to harvest his domestic prairie. When a tractor is being repaired at the Land, or a solar heating system designed, Jackson usually has his hand in. In his love of tinkering he is very like the New Alchemists—and very like a Kansas farmer.

Jackson grew up on a farm of forty acres in the Kansas River valley. He was the youngest of six children—a fair number for his parents to be raising during the Great Depression. But then, as Jackson says, "it wasn't six mouths to feed so much as twelve hands to work." The family never went on the dole. "If you're on a farm, even though the money economy may have collapsed all around you, if you have land you can eat." That might surprise some modern farm families who put all their acreage in one cash crop, get their food at the supermarket, and suffer malnutrition while the national economy swells. But the Jacksons were old-fashioned. They hedged their bets, growing crops they could eat, crops they could sell, and crops they could sell if the others stopped selling. "Now, get this inventory of things that they grew," Jackson said to me. "Watermelons. Sweet potatoes. Cantaloupes. Phlox. Peonies. Strawberries—they had as many as twenty acres of strawberries once; people would come from Topeka and pick their own. They had hay, alfalfa hay. Potatoes. Milk cows, horses, hogs, chickens, turkeys. They grew corn, and wheat, and soybeans—I think soybeans. They even had gladiolas—we'd go out and dig gladiola bulbs. Rhubarb. Asparagus. Now, they didn't have all these things at once, but things would fade in and fade out. I don't think I've known of a place that had more diversity."

Three of the Jackson children became farmers. Wes, however, began to suspect that till farming was not the only way, maybe not even the best way, of living on the land. When he was fifteen or sixteen, he spent a summer working on a cousin's ranch in South Dakota. "And I've only lately come to think that that's where I got a psychological fix on grassland. I didn't get it, I know, hoeing. Grass was something to get out of the row, and it was particularly bad in strawberries. But on that grassland I lived in a little shack. I had two horses. I'd go out and count cattle, fix fence, put up hay. I'd go along on a horse in hot sun, and come by a rattlesnake den with snakes all coiled up in there, and look over and see the prairie dogs and the burrowing owl. I think I got some kind of psychological fix on the *rightness* of that, compared with hoeing fifteen acres of strawberries."

If that rightness had something to do with gratified sloth (not always a bad adviser), it also had to do with something deeper. Jackson was doing what he liked, the prairie was doing what it liked, and both throve. Still, things were not as right as they might have been. In the rustle of the midgrass Jackson heard rumors of a day when the prairie had truly been free to do what it liked: before the fences hemmed it in, before the thunder of the bison was stifled. "On some Sunday afternoons I would ride with some Indian kids that were half-breeds, and they told stories about how their grandfather used to trap eagles on that little mound that was now Andrew and Ina's ranch, how it was once all their land. I made the mistake once of coming back and telling Andrew and Ina that. The fury, the blaze in Andrew's eyes. *The Indians weren't doing anything with the land, it wasn't theirs,* obscenities shouted, and so on. Well, you know, here was the grass, which was now fenced, and fewer than eighty years before—one person's lifetime—the Great Herd had migrated over that ranch."

Jackson leaned forward, hushed, intense as a tracker. "And I think that subconsciously, in that white culture, they know"—a bark, feral, then again his hushed voice—"they know that they, in their creation of Eden, created the fallen world." He gave a horsey snort, half laugh, half self-deflation. "I mean, the fences . . ."

Two Visions

Jackson's vision is more than a technical fix. It proposes a radical revision of the way human beings live—radical but oddly familiar. The best way to see this is to compare Jackson with his friend Wendell Berry.

Doctoral students in the next century may chew over the friendship of Jackson and Berry as thoroughly as today's students do that of Coleridge and Wordsworth, Darwin and Huxley, or Emerson and Thoreau. If either of the pair had not existed, the other would have had to invent him. Berry gives Jackson the sustenance of the past. And Jackson, standing on Berry's shoulders, gives him a vision of the future.

With the help of draft horses, Berry works a small hill farm near the Kentucky River. Although he happens to think that horses are both efficient (tractors can neither run on grass nor leave manure) and a joy to work with, he does not insist that everyone go back to farming with horses. But on the grounds that a countryside full of small farms, and the communities that grow from them, is healthier in every way than a countryside dominated by a few factory farms, he does insist that we go back to small farming.

Berry is taller than Jackson but so lean and gangly that he seems to hang from his shoulders. Instead of Jackson's sturdiness, Berry appears to have the Lincolnesque, rail-splitting strength that comes from leverage. He lives in a private, family-centered way,"on principle,"and dislikes giving interviews. When you ask him a question, he does not, like Jackson, take a deep breath and let fly; he stares at his knees (when he's sitting, they are very close to his face) for a while and then, in his soft drawl, delivers an epigram or a paragraph that responds exactly to what you have asked. Severe at first, when he warms up and starts to laugh he has a helpless joy in his eyes.

The Unsettling of America, Berry's study of industrial agriculture, the culture that drives it, and the way it has scarred the body and soul of the republic, is a very great book. Its heart can be found in these lines:

> The "drudgery" of growing one's own food . . . is not drudgery at all. (If we make the growing of food a drudgery, which is what"agribusiness"does make of it, then we also make a drudgery of eating and of living.) It is . . . a sacrament, as eating is also, by which we enact and understand our oneness with the Creation, the conviviality of one body with all bodies. . . . We are working well when we use ourselves as the fellow creatures of the plants, animals, materials, and other people we are working with. . . . Such work is unifying, healing. It brings us home from pride and from despair, and places us responsibly within the human estate. It defines us as we are; not too good to work with our bodies, but too good to work poorly or joylessly or selfishly or alone.

Berry is the moral equivalent of magnetic north. Yet his agrarian declination prevents him from serving as a sure guide for the rest of us. By"the rest of us"I mean city-dwellers, cosmopolitans, people whose roots are not in the alluvium of a specific river valley but in the world-wide industrial culture that Berry cannot stand.

Although Jackson is less urbane than Berry—he likes to play the bumpkin, and looks the part—he is easier for an urban mind to approach. That has to do, first of all, with the length and breadth of his view, whose increments are not centuries but eons. Berry thinks Renaissance, Enlightenment, Romantic: Jackson thinks Pliocene, Pleistocene, Holocene. He sees that human beings have lived in various ways, of which farming is one of the more newfangled and problematical. (Cities have been around almost as long as farms—or longer, if we believe Jane Jacobs.)

It has to do, too, with his impatience. At administrative meetings he hunches like a benched athlete and his long arms trawl, swishing against the leg of his chair. He must have been impatient hoeing fifteen acres of strawberries, when his bodily energy had an outlet but his mind was strapped down. It's hard to picture him as a till farmer.

It is not so hard to picture him as a farmer of the future he imagines, a ranger of the domestic prairie. That farmer will be more like "an old-fashioned English naturalist," he says—a Gilbert White roving the sheepwalks of Selborne, a Charles Darwin taking in, in the same mental glance, the finches of the Galápagos and the pigeons in his dovecote—than like the till farmer we know. By the same token, he will be a lot like a Paleolithic hunter-gatherer: the canniest naturalist of all, if the few surviving tribes are representative. And that will be just as well, since a hunter-gatherer is what his chromosomes want him to be. In an odd way, he will be more like a city-dweller.

"The romantic back-to-the-land movement," Jackson writes, as minimal as it is thus far, is a sign of something desirable. But what happens when we all get there, after the first generation of back-to-the-land romantics have been buried organically in their gardens? Will their children maintain the back-breaking work most humans have sought to avoid over the centuries? Isn't this one of the components of the human condition? It has yet to sink into our culture that we are still basically gatherers and hunters, and that the era of agriculture is but a thin veneer over an evolutionary past which tolerated a great deal of leisure. The appeal of the countryside is the appeal open space has always had for us gatherer-hunters. The appeal of the city is that it at least faintly suggests a mixture of leisure and stimulation most of us need. R. B. Lee has reported that only 65% of 31 !Kung Bushmen in Botswana spent two to three days each week gathering food while "35% of the people did no work at all."

Like Berry, Jackson thinks that many of the four million American farmers who have left the land in the past fifty years have been pushed off it by economic forces: forces applied knowingly, as often as not, by businessmen and politicians with a stake in cheap food and cheap urban labor. Unlike Berry, Jackson thinks a good many may not have been pushed so much as pulled by the easier and jazzier life of the cities. How *are* you going to keep them down on the farm after they've seen Paree, or Chicago, or Topeka?

Jackson is not lazy. He has energy enough for two men and has accomplished enough in his fifty-odd years to make both of them proud. When he jumped up at the Prairie Festival symposium and said, "When the oil runs out, we're going to need the work ethic more than ever," he was not just being ornery. The transition to the world he envisions will not be easy. Living in it may not be easy either—farmers, in particular, will have to be endlessly watchful, thoughtful, and inventive—but it could be more satisfying than anything our society has known.

God's Own Lawn

A radical is someone who traces the tree of culture back to its roots in human nature to see where it got twisted. Wendell Berry goes back a few hundred years and finds a peaceable agrarian animal. His proposal that many of us go back to being that animal strikes most of us as radical in the sense of being impossibly stringent. Jackson's proposal is at once less radical because it is more forgiving, and more radical because it goes deeper. Berry's vision seems reactionary at times because he harks back to a past we remember. Jackson's seems forward-looking for the ironic reason that he has dug all the way back to a past we can't remember but can feel inside us.

It took New England two hundred years, Jackson has said, to produce Thoreau, who could see New England for what it was rather than through the eyes of a European. It took Kentucky about two hundred years to produce Wendell Berry, who can see Kentucky for what it is. He does not go on to say that Kansas, a precocious youngster, seems to have skipped a century.

So here is the sequence: Thoreau finds the soul of the eastern forest, but never figures out how farms fit into it. He advises foresters to take nature as their model—"for she is the most extensive and experienced planter of us all"—but knows that applied to farming that advice would seem plain folly. Berry, discovering the wooded hills, plains, and bottomlands of the interior, of the Middle Border, sees clearly how farms should fit them: gently, respectfully, as a husband embraces a wife. And finally Jackson, intoxicated by the prairie—seeing it as if for the first time, like a child who has wandered onto God's own lawn—sees that farming as we know it is unnatural, a limp resulting from Adam's fall; that even the gentlest husbandman is still too much the master, too separate, not in truth one flesh; but that Eden is not forever closed to us. Though the angel with the fiery sword is still at his post, with patience and luck we may catch him napping.

Jackson says,"Thomas Wolfe wrote somewhere that the discovery of America lies before us. I think that's right. So far we've only colonized it. So what each of us has as our mission is to assist one another in seeing where we are, not only in the sweep of history but also in the history of the landscape. I don't think that we will have discovered America until we have some sense of the feelings of the natives that walked over this place. And the people that were the furthest from understanding the sense of the natives that walked over this place"—he thumps his desk like a tom-tom—"were the conquerors of those natives. As Wendell said, we came through and plowed the prairies and we never knew what we were doing because we never knew what we were undoing. What we are about is the discovery of what we have undone, and then we can begin to talk about what to do."

MAKING PEACE WITH THE EARTH:

Indigenous Agriculture and the Green Revolution

DEANE CURTIN*

Since its inception in the years following World War II, the green revolution has been defended, not just as a technical program designed to alleviate world hunger, but on moral grounds as a program to achieve world peace. In this paper, I dispute the moral claim to a politics of peace, arguing instead that the green revolution is warist in its treatment of the environment and indigenous communities, and that the agricultural practices that the green revolution was designed to supplant—principally indigenous women's agriculture—are forms of ecological peacemaking, akin to pacifism. I argue, as well, that the warist intentions of the green revolution are characteristic of a form of domination called developmentalism. A complete understanding of domination necessitates linking developmentalism with other forms of domination such as racism, sexism, and naturism.

I. Introduction

In 1970 Norman Borlaug won the Nobel Prize recognizing his role as father of the green revolution. While Borlaug's work was on plant genetics, he did not win for biology. He won the prize for *peace*. Thus culminated the effort of two decades and more to represent the green revolution as a peace program. According to narrowly defined criteria, the green revolution has been proclaimed a success: yields on all the major seeds produced by green revolution research have at least doubled since World War II. Yet, the claim that it is a peace program demands a broader assessment in terms of its social, political, and environmental impact.

My analysis has two dimensions. Both involve assessments of the characteristic *violence* of the green revolution despite its explicit peace agenda. The green revolution's violence is, I believe, enabled by a set of "First World" attitudes toward the so-called "Third World" that puts the First World at the center ethically and conceptually, spinning out other "worlds" from that center to the ethical and conceptual margins.[1] This system of marginalization, which I call *developmentalism,* is a system of domination that connects with, and benefits from, other forms of domination such as racism, sexism, classism, castism, and naturism. Any analysis of domination that is not linked to developmentalism risks a conceptual bias toward the First World.

Second, to address the peace claims of the green revolution specifically, I characterize developmentalism as an ideology that is akin to what Duane Cady has called *warism.*[2] It is structurally committed to violence in its treatment of indigenous communities and the Earth, violence that is justified as a moral good. If the green revolution is a peace program, therefore, it must defend the claim that peace comes through violence. This claim, I believe, confuses peace with pacification.[3]

Because women have been the Third World's principal food producers, the green revolution was designed, in effect, to displace women's agriculture. These agricultural practices, ironically, tend to be forms of ecological peacemaking, akin to pacifism. I do not intend these remarks to romanticize indigenous women's agriculture. Violence does exist within indigenous communities. This violence, however, is immeasurably worsened by the effects of the green revolution. Furthermore, in speaking of "women's agriculture," I am not essentializing their typical practices. My approach is nonessentialist. Women are not inherently closer to nature than men. Men are not the sole perpetrators of environmental violence. For complex reasons, women's practices do tend toward peacemaking with the Earth, but in dysfunctional cultures, all practices are distorted.

I am concerned, as well, about overgeneralization. Any generalization, however nuanced, risks losing sight of the diversity of women's lives.[4] Yet, this concern generates a dilemma: women's practices are often invisible to development experts. Without some degree of generalization, they will remain invisible. I focus on the effects of the green revolution in India, therefore, because of personal experience with Indian agriculture, and because India was one of the original test sites for the success of the green revolution. Even within India, generalization can be difficult. The state of Kerala, for example, has a matrilineal social structure that may be seen as having affected agricultural practice (and success) there. While granting these important points of diversity, I believe my generalizations—as qualified as they must be even limited to a country like India—can be useful in understanding the condition of women's agriculture elsewhere. Here I am content to suggest some of the connections that might be taken up at a later time.

II. Developmentalism and the Green Revolution

In *Principles of Political Economy,* John Stuart Mill describes the British Empire's colonies as

> . . . hardly to be looked upon as countries, . . . but more properly as outlying agricultural or manufacturing estates belonging to a larger community. Our West Indian colonies, for example, cannot be regarded as countries with a productive capital of their own. . . [but are rather] the place where England finds it convenient to carry on the production of sugar, coffee and a few other tropical commodities.[5]

Mill's colonialist ideas have hardly become outmoded in an age when agricultural multinationals have replaced the East India Company. The sense that the Third World is "outlying," distant from, dependent on, and defined by, the center is at the heart of contemporary developmentalism.

Richard Levins captures the dynamic of center and margin through "seven developmentalist myths in agriculture":

1. *Backward is labor-intensive, modern is capital intensive agriculture.*
2. *Diversity is backward, uniform monoculture is modern.*
3. *Small scale is backward, large scale is modern.*
4. *Backward is subjection to nature, modern implies increasingly complete control over everything that happens in the field or orchard or pasture.*
5. *Folk knowledge is backward, scientific knowledge is modern.*
6. *Specialists are modern, generalists backward.*
7. *The smaller the object of study, the more modern.*[6]

Developmentalism is defined by a complex set of attitudes including issues of economic scale, reductionism, relationship of the person and community to the land, and the epistemic status of the outside "expert" as against the insider to a practice. In this section, I characterize the green revolution historically as form of developmentalism.

The green revolution began in 1944 when the Rockefeller Foundation invited Borlaug to leave his wartime job in a Dupont laboratory to direct the wheat breeding program at the International Maize and Wheat Improvement Center in Mexico. Under Borlaug's direction, the center produced the so-called high yielding variety (HYV) of wheat that was initially targeted for two areas, northwest Mexico and the Punjab region of India and Pakistan.

In principle, Borlaug's accomplishment is simple. Inorganic nitrogen fertilizer, when applied to traditional varieties of wheat, makes the whole plant grow larger. Tall varieties, top-heavy with grain, have the tendency to topple over (called "lodging"), thus reducing the yield. Borlaug genetically engineered a dwarf variety of wheat to concentrate fertilizer in the grain, the profit-producing part of the plant. The green revolution is possible because HYVs can accept very high doses of fertilizer.

Although the core of the green revolution is simple, the broader impact is complex. As Borlaug himself explained, a whole package of technologies was transferred to the Punjab, including ". . . seeds, fertilizers, insecticides, weed killers, and machinery—and the credit with which to buy them."[7]

High doses of fertilizer require much greater reserves of water to be effective. Green revolution crops, therefore, require massive irrigation.[8] Unlike traditional agriculture, which is based on crop rotation, the green revolution involves plant monoculture. This change makes HYVs especially susceptible to pests and diseases, thus, making the green revolution dependent on pesticides.[9] In contrast to agriculture based on the principle of recycling inputs, the green revolution depends on external sources for seeds, chemicals, and machinery. Because green revolution hybrid seeds are not self-pollinating, they must be purchased each year from a seed company. The biotechnology revolution now occurring in agriculture carries plant monoculture further, to the level of individual brand names. Agrochemical companies are engineering seeds to respond to (or not to respond to) particular commercial brands of fertilizer, herbicide, or pesticide.[10]

Though brief, this summary of green revolution technology is sufficient to illustrate the impact that it has had on indigenous women's agriculture. The green revolution does not depend on replenishing the soil, the peasant's traditional source of security. It does not depend on conserving water; rather, it assumes that access to water is unlimited. It demands political and economic dependence on a small number of seed and agrochemical companies rather than building on local self-reliance. Because its economy of scale depends on bigger being better, it depends on loans—and the ability of the debtors to repay loans—from Western dominated international agencies, such as the World Bank.

The ideology of the green revolution is replete with developmentalist myth. Everything that characterizes traditional agriculture—a type of agriculture that is labor-intensive, diverse, small-scale, and responsive to the cycles of nature—is represented as backward and primitive. Indigenous knowledge is reduced to folk tale or superstition. Only the outside expert "knows."

III. Indigenous Women's Agriculture

What are the agricultural practices that need "development"? Anthropological research helps to explain the gendered roles of traditional farmers. The hunter/gatherer paradigm, according to which the hunters provided most of the food, has been dismissed as a sexist myth. In such societies, up to eighty percent of the food was gathered by women. Women's roles as gatherers, in turn, led to the invention of agriculture as predominantly a women's practice. Assuming that "the workers invented their tools," most anthropologists now agree that women invented the important agricultural tools. Because women have been the traditional plant breeders, they must have also bred most of the world's grains including wheat, rice, maize, barley, oats, sorghum, millet, and rye. These cereals still supply seventy-five percent of all human food energy.[11]

In Africa and many parts of Asia, agriculture is still disproportionately women's work. According to a recent Worldwatch paper, "Gender Bias: Roadblock to Sustainable Development," women in sub-Saharan Africa grow eighty percent of all household food. Women's labor produces seventy to eighty percent of food on the Indian subcontinent, and fifty percent of food consumed in Latin America and the Caribbean.[12]

Sir Albert Howard encountered this state of affairs in the late 1800s when he was dispatched to India by the British government to investigate methods of improving Indian agriculture. He found, much to his surprise, that their crops were free of pests, and that insecticides and fungicides had no place in their system of agriculture. Impressed, he decided" . . . I could not do better than watch the operations of these peasants, and acquire their traditional knowledge as rapidly as possible. For the time being, therefore, I regarded them as my professor of agriculture. Another group of instructors were obviously the insects and fungi themselves."[13] The best agriculture, Howard later wrote, is modeled on "nature's agriculture":

> The main characteristic of Nature's farming can therefore be summed up in a few words. Mother earth never attempts to farm without live stock: she always raises mixed crops; great pains are taken to preserve the soil and to prevent erosion; the mixed vegetable and animal wastes are converted into humus; there is no waste; the processes of growth and the processes of decay balance one another; ample provision is made to maintain large reserves of fertility; the greatest care is taken to store the rainfall; both plants and animals are left to protect themselves against disease.[14]

The cyclical principle made Howard's book, *An Agricultural Testament,* a classic in the organic, sustainable agriculture movement. It is not often recognized, however, that this movement owes a great debt to Third World women's agricultural knowledge.[15]

The foremost commitment of cyclical agriculture is to biological diversity. In a healthy ecosystem in which decay balances growth, soil regenerates through the recycling of plant and animal matter. Plant agriculture, thus, requires animals for fertilizer. Crops that use nitrogen, such as wheat and rice, must rotate with plants that return nitrogen to the soil, such as pulses (beans, peas, and lentils).

The cyclical principle also embraces water conservation, which begins with healthy soil that retains moisture. Water depends on mixed land use as well. Agricultural land must be mixed with forested land. Trees prevent flooding, and are a self-regenerating resource for fodder that retains water. Forests are also necessary for food, building materials, and traditional medicines.

Traditional agriculture does not depend on pesticides. It lets biological diversity do the work. As any agriculturist knows, it is not true that everything is connected in nature. If that were the case, Dutch Elm disease would have wiped out oak and other species. Traditional agriculture takes advantage of the disjunctions in nature to control pests. Because corn rootworm do not eat soybean roots, for example, crop rotation removes the rootworm's source of food.

Traditional agriculture is also cyclical and localized in the genetics of its seeds. It is typical for each family to maintain its own stock of seeds from year to year. Women are usually responsible for selecting the best seeds for cultivation the next season. Village-based genetic breeding means that seeds are developed to meet the needs of widely diverse growing conditions. Seeds are bred to "meet the expectations of the land," in the words of Wes Jackson.[16] Land is not altered to meet the demands of the seed. The genetic diversity of traditional agriculture is a safeguard against widespread devastation of crops due to climatic change or pests. Village plant breeding is also a political issue. When women control plant breeding, they are at the nexus of activities that most determine family and community survival. It is just this sense of what is appropriate to a place, the refusal to dominate a place, I contend, that developmentalism reads as backward.

IV. Warism, Women, and the Environment

I do not argue directly against warism. Rather, my purpose here is to clarify what would have to go into a defense of a warist environmental ideology. In keeping the argument "close to the ground," in focusing, that is, on the actual impact of the green revolution on Third World peoples, I hope to show how *difficult* it would be to defend the violence of the green revolution as being moral.

Duane Cady has coined the term *warism* as a parallel term to *pacifism.* Warism is the view "... that war is both morally justifiable in principle and often morally justified in fact." Pacifism is the view that "... war, by its very nature, is morally wrong and ... humans should work for peaceful resolution of conflict." When he coined the word, Cady was thinking in terms of explicit military violence rather than the covert political violence of the green revolution. Recently, however, Cady has connected warism with other forms of domination: sexism, racism, and classism.[17] As a reminder that I am stretching Cady's meaning beyond military violence, I claim here only that the ideology of the green revolution is *akin* to warism—that developmentalism is warist.

In connecting warism with other forms of domination, Cady draws on Marilyn Frye's insight that "arrogant perception" underlies the structures of oppression. According to Frye, Western civilization's answer to the question of man's place in nature is that "... everything that is is resource for man's exploitation. With this world view, men see with arrogant eyes which organize everything seen with reference to themselves and their own interests."[18]

Developmentalism and warism both perceive the world arrogantly. Like racism, sexism, classism, and naturism, they construct a moral pecking order. Those at the top "know," "act," "develop," and "enjoy" the moral prerogative. Those at the bottom are ignorant, passive (and deserving) recipients of the actions of others. For example, Lawrence Summers, former Harvard economist and World Bank official, now Assistant Secretary of the Treasury, argued in a World Bank memo that the Third World may be "underpolluted" and that it would make sense to shift highly polluting industries from the First World to the Third World.[19] The arrogance of this suggestion—reminiscent of John Stuart Mill—is unusual today only for its candor.

While the dominations of women and nature are connected in the green revolution, they are not identical. Nature is entirely outside the moral sphere for the developmentalist; it is something to be controlled as a "resource for man's exploitation." Since it lacks moral standing, the developmentalist's impact on nature is regarded as nonmoral. While developmentalism itself is a moral crusade, applications of science to nature are understood as morally neutral technical improvements.

Women and indigenous cultures are morally unlike nature in this respect. They do have moral standing for the developmentalist. This standing is revealed in the very fact that they are constructed as needing "development." The connection between the domination of women and the domination of nature is that developmentalism operates in a mode of "control over" indigenous communities by controlling their natural "resources." I consider, first, the warist domination of women and indigenous communities in the green revolution, then the warist domination of nature.

Warism and Indigenous Peoples

It is inviting to think that the effects of the Green revolution on women, their communities, and their environments, are unfortunate side effects of well-intentioned development programs. However, this claim is not true. Before the green revolution, there were important research programs in both Mexico[20] and India on a wide variety of self-pollinating seeds traditionally cultivated by peasants. These programs were intentionally undermined by the politics of the green revolution.

In India, the Cuttack Institute was investigating techniques to increase yields on traditional varieties of rice based on the indigenous knowledge of tribal peoples. The institute had collected and preserved 20,000 indigenous varieties of rice, a storehouse of genetic information patiently developed over centuries by village women. Unlike the genetically uniform rice developed by the IRRI in the Philippines—another Rockefeller/Ford "food for peace" program—these indigenous species were genetically diverse, having been bred over the entire Indian subcontinent. Nevertheless. for political reasons, the Indian minister of agriculture, who had been trained in Mexico, demanded that the Cuttack Institute turn its germ plasm over to IRRI. When the director resisted, he was fired.[21]

There were elements in Mexico and India that resisted the green revolution as an assault against national autonomy. They were often silenced when food was used as a political weapon. In 1966, for example, Lyndon Johnson refused to commit food aid to India until it adopted the green revolution as national agricultural policy.[22] U.S. AID was also involved in subverting the Allende administration in Chile—by withholding food aid.[23]

There was no scientific reason for directing research exclusively toward improving seeds that undercut the political control of the world's poor. More appropriate programs already existed, as I have noted; Richard Lewontin has argued that open-pollinated varieties of seed could have been as productive as HYV hybrids.[24] In fact, under less than ideal conditions, traditional seeds are often *more* productive than HYVs. Since most Third World farmers work small plots of marginal land with irregular access to water, less than ideal conditions are the norm, not the exception. The fact that HYVs were chosen for development, therefore, while programs involving open-pollinated varieties were forcibly shut down, reflects the class and gender interests of the green revolution. HYVs are genetically engineered to be a privately owned commodity.

In addition to the attempt by multinationals to control indigenous communities through genetic research, control is also increasing as courts in the United States and its industrialized allies are now recognizing the legal right to patent genetic information, thus privatizing the seed. American companies often claim that the Third World steals intellectual property from the First World. However, this conveniently narrow argument only applies to items that are defined by the First World as commodities. Resources taken from the Third World are defined (by the First World) as "the common heritage of mankind."

Most of the world's sources of genetic diversity, upon which First World research depends, are in the Third World. Colonialism has always depended on exploitation of this diversity to produce commodities that are sold back to the Third World on credit provided by the First World. As early as 1848, for example, the East India Company was collecting plant species in India. In 1853, when Admiral Perry colonized Japan, he collected plant species, including rice.[25] That this kind of treatment is warist is not news to many residents of the Third World. Recently, for example, five hundred farmers in Bangalore, India stormed and ransacked the office of Cargill, Inc. to protest the intellectual property provisions of the General Agreement on Tariffs and Trades.[26]

The effects of this "top-down" ideology on indigenous people are both predictable and tragic. Vandana Shiva and Maria Mies have challenged the claim that so-called "ethnic conflict" is responsible for recent violence in India's Punjab region.[27] Religious conflict, they argue, is the effect. The cause is the green revolution. The same system of irrigation that caused the Punjab to be chosen in the first place as the test site for green revolution techniques is now causing violence over water rights. Farmers who have the means to mechanize and repay foreign loans have benefited; their acreages are increasing rapidly as small farmers are driven out of business. Small farmers—women and men—have been displaced and cut off from their traditional source of security. Men who are displaced by this process seek wage-labor jobs on the large farms or they migrate to the cities, leaving their families behind. Women who are displaced often are expected to work as unpaid adjuncts to their husbands' wage labor. As is almost always the case in such situations, violence against women has increased in the form of *sati*, "kitchen accidents," and female infanticide.[28] In many cultures, the same women who grow and prepare the food are the last to eat, even in the best of times.

Warism and Nature

The green revolution's vision of agriculture is the violent conquest of nature. This begins with a reductionist, non-contextual conception of science. A Harvard botanist said, "We now operationally have a kind of world gene pool. . . . Darwin aside, speciation aside, we can now envision moving any gene, in principal at least, out of any organism and into any organism." Nobel laureate David Baltimore put it even more succinctly:"We can outdo evolution."[29] The desire to outdo evolution in the control of nature reveals the dualistic intentions of green revolution ideology. Science is not a romance *with* nature; rather, it is a race *against* nature, and science is winning.

Most revealing are the conceptual and empirical connections between green revolution techniques and the technology of war. Vandana Shiva has pointed out that

> Violence was part of the very context of discovery of pesticides during World War I. The manufacture of explosives had a direct spin-off on the development of synthetic insecticides. The tear gas, chloropicrin, was found to be insecticidal in 1916 and thus changed from a wartime product to a peacetime one.[30]

Organophosphates, such as malathion, are designed to destroy the central nervous system. Insecticides and herbicides are tested by their "kill ratios." Broad-spectrum herbicides kill all vegetation leaving the Earth biologically neutral. Agent Orange, used as a defoliant during the war in Vietnam, is one such product.

The infamous disaster in Bhopal, India in 1984 summarizes the impact of developmentalism. It was caused by a leak of methyl-isocyanate from a Union Carbide plant that produced nitrogen fertilizer. More than 3,000 people died; roughly 500,000 were injured. A 1989 survey found that fifty percent of the women in the affected area were experiencing menstrual problems, and that there had been a marked increase in stillbirths, spontaneous abortions, and genetic defects in infants. Fifty-seven percent of those examined suffered from posttraumatic stress disorder, the same disorder that affects many veterans of war. Union Carbide has refused to admit any liability, claiming it was a terrorist act by a disgruntled employee.[31]

It is often claimed that science and technology are morally neutral and that only political judgments about the applications of science are open to moral scrutiny. The green revolution makes this distinction between technology and gender bias difficult to sustain. Class, race, and gender biases have been genetically engineered into the "miracle seeds" of the green revolution. Research that is sensitive to local traditions, for example, the Cuttack Institute's, shows that women's agriculture is not at all opposed, in principle, to culturally sensitive development or to scientific research. It is only opposed to the political genetics of the green revolution.

I do not doubt the sincerity of Borlaug and others who believed that the green revolution would bring world peace. Nevertheless, the green revolution succeeded to a much greater degree than any overt form of militarism in reducing Third World peasants to political and economic dependence on Northern industrial and scientific powers. A recent survey of sustainable Third World development concluded that "There is much merit to the argument that if given the choice between the present system of international assistance and no assistance at all, the Third World's poor would be better off with none."[32]

V. Women's Agriculture as Ecological Pacifism

In stark contrast to the ideology of the green revolution, Third World women's agriculture is structurally akin to pacifism in its commitment to ecological peacemaking. Its mode is "collaboration with" nature. Unlike the universalism of the developmentalist, which regards everything that is different as needing "development" in its own image, women's cyclical practices reveal a sense of what is appropriate to and sustainable in a particular place. Nature is not on the far side of the moral divide. The community includes the land as much as its people. Treatment of the land, therefore, reveals the moral self.

The sense of the deep relation of person and place is often expressed in moral or mythological terms. Consider, for example, the arrogant perception in the following passage from the *Journal of the Indian Pesticide Industry* as the author struggles to explain the peasant's relationship to the land:

> [A] more important [difficulty in marketing products in India] is the mental attitude of the agriculturalist about killing. Pesticides spell killing, maybe small and perhaps invisible insects. But it is killing that they are used for. This killing is anathema to the majority of the agriculturalist, be they Hindu, Jain or others. By nature, the agriculturalist is generous, wanting to bestow on others what he reaps out of Mother Earth. He [sic] does not think that he alone should enjoy the fruits of his labor . . . to kill those unseen and unknown lives, though they were thriving on what Mother Earth yields, is foreign to his nature. . . . It takes some time for the simple folk to get acclimatized to the very conception of killing tiny helpless and unarmed creatures.[33]

The Indian idea of *ahimsa,* or nonharming, puts the violence of the green revolution, and its cultural specificity, into sharp relief. *Ahimsa* requires a moral universe that includes insects in the moral cycle of life and death. Nature is seen in moral relationship to human beings. We are defined morally by our conduct toward nature.

It is telling how often, in practice, development experts fail to perceive this deep relationship between women, indigenous cultures, and place. Third World women, for example, have long depended on grasses that grow along the borders of fields to make baskets and mats. When development experts decide that these grasses have no market value and plan and carry out programs to kill them with herbicide, they rely on an arrogant perception of their relationship to nature. When these same experts decide that public, forested land is "undeveloped," and only has value when it is privatized and "developed" for profit, this decision too involves arrogant perception.

Of Richard Levins' seven developmentalist myths, one seems most powerful in explaining the developmentalist's inability to appreciate place: "Backward is subjection to nature, modern implies increasingly complete control over everything that happens in the field or orchard or pasture." In the developmentalist's moral pecking order, one is either dominant over nature or subject to it, either master or slave. The sense of living with nature in a particular place, which is neither subjugation nor dominance, is misread by the developmentalist as backwardness, as a life fit for a slave.

Far from defending the world's poor, the green revolution sought to defeat communism by destroying the world's peasant class. It did so by dividing the peasants economically, politically, and spiritually from their sense of place. Conversely, one could understand much about peasant resistance movements by considering them to be localized defenses of the connection between person and place.[34] It is no accident that liberation theology in Central America or Dalit theology[35] in India are perspectives that express the particular conditions of specific peoples. They reject the universalizing tendencies of traditional theology.

Finally, it cannot be insisted often enough that pacifism is not passivism. Characterizing typical women's farming as pacifist does not mean that women must simply accept the violence perpetrated against them and their communities. To the contrary, pacifism is a form of resistance. The Chipko Movement in India, the Green Belt Movement in Africa,[36] and countless other small-scale women's movements the world over testify to the fact that women are not passive in defending the integrity of their communities and their environments.

VI. Conclusion

Out of a concern to recognize the diversity of women's labor properly, I have confined the majority of my remarks to women's agriculture in India. I believe, however, that these insights can be expanded to other contexts. Having begun with the 1970 Nobel Peace Prize, I want to mention the recipient of the 1992 prize in closing as a way of indicating how these connections might be made.

Rigoberta Menchu was born to a Mayan family in the highlands of Guatemala. She is the sixth of nine children. At the age of eight she began working on a plantation picking coffee and cotton. In her autobiography, she describes a system of near slavery in which she and her family were forced to buy food and medicines from the landowner's store and work the fields for endless hours, only to find at the end of the growing season that they owed the landowner more than they had made.

Menchu's father and later other family members joined the Peasant Unity Committee in their struggle to prevent these plantation owners from removing Mayan families from their homelands. In 1979, Menchu's brother was detained by government troops supporting the wealthy landowners; he was publicly tortured and executed. Two months later her father joined a march that occupied the Spanish embassy to protest abuses of indigenous peoples. He and thirty-seven others died when security forces set fire to the embassy. Menchu's mother was later kidnapped, raped, mutilated, and killed. Roughly 40,000 people, mostly indigenous people, have been killed or have "disappeared" in Guatemala, and 45,000 others live in Mexico in refugee camps where Menchu herself has lived since 1981.

The green revolution is inscribed in the details of Menchu's life. Because it depends on expensive technologies that increase political and economic dependence, and because its economic scale requires control of enormous numbers of acres, the wealthy, rather than indigenous owners of land, have benefited. As in Guatemala, local governments are generally all too happy to side with wealthy interests who promise to increase exports—such as coffee and cotton.

Since the end of World War II, peace has been understood as the absence of military violence. Its language has been the language of nuclear deterrence, and "mutually assured destruction." Perhaps in keeping our eyes trained on the skies we have neglected to see the Earth. Menchu's life, and the lives of other women of the Third World, argue for a frankly agrarian vision of peace.

---

* Department of Philosophy, Gustavus Adolphus College, St. Peter, MN 56082. Curtin's current research interests are focused on environmental conflicts between the First and Third Worlds. He is coeditor of *Cooking, Eating, Thinking: Transformative Philosophies of Food* (Bloomington: Indiana University Press, 1992), as well as "Dogen, Deep Ecology, and the Ecological Self," *Environmental Ethics* 16 (1994): 195–213 and "Toward an Ecological Ethic of Care," *Hypatia* 1 (1991): 60–74.

Endnotes

1. The First/Third distinction marks this marginalization, and is, of course, part of the problem. I do not address the linguistic issue here.
2. See Duane Cady, *From Warism to Pacifism: A Moral Continuum* (Philadelphia: Temple University Press, 1989).
3. It is significant historically that the green revolution was designed by the First World to block the expansion of communism among Third World peasants by increasing crop yields. See Vandana Shiva's discussion of the 1952 Columbo Plan, in *The Violence of the Green Revolution: Third World Agriculture, Ecology and Politics* (London and Penang: Zed Books and Third World Network, 1991), p. 14.
4. In "Women's Knowledge as Expert Knowledge: Indian Women and Ecodevelopment," in Karen Warren, ed., *Ecological Feminism* (Bloomington: Indiana University Press, forthcoming). I argue for a nonessentialist account of women's agriculture in terms of the actual practices in which women engage and the forms of knowledge that are generated by those practices. This approach provides a context-sensitive way of making women's practices visible, and reclaims the concept of expertise for those who have been defined in opposition to that concept.
5. John Stuart Mill, *Principles of Political Economy,* vol. 3, ed. J. M. Robson (Toronto: University of Toronto Press, 1965), p. 693.
6. Richard Levins, "Science and Progress: Seven Developmentalist Myths in Agriculture," *Monthly Review* 38, no. 3 (1986): 13–20 (emphasis in original).
7. Norman Borlaug, "The Green Revolution, Peace, and Humanity," in *Les Prix Nobel en 1970* (Stockholm: Imprimerieal Royal P. A. Norstedt and Söner, 1971), p. 231.
8. One reason the Punjab was chosen as the site for transfer of green revolution technology is that it already had a system of irrigation based on partial diversion of rivers. See Shiva, *The Violence of the Green Revolution.* p. 122.
9. F. Chabousson, "How Pesticides Increase Pests," *Ecologist* 16, no. 1 (1986): 29–36.
10. See Jack Ralph Kloppenburg, *First the Seed: The Political Economy of Plant Biotechnology: 1492—2000* (Cambridge: Cambridge University Press, 1988). The impact of these technolo-

gies is indicated by the juxtaposition of two facts: in the latter half of the 1960s, consumption of inorganic nitrogen fertilizer in India increased from 58,000 metric tons to 1.2 million metric tons. During that decade, India's currency was devalued by 37.5 percent as a result of its rising foreign debt. See Borlaug, "The Green Revolution, Peace, and Humanity," p. 233, and Shiva, *The Violence of the Green Revolution,* p. 30.

11. Autumn Stanley, "Daughters of Isis, Daughters of Demeter: When Women Sowed and Reaped," in *Women, Technology and Innovation,* ed. Joan Rothschild (New York: Pergamon, 1982), pp. 293–94.
12. Jodi L. Jacobson, *Gender Bias: Roadblock to Sustainable Development.,* vol. 110 (Washington, D.C.: Worldwatch Institute, 1992), p. 19. Also see J. B. Bhati and D. V. Singh, "Women's Contribution to Agricultural Economy in Hill Regions of North-West India," *Economic and Political Weekly* 22, no. 17 (1987): 7–11 (special women's studies supplement); Ester Boserup, *Woman's Role in Economic Development* (London: Allen and Unwin, 1970); Lynne Brydon and Sylvia Chant, *Women in the Third World: Gender Issues in Rural and Urban Areas* (New Brunswick: Rutgers University Press, 1989); Sue Ellen Charlton, *Women in Third World Development* (Boulder: Westview Press, 1984); Irene Dankelman and Joan Davidson, eds. *Women and Environment in the Third World* (London: Earthscan, 1988); Donald Michelwait, Mary Ann Riegelman, and Charles Sweet, eds. *Women in Rural Development* (Boulder: Westview, 1976); Maria Mies, "The Dynamics of the Sexual Division of Labor and Integration of Rural Women into the World Market," in *Women and Development,* ed. Lourdes Benería (New York: Praeger, 1982); Ingrid Palmer, "New Official Ideas on Women and Development," *Bulletin Institute of Development Studies, University of Sussex* 10, no. 3 (1979): 42–52; Vandana Shiva, *Staying Alive: Women, Ecology and Development* (London: Zed Books, 1988).
13. Sir Albert Howard, *An Agricultural Testament* (1943; special ed., Emmaus, Pa.: Rodale Press Edition, 1972), p. 160. It is clear from the history of Indian agriculture that Howard had placed his education primarily in the hands of Indian peasant women.
14. Ibid., pp. 23 and 14.
15. The Rodale system of organic gardening is a direct result of peasant women's farming. Rodale was a student of Howard's.
16. Wes Jackson. "Meeting the Expectations of the Land," in *Altars of Unhewn Stone: Science and the Earth* (San Francisco: North Point Press, 1987).
17. Cady, *From Warism to Pacifism.* pp. 3–4. and Cady, "War, Gender, Race and Class." *Concerned Philosophers for Peace Newsletter* 11, no. 2 (1991): 4–10.
18. Marilyn Frye, *The Politics of Reality: Essays in Feminist Theory* (Trumansburg, N.Y.: Crossings Press, 1983), p. 67.
19. *Minneapolis Star Tribune,* 11 January 1993, p. 4A.
20. For the history of the issue in Mexico, see Frances Moore Lappe and Joseph Collins, *Food First: Beyond the Myth of Scarcity* (New York: Ballantine, 1978), pp. 112–16, and Andrew Pearse, *Seeds of Plenty, Seeds of Want: Social and Economic Implications of the Green Revolution* (Oxford: Oxford University Press, 1980), pp. 33–37.
21. Shiva, *The Violence of the Green Revolution,* pp. 43–44. The fragility of such genetic information cannot be overstated. As one observer commented, "The genetic heritage of a millennium in a particular valley can disappear in a single bowl of porridge." Once lost, it is lost forever. This has, in fact, happened. A cache of wheat germ plasm stored at the Maize and Wheat Institute in Mexico was destroyed when an institute refrigerator shut down during a power outage. (Lappé and Collins, *Food First,* p. 174.)
22. Shiva, *The Violence of the Green Revolution,* p. 31.
23. Lappé and Collins, *Food First,* pp. 357–60.

24. Richard Lewontin, "Agricultural Research and the Penetration of Capital," *Science for the People* 14, no. 1 (1982): 16.

25. Kloppenburg, *First the Seed,* pp. 14 and 55.

26. *Minneapolis Star Tribune,* 1 January 1993.

27. Maria Mies, Veronika Bennholdt-Thomsen, and Claudia von Werihof, *Women: The Last Colony* (London: Zed Books, 1988), p. 134, and Shiva, *The Violence of the Green Revolution,* pp. 189–92. See also Maria Mies, *Patriarchy and Accumulation on a World Scale* (London: Zed Books, 1986).

28. *Sati* is the practice of burning the widow on the husband's funeral pyre. The euphemism, "kitchen accident," refers to situations in which women are doused with kerosene in the family kitchen and burned to death, often by the husband's relatives. The cause is often disappointment over a dowry. Female infanticide has become a subject of national debate in India with the advent of genetic testing for sex before birth.

29. Kloppenburg, *First the Seed,* p. 3.

30. Shiva, *Staying Alive* (London: Zed Books, 1988), p. 156. At a political level, Green revolution techniques marginalize Third World countries in other ways. Products such as DDT that have long been banned in the First World, because of their known damage to human health and the environment, are still widely sold in the Third World. Pesticide accidents among people who are illiterate and cannot read directions for recommended use are responsible for 40,000 deaths each year.

31. Andy Crump, *Dictionary of Environment and Development* (London: Earthscan Publications, 1991), p. 32.

32. David C. Korten, "Sustainable Development," *World Policy Journal* 9 (1991): 177.

33. Lappé and Collins, *Food First,* p. 61.

34. See Ramachandra Guha, *The Unquiet Woods: Ecological Change and Peasant Resistance in the Himalaya* (Berkeley: University of California Press, 1990) for a detailed account of the relation of person and place in the Himalaya foothills.

35. Dalit means "the oppressed." It is often preferred as a more descriptive term than either Untouchable or *Harajin* (Gandhi's term, meaning "children of God").

36. Founded by Wangari Maathi, the Green Belt Movement is run by women. It provides tree seedlings for planting in green belts in rural and urban areas. The women who plant the trees receive twenty-five cents for each tree that survives for at least three months. This generates income for women, who are generally excluded from the wage labor market. It also generates self-esteem: these plantings have a success rate of over eighty percent. 50,000 women and children have planted over 10 million trees since the movement began. Maathi recently won the alternative Nobel Prize. The Chipko Movement is a movement of indigenous women in the Himalayan foothills which began when they hugged trees to prevent deforestation at the hands of lumber companies. The movement has grown into a comprehensive development program that addresses the environment, health, education, and political justice.

BIOTECHNOLOGY TO THE RESCUE?

Twelve Reasons Why Biotechnology is Incompatible with Sustainable Agriculture

JACK KLOPPENBURG, JR.
AND BETH BURROWS

Biotechnology is a commodity being researched and paid for by profit-oriented corporations. Yet even amongst those who are opposed to corporate control of biotechnology, there are many who argue that genetic engineering has a role to play in developing sustainable agriculture. Biotechnology, however, is a technology that has been shaped by a narrow range of private interests—interests that are incompatible with the demands of an ecologically-sound and socially-just agriculture.

The transnational corporations which are the principal engines of"globalization" foster and promulgate a vision of their activities as both inevitable and progressive., On 15 April 1994, the *New York Times* carried a six-page advertisement that began:

> *1944 Bretton Woods: The IMF and the World Bank*
> *1945 San Francisco: The United Nations*
> *1994 Marrakesh: The World Trade Organization*
> *History knows where it is going.*
> *It's in Morocco that 124 countries are signing*
> *the GATT agreement.*[1]

The business interests that placed the advertisement want the public to believe that history has an immanent direction and purpose, that the International Monetary Fund and the General Agreement on Tariffs and Trade (GATT) are part of a natural evolutionary flow and that the business view of the future is the only possible view.

In the same vein, proponents of biotechnology increasingly portray genetic engineering as both "natural" and inevitable: by doing so, these industrial *conquistadors* effectively plant biotechnology's flag on the future—on everyone's future—claiming it as their own and denying the space for other visions to be articulated. For example, in a 1990 article entitled"Planetary Patriotism", two top executives of the Monsanto Corporation maintained that caring for the environment while meeting a growing demand for food "requires sustainable agriculture" and that "sustainable agriculture is possible only with biotechnology and imaginative chemistry,."[2] To oppose biotechnology is thus construed as opposition to sustainable agriculture. It is to be a know-nothing, backward looking, knee-jerk opponent of technical progress.

But there is nothing inevitable about biotechnology, and to see it as such is to succumb to a determinism that is disempowering to many who might otherwise oppose the transformations that capital is so busy engineering. Whether biotechnology is widely adopted or not will depend on choices made by people in all walks of life, not least the choice of whether or not to resist its application and further development.

Yet, even among those whose knowledge of history renders them sceptical of corporate claims as to the benefits of biotechnology, there is a reluctance to throw the biotechnology baby out with the corporate bath water. Might there not be *some* truth in the assertion that "sustainable agriculture is possible only with biotechnology and imaginative chemistry?" Could judicious use of biotechnology's powerful capabilities by people of goodwill not help facilitate a transition to sustainable agriculture? Surely biotechnology must in some way or another be compatible with sustainable agriculture? Following are 12 reasons why it cannot.

1. The Commodity Form

Commodification is the central tendency in contemporary society. The principle on which businesses operate is that everything should be bought and sold, that nothing should be beyond the reach of the market. Neither knowledge nor the production of knowledge have escaped this commodification. By definition, commodities are available to those who can purchase them, and since purchasing power is a function of the distribution of wealth and income, business interests are the ones today in a position to dominate the purchase of knowledge and thus to control its development and deployment.

Those business interests have a record of gross environmental abuse. Despite its executives' pretensions to the status of "planetary patriots," Monsanto remains one of the largest polluters in the United States. The company was responsible for five per cent of the 5.7 billion pounds of toxic chemicals released into the US environment in 1992. Moreover, Monsanto has played a central role in promoting the type of high-input, homogenous, industrial agriculture that environmentalists have long opposed as non-sustainable, non-regenerative and socially inequitable.

Not surprisingly, those who are familiar with the dismal record of Monsanto and other major corporations promoting biotechnology—ICI, DuPont and Bayer to name but a few—are rightly sceptical of claims that biotechnology can be a force for good whilst it remains in corporate hands. In the United States, some 75 per cent of all research and development is now undertaken by private industry. Worldwide, companies such as Monsanto, Ciba-Geigy, American Cyanamid, DuPont, Hoechst, ICI, Unilever and Elf-Aquitaine now have the almost exclusive capacity to develop and introduce biotechnologies.

Given the increasing commodification of science and technology and the reality of where purchasing power lies, it is naive to expect a new tool such as biotechnology to remain outside of corporate control, let alone to deliver an ecologically-sustainable and socially-just agriculture.

2. You Can't Take the Baby Out of the Bath Water

Technological utopianism, even among progressive people, manifests itself in the hope that, if removed from its corporate entanglement, biotechnology could be used for sustainable ends. But the baby of modern biotechnology is not so easily separated from the corporate bath water. The new genetic technologies are being developed in corporate and academic laboratories in the industrial, commoditized, capitalist North to serve industrial, commoditized and capitalist ends. It may be possible to use genetic engineering technologies for other ends, but they will best serve those purposes for which they were originally and expressly designed.

Even if biotechnologies developed in corporate laboratories could be tactically modified into derivative "appropriate" or "indigenous" or "peoples'" biotechnology, any successes—and there would certainly be some—would be relatively insignificant compared to the social effects of deploying commercial biotechnologies. Some derivative biotechnology may help the poor, but it will not do so as well as research that starts with the needs of the poor as its point of departure.

Furthermore, any successes would be used to justify the unrestricted advance of genetic innovation by providing evidence that biotechnology can indeed serve (a few of) the poor or (a few of) the goals of agricultural sustainability. Indeed, Monsanto, USAID and other organizations have already started distributing surplus biotechnology (non-critical techniques and knowledge) just as Public Law 480 programmes were used to distribute surplus US wheat in the 1950s and 1960s. The Rockefeller and Ford Foundations—as they did with "Green Revolution" technologies—are financing the development of biotechnology expertise in the network of international agricultural research centres associated with the CGIAR system and in selected developing nations.

Such technical "aid"—like the wheat in earlier years—will bind recipients more closely to an expanding global market for these technologies and to a commitment to a particular approach to the production of knowledge.[3] Scientific and technical cadres from developing nations are already trained in the use of techniques and the identification of problems as they are defined by the techno-scientific community in the industrial North, implying continued technical, economic and epistemic dependency. Following someone else's leading edge means foreclosing opportunities to take a fundamentally different path to knowledge production.

3. Against Technological Determinism

Knowledge and technologies are social creations. Of course, neither scientists nor engineers make technology any way they like (just as people make history but not in any way they please). Nature presents certain constraints, but which elements of nature are explored and how nature is interpreted are determined largely by those who have the power to fund research.

The results of research can be unexpected; new knowledge can open up multiple possibilities for future application. But how such new knowledge or technologies are deployed depends in large measure on social rather than technical parameters. Existing agricultural technologies have not so far been used in an environmentally-responsible and socially-just manner. Why, then, should one expect that the new biotechnologies will be so used without any changes in the social institutions that manage the creation and use of knowledge?

Ecological and social sustainability follow from social arrangements, not from the technologies developed.[4] A focus on technologies as causal agents is a deflection from the hard but necessary task of social transformation, which is precisely what corporations intend. In their advertising, annual reports and public presentations, companies researching biotechnology continue to speak of technical miracles as both necessary and sufficient solutions to all the problems we face.[5]

The world's poor and hungry find a prominent place in these fables as the ultimate beneficiaries of biotechnological research and development. Companies such as ICI Seeds, Britain's largest seeds merchant, proclaim that biotechnology will be the most reliable and environmentally acceptable way to secure the world's food supplies. Similarly, an advertisement from Monsanto depicts maize growing in the desert with the caption:"Will it take a miracle to solve the world's hunger problems?" The implicit message is that to regard biotechnology as anything but an unalloyed good is not only irrational but downright inhumane: to reject biotechnology is to reject the best hope for a solution to world hunger and is thus to perpetuate the suffering of starving children.

But the poor and hungry are poor and hungry because they have little money with which to purchase agricultural inputs (not to mention land) or food. And companies are not in business to give away their products; they sell to those who have the money to buy them. Increasing agricultural production (even assuming this is possible through biotechnology) whilst leaving the structural causes of poverty and hunger unaddressed is a recipe not for feeding the world but for continuing to starve sizeable numbers within it.

4. Structural Limits to Corporate Research

The scope of current biotechnology research is in fact fairly restricted. Even assuming that industry's intentions are as green as it claims, companies still need to sell products and therefore need to develop saleable technologies as quickly as they can. This imperative imposes structural limits on how "alternative" corporate research can be. Private research money is invested in those areas of research that are commodifiable; they are not invested in those areas that cannot be quickly or easily packaged and sold or which are antagonistic to commodification because they are practices rather than discrete technologies (crop rotations for insect control, for instance). Industry will not develop "management technologies" (practices which involve manipulation of multiple variables in an agroecosystem and are therefore systemic) because they cannot be easily sold.

Thus industry tends to develop technical products which redress symptoms rather than solve underlying problems. For instance, against the depredations of the Colorado Potato Beetle, Monsanto offers not better methods of crop rotation but potato plants engineered to produce bacterial toxins in their tissues which will kill the beetle,[6] Monsanto has constructed the problem as the Potato Beetle, not as potato monoculture. Existing systems of agricultural production are left intact while the root causes of unsustainability are left untouched. Bovine growth hormone, the Flavr Savr tomato and herbicide resistant crop varieties are additional examples of this tendency.

5. By Their Fruits Shall You Know Them

What is there to show for 20 years of research into biotechnology, a billion dollars of research expenditure and countless hours of scientific labour? There is recombinantly-derived bovine growth hormone (rBGH or BST), a product which cost more than $500 million to develop and is so socially unsustainable that the European Union took the unprecedented step of banning it on economic grounds. There is the Flavr Savr, Calgene's $25-million tomato with an added "anti-sense" gene that confers an additional three to five days' shelf-life, its sole supposed advantage over existing commercial varieties. Unfortunately, Flavr Savr tomatoes are also unusually susceptible to bruising and damage in shipment.[7] The results of blind taste tests have been mixed. Whether or not the Flavr Savr is something to savour, its economic palatability may be compromised by its price, about double that of existing tomatoes.

6. It Hasn't Paid Off

Indeed, perhaps the most striking thing about the products of biotechnology is that there are so few of them. Beyond rBGH and the Flavr Savr, there are only a handful of additional products available. A few corn and potato varieties engineered to contain *Bacillus thuringiensis* toxin are soon to be marketed, and soybean and cotton plants resistant to herbicides can also be purchased. But despite frequent media reports of"a revolution on the farm", there has been no flood of agricultural products from either commercial or academic laboratories.

Living organisms are simply not as easy to transform or to recode as a computer programme; biology is far more complex than the genetic engineers anticipated. Developing more biotechnological products will be far more difficult and orders of magnitude more expensive than companies had expected.

Researchers maintain a public optimism that breakthroughs are just around the corner, if only the research monies continue to flow.[8] Difficulties in actually developing and bringing genetic engineering to the market are seen, not as a reason to consider other strategies, but as evidence of the need to commit further resources to the effort. An enormous amount of time, money and labour has already been lavished on biotechnology with uncertain benefits. Monsanto alone has spent $500 million over the last decade in order to develop a plant resistant to its Roundup herbicide. If genetic engineers cannot get the Flavr Savr to market unbruised unless it travels in special packaging, can they be expected to produce anything of benefit to campesinos in Guatemala?

7. The Road Not Taken

In the United States as elsewhere, there are all manner of traditional and visionary farmers at the margins and in the interstices between technological convention and scientific orthodoxy—Amish, Mennonites, Native Americans, organic farmers, perennial polyculturists, low input producers, seed savers and horse farmers—who continually produce and reproduce a landscape of alternative agricultural possibilities. Despite an agricultural policy environment actively hostile to their interests, many of them have managed not just to survive but to thrive. Would a sustainable agriculture be more realistic if the $300 million Monsanto spent in developing bovine growth hormone had been spent on gaining a better understanding of rotational grazing of dairy herds or the structure of Amish Farming?

Compared with the resources directed to biotechnology, little time, money or scientific brainpower has gone into pursuing sustainable agriculture through means other than genetic technologies. Little effort has been made to understand the farming systems developed by those who—by their very survival outside conventional agriculture—have demonstrated useful and workable alternatives. The University of Wisconsin, for example, raised $27 million for a new Biotechnology Center. Meanwhile, it houses both its Center for Integrated Agricultural Systems (a phrase that intimates sustainability without offending corporate sensibilities) and its Agricultural Technology and Family Farm Institute in a tiny, remodeled furnace building on the university campus.

There are, however, various progressive people of goodwill scattered throughout the biotechnology community who try to bring a commitment to peasants and small farmers or, more frequently, to an environmentally-benign agricultural biotechnology to their work. But their work is poorly funded or embedded within other research considered more interesting by conventional agribusiness or agriscientific interests.

An example of this is the work of CAMBIA (the Centre for the Application of Molecular Biology to International Agriculture), based in Canberra, Australia where researchers are trying to use sophisticated genetic engineering techniques to serve small and peasant farmers. They are attempting to develop apomixis—the ability to develop seeds through asexual reproduction—to allow farmers to propagate and maintain superior cultivars through seed. They are also attempting to develop "marker gene codes" with which to signal plant reactions to environmental changes, enabling farmers to improve observation and selection of cultivars.[9]

However useful apomixis might be to farmers, it is a threat to the seed industry, and thus despite the potential of its work. CAMBIA maintains its handful of scientists with difficulty. An argument put forward for not supporting them is that no one has enough knowledge to attempt what CAMBIA proposes. But meanwhile, the US National Science Foundation, the Department of Energy and the National Institutes of Health are providing $100 million over a ten-year period for researchers to map the genome of a small, mustard-like plant, *Arabidopsis thaliana,* which is emerging as a key model for genetic engineering.

8. Our Compromised Universities

If companies cannot or will not carry out alternative research agendas, could publicly-funded agricultural universities do so? Universities have long been the servants of agribusiness, and industry's influence on the direction of university research is growing rather than diminishing. Companies engaged in biotechnology are now reaching into universities in an unprecedented fashion, establishing a wide variety of arrangements to draw knowledge out of academia and into the corporate labs. Popular mechanisms include consulting, participation of university researchers on corporate scientific boards, university/industrial consortiums, and direct, sometimes very large, industry contracts for research to be undertaken by universities. And the companies do not make the grants out of the goodness of their corporate hearts. Hoechst, for instance, expects something in return for the $70 million it has provided to Harvard; Monsanto expects something out of the $62 million it has given to Washington University. What the companies get is exclusive licences, patent rights, placement of corporate scientists as observers in university labs and early access to information.

Such arrangements present several challenges to academic and scientific integrity; conflicts of interest may arise; the free flow of information may be constrained; and the public may lose a pool of experts who might have been relied upon for a disinterested analysis of scientific issues and options since the best and the brightest scientists are often most closely associated with industry.

But most problematic is the influence of corporate support on the research agenda itself. American Cyanamid can have research on rBGH undertaken at the University of Wisconsin because it has given more than $300,000 to the Department of Dairy Science. As public funding shrinks, the university is turned into a marketplace in which its intellectual labour is sold to the highest bidder. Public science, because of its close connections to industry, is constrained in its capacity to provide alternatives to industrial modes of developing biotechnologies. It has become complementary to, rather than competitive with, corporate biotechnology.

9. Pascal's Wager: Betting on the Wrong Side

Does the existence of possible useful applications of genetic engineering in agriculture alter this critique of biotechnology? Consider the application of Pascal's Wager to the possible effects of modern biotechnology.[10] Pascal decided that if he bet against the existence of God, lived a life of "poisonous pleasures" and was wrong, he condemned himself to eternal misery. But if he bet on the existence of God and lived an honest, generous life and was wrong, he sacrificed only a few fleeting pleasures.

Following Pascal, it would be prudent to bet against the free deployment of genetically-engineered organisms unless we were more than reasonably certain that their effects and proliferation could be known and controlled.

Corporate and academic biotechnologists argue that they are or can be reasonably certain of the environmental effects. Indeed, they maintain, restrictions on the release of genetically-engineered organisms into the environment should be reduced on the grounds that gene transfer is nothing more than what nature itself does. Yet techniques such as recombinant DNA transfer and protoplast fusion are qualitatively different from sexual techniques of reproduction. To suggest that splicing a gene into a plant or animal is the same as making Roquefort cheese is deeply problematic. It is like saying that the atomic bomb is nature's way of making war and is not fundamentally different from napalm.

Attempts to legitimize local and indigenous knowledge by calling it "people's biotechnology" merely play into the hands of corporations by making the very point that they are trying to get across; that there is nothing particularly new about biotechnology. Given biotechnology's power to create wholly new organisms that could not be produced without the intervention of a scientist, and a corporately-funded one at that, it represents a qualitatively new threat to the environment from the possible escape and proliferation of recombinant organisms, from unintended gene interaction which no one foresaw or intended.

The degree of the threat is uncertain, but it is possibly very large and certainly, given our track record with exotic species (for example, kudzu and zebra mussels introduced into the US and rabbits released for hunting in Australia), the risks are far from zero.[11] This is especially true of organisms such as insects and viruses now being designed as "transgenic predators" for biological control programmes.[12]

That there is a relative lack of ecologists and ecological data to gainsay or confirm claims about the environmental effects of genetically engineered organisms reflects more the paucity of funding for ecological studies than it does the robustness of claims of low potential risk. Apparent lack of evidence of hazards should not be taken to imply safety but to reflect lack of testing for such hazards.[13]

Nevertheless, a variety of studies have shown that genetically engineered organisms may act in entirely unexpected ways.[14] Danish researchers recently confirmed that genes inserted into oilseed rape are capable not only of moving to related weeds but of doing so very rapidly.[15] University of Nebraska scientists have found that the Brazil nut genes transferred into a soybean plant not only produced the desired proteins but also produced other proteins potentially fatal to people allergic to Brazil nuts.[16]

The very difficulty that biologists have often had in getting transgenes to function is evidence of how imperfect their understanding is not only of the genome itself but also of genome-environment interaction. The environmental inactivation of newly-introduced genes means that one should be particularly wary of efforts to use transposable genetic elements to confer pesticide resistance on predatory mites, for instance. The very transposable genetic elements that permit genetic engineers to make cross species transfers may facilitate unanticipated additional transfers.[17]

Activist groups opposed to the development of biotechnology have emphasized potential dangers to human health and the environmental impacts of releasing genetically-engineered organisms. Claims of imminent danger are symmetrical with claims of imminent benefit, and people do respond to obvious or apparently obvious threats. But emphasis on possible threats to human health may not provide a broad-enough basis for mobilizing wider public concern. Just as the benefits of agricultural biotechnology may be slower in coming than many had anticipated, so too some real health problems associated with genetically-engineered food may be rarer than opponents suggest. Health and environmental problems associated with biotechnology will doubtless increase relatively slowly over time but it will be difficult, if not impossible, to partition the effect of genetic engineering in causing the problem from other variables.

Each of those variables acting in isolation may be relatively benign. An increase in cancers, for example, might not be attributable just to atrazine or just to a genetically-engineered tomato. But individual benign components may have adverse synergistic effects. Again, Pascal's Wager seems useful. Why add to the chemical and biological burden our bodies are already bearing?

10. Against Reductionism

It is also possible to oppose biotechnology on epistemological grounds. What is at issue is not simply the nature of biotechnology, but the nature of Western science.[18] In this critique, science is regarded not as a neutral tool, but as an undertaking stamped with the character of the society which produced it. Our society has been characterized by domination and control (of women, of classes, of ethnic or cultural groups, of other species and of natural resources) and the frames of reference we use to analyze and interpret the physical world reflect that character.

Hence, even if biotechnology were to come under broader social guidance, simple redirection of objectives might not be enough to develop sustainable technologies in agriculture. Agricultural scientists are, after all, working within a dominant epistemological paradigm, commonly described as reductionist. That is, the methodological principle of scientific research is to break a problem down into manageable pieces and look for those elements which provide the greatest amount of control over the phenomenon of interest. This method of research is powerful but fundamentally anti-systemic. When a few variables are selected for detailed analysis, scientists tend to lose sight of how those variables are articulated to the rest of the problem under scrutiny, with unanticipated and unfortunate results.

For example, in trying to increase yields in corn, agricultural scientists in effect reduced a cornfield to corn genes and bags of nitrogen. The development of genetically-uniform hybrid corn, needing high applications of inorganic fertilizer, produced high yields (and created the fertilizer industry and the seedcorn industry). Because they have inorganic fertilizer, most corn farmers no longer worry about soil fertility and thus, instead of planting rotations of leguminous crops to maintain fertility, plant corn continuously. With apparently unlimited fertility, they plant the plants closer together, for which they need to get rid of weeds and thus use herbicide. When insects and weeds become resistant to the herbicide, farmers use more and more pesticides which flow into the groundwater. Because nothing but corn is grown, soil begins to wash away; in addition, the pesticides kill off micro-organisms in the soil and organic additions are reduced. Using heavy mechanical equipment, the soil is compacted, compounding erosion problems. Costs continue to rise because the equipment runs on oil and the chemicals are petroleum-based. Farmers end up paying more and more for inputs that eventually degrade and destroy the environment into which they are being ploughed.

Such effects of reductionistic and linear thinking take a while to emerge. By the time they do and are recognized as such, a powerful set of interests has built up to sell the various piece-meal solutions to the different problems and thus sustain the reductionist system—interests which want to carry on selling their solutions regardless. In California, for example, companies did not stop promoting methyl bromide, even though they acknowledged it was destructive to the ozone layer.

Biotechnology merely slots into this approach rather than challenging it. Private industry thrives on the reduction of complex systems to component parts, a simplification which creates space for perpetual rounds of product development. The very nature of modern biotechnology, operating as it does at the molecular and cellular levels, makes the reductionistic, piecemeal route especially tempting.

Rather than drawing attention to a systemic understanding of the interactions in agroecosystems such as a cornfield, apple orchard or cow pasture, biotechnology focuses its vision in the opposite direction, downward and inward rather than upward and outward. Recombinant DNA technology may provide an atrazine-resistant soybean plant, but it will not yield information about crop rotations, cultivation, natural predators, allopathy, cover crops, intercropping, timing of planting—all of which might be alternate and systemic routes to weed control.

11. Biotechnology Distances Us From Other Species

Biotechnologists are not limiting their activities to micro-organisms. Scorpions, salmon, pigs and cattle are all slated for "improvements" of various sorts. In what has been dubbed "pharming." sheep, goats and cows are being genetically engineered to produce human proteins in their milk.[19] The intention is to make animal milk a perfect substitute for human milk. This will doubtless be marketed as better than human since it will be subject to rigorous quality control and will not contain the toxins now found in all humans.[20]

Such a wholesale loss of respect for the integrity of other species is endorsed even by many conservation biologists captivated by the "price it to save it" approach. In writing of "unmined riches", E. O. Wilson describes what he calls "The New Environmentalism":

> "The race is on to develop methods, to draw more income from the wildlands without killing them, and so to give the invisible hand of free-market economics a green thumb."[21]

All species come to be viewed as little more than sets of genetic codes.[22] The UN Convention on Biological Diversity comes to be little more than a means to legitimate a market for genes.[23]

12. Biotechnology Distances Us From Each Other

If we regard other species as mere commodities, can we fail to see each other—or parts of each other—as commodities as well? Corporate and academic biotechnologists have recently begun to focus on human genetic information as a raw material. The leading journal *Science* reports on a programme of bioprospecting that targets our own species: "Geneticists want to collect DNA from such groups as the Arewete. Just 130 members of this tribe remain on the Xingu River in Brazil."[24] Reading that, one wonders what kind of sensibility would rather have the genes than the people? A clue may lie in the $70 million deal that Hoffman la Roche recently struck with the Millennium company to work on mapping the human genome. If species, including our own, are treated as commodities, our treatment of each other will increasingly come to reflect differential values in the market.

Looking Elsewhere

For these 12 reasons and in the absence of dramatic social, political and economic changes, we should not expect companies, universities, scientists and technocrats to use biotechnology to develop a sustainable, regenerative, low-input and diversified agriculture. There is little reason to believe that rBHG, herbicide resistant corn or cloned Douglas-firs assist a move towards sustainable agriculture. Biotechnology for the foreseeable future will continue to be dominated by and respond principally to the needs of industry.

Technology and social power are not separable. The way knowledge is developed and the uses to which knowledge are put reflects the distribution of social power. The development of modern biotechnology is being shaped by a narrow range of private interests which considerably influence the agendas of our public universities. There are no reasons to believe the Planetary Patriots who tell us otherwise or support the technical applications they are promulgating. We should look elsewhere for approaches towards sustainable agriculture.

This article is drawn from *Does Technology Know Where it is Going? 12 Reasons to Stop Expecting Modern Biotechnology to Create a Sustainable Agriculture and What to do After the Expectation Has Ceased,* available from the Edmonds Institute, 20319-92nd Avenue West, Edmonds, WA 98020, USA.

Notes and References

1. *New York Times,* 15 April 1994, p. A19.
2. Schneiderman, H. A. and Carpenter, W. D., "Planetary Patriotism: Sustainable Agriculture for the Future," *Environmental Science and Technology,* Vol. 24, No. 4, April, 1990, p. 472.
3. For discussion of whether indigenous farmers receive appropriate compensation for patented materials derived from seed they have been induced to store in international germplasm collections, see Rural Advance Foundation International, *Conserving Indigenous Knowledge: Integrating Two Systems of Innovation,* commissioned by the United Nations Development Programme, 1995. See also Mackenzie, D. F., "Battle for the World Seed Banks", *New Scientist,* July 2, 1994, p. 4.
4. See the essays in Smith, M. R. and Marx, L., (eds.), *Does Technology Drive History?* MIT Press, Cambridge, 1994.
5. See Kloppenburg, J., Jr., "Sustainable Agriculture and the New Biotechnologies," *Science as Culture,* Vol. 2, p. 4, No. 13, 1991, pp. 482–506; Kleinman, D. L. and Kloppenburg, J., Jr., "Taking the Discursive High Ground: Monsanto and the Biotechnology Controversy," *Sociological Forum,* Vol. 6, No. 3, September 1991; *From Field to Plate and Biotechnology: Agriculture and Your Food,* Ag-West Biotech Inc., Saskatoon, Saskatchewan, Canada.
6. Galloway, J. A., "Potato Blight has Built-in Insecticide," *The Wisconsin State Journal,* May 1995.
7. King, T. T., Jr., "Low-Tech Woes Keep Hindering the Success of High-Tech Tomato," *The Capital Times,* 12 April 1995, p.5B.
8. See, for example, Moffat, A. S., "Mapping the Sequence of Disease Resistance," *Science,* No. 265, September 23, 1994, pp. 1804–1805.
9. Jefferson, R. A., "Beyond Model Systems: New Strategies, Methods, and Mechanisms for Agricultural Research," *Biotechnology R & D Trends* in the Annals of the New York Academy of Sciences, Vol. 700, New York, 1993.

10. See Orr, D.,"Pascal's Wager and Economics in a Hotter Time," *The Ecologist*, Vol. 22, No. 2, March/April, 1992, pp. 42–43.

11. For the latest problem, see Shenon, P.,"In Vietnam, the Snails Dine Regally," *The New York Times*, 17 May 1995, p. A9. See also Ingham, E., *Biosafety Regulation: Why We Need It*, The Edmonds Institute, Edmonds, Washington, 1995.

12. See Goodman, B.,"Debating the Use of Transgenic Predators," *Science*, No. 262, 3 December 1993, p. 507; Morell, V.,"Australian Pest Control by Virus Causes Concern," *Science*, No. 261, August 6, 1993, pp. 683–684.

13. See Doyle, J. et al.,"Effects of Genetically Engineered Microorganisms on Microbial Population and Processes in Natural Habitats," *Advances in Applied Microbiology*, Vol. 40, 1995, p. 237ff.

14. For examples, see Hill, R., "OSU Study Finds Genetic Altering of Bacterium Upsets the Natural Order," *The Oregonian*, 8 August 1994; Holmes, M. T. and Ingham, E. R.,"The Effects of Genetically Engineered Microbes on Soil Food Webs," *Abstracts*, Ecological Society of America, Vol. 75, No. 2, June 1994; Finnegan, J. and McElroy, D.,"Transgene Inaction: Plant Fights Back," *Bio/Technology 12*, pp. 883–888.

15. Leary, W., "Gene Inserted in a Crop Plant Quickly Spreads to Weeds, a Study Shows", *The New York Times*, 7 March, 1996, p. A8.

16. Nordlee, J. A., et al., "Identification of a Brazil-Nut Allergen in Transgenic Soybeans," *The New England Journal of Medicine*, 14 March 1996, pp. 688–692.

17. Goodman, B., op. cit. 12.

18. A different criticism of biotechnology—as outmoded reductionist science—has been made by Mae Won Ho, Brian Goodwin and others. Much of what we have labelled "science" they might term "technology" parading as science. For their criticism of biotechnology, see, for example, Ho, M. W., "Genetic Engineering: Hope or Hoax?", *Third World Resurgence*, no. 53/54, p. 28ff.

19. Mullaney, T. J., "Martek sells its First Product," *The Sun*, December 14, 1994, p. 9c.

20. See Crouch, M., "Like Mother Used to Make?", *The Women's Review of Books*, Vol. 12, No. 5, February, 1995, pp. 31–33.

21. Wilson, E. O., *The Diversity of Life*, W. W. Norton & Co., New York, 1992, p. 283.

22. See, for example, "From Wolf to Poodle: Taming and Changing Animals", *Your World, Biotechnology & You*, Vol. 5, Issue 1, 1995 developed by the Pennsylvania Biotechnology Association "for seventh and eighth grade classroom use".

23. Note, for example, the argument for ratification made by US Secretary of the Interior Bruce Babbitt, Secretary of State Warren Christopher, and Secretary of Agriculture Mike Espy in a 16 August 1994 letter and Accompanying Memorandum of Record to George Mitchell, Senate Majority Leader, and Robert Dole, Senate Minority Leader:

> "US ratification of the Convention benefits US agriculture by safeguarding US access to agricultural genetic resources, and encouraging conservation of such resources in other countries . . . The majority of important US agricultural crops and livestock originated in other parts of the world, and the major sources of the variation essential to future improvements, through traditional breeding and biotechnology, are located outside US boundaries. Access to this germplasm is essential to continued improvement in the productivity of US crops . . . The US depends on access to foreign germplasm for plant breeding programs of such key crops as corn, wheat, soybeans, potatoes, cotton, and most vegetables . . . By becoming a

party to the Biodiversity Convention, the US will ensure continued access to genetic resources."

24. Roberts, L., "A Genetic Survey of Vanishing Peoples," *Science*, Vol. 252, 21 June 1991, pp. 1614–1617.

ENVIRONMENTAL ETHICS AND TROPICAL RAIN FORESTS:

Should Greens Have Standing?

ALASTAIR S. GUNN*

Almost everyone in the developed world wants the logging of tropical rain forests to stop. Like Antarctica, they are said to be much too important and much too valuable to be utilized just for development and are said to be part of a global heritage. However, it is not that simple. People in the developing world consider our criticisms to be ill-informed, patronizing, and self-serving. We are seen as having "dirty hands." They hold that we neither have nor deserve moral standing as critics until we change our trade policies, rhetoric, and extravagant lifestyles.

Introduction

Almost everyone in the developed world (hereafter "the North") wants the logging of tropical rain forests to stop. Environmentalists argue that at present rates of felling there will be no major tropical forests left within a few decades, that these forests contain most of the world's biological diversity, that they are the sources of many existing and potential foods, drugs, and other products, that millions of species are becoming extinct before they are scientifically described, that governments corruptly hand out logging concessions to relatives and friends, that the cultures of traditional peoples are being destroyed, and their standard of living greatly reduced, that the economic benefits of logging go primarily to elites, that the agriculture created on deforested land is unsustainable, and that tropical forests are needed to slow down climate change. Like

Antarctica, they are said to be too important and valuable to be utilized just for development and are said to be part of a global heritage.

Brazil, Malaysia, and Indonesia especially have been the targets of criticism and, sometimes, protest. The Penan people of Sarawak, Eastern Malaysia, have become a symbol of the worldwide situation of hunter-gatherers. The Swiss environmentalist Bruno Manser, star of the movie *Blowpipes and Bulldozers,* [1] has been instrumental in bringing the Penan to world attention; in addition, pop star Sting has toured the world with an Amazonian tribal leader raising money and support to save Amazonian forests. The case against continued logging seems overwhelming to many in the North.

However, it's not that simple. For instance, much of the criticism presented by environmentalists is irrelevant to the lives and aspirations of people in the South, the vast majority of whom would be more at home living in San Diego or Sydney than in a rain forest: there are probably only a few hundred hunter-gatherers remaining in Asia. In this paper, I am concerned not so much with the truth of the claims made in the opening paragraph. True or not, it doesn't follow that publicizing them will change the world, largely because the governments and individuals in the North have "dirty hands" and neither have nor deserve standing as critics of the development process in the South until they change their trade policies, their rhetoric, and their extravagant lifestyles.

Ignoring the South

In late 1991, I spent three months in Southeast Asia, including a three-week visit to Malaysia; I concentrate in this paper on Sarawak because it has been a focus of anti-logging protest.[2]

Part of what I mean by "ignoring the South" is the fact that during a six-week visit to the United States in spring 1992 I presented versions of this paper several times and discussed the ideas in it with many people, all of whom were very opposed to the logging of tropical rain forests. When I asked them to specify *which* areas they were especially concerned about, apart from Amazonia, many mentioned Sarawak, yet very few had any idea at all of the location of Sarawak or any other areas of Southeast Asia where rain forest logging is taking place.

The view from Southeast Asia is very different from the view of the North, which to Asian people often seems ignorant, biased, hypocritical, patronizing and self-serving. We[3] are seen as having no concern for their affairs except when our self-interest is affected.

To illustrate, in the second half of 1991, a thick haze developed in areas of Sarawak and Sumatra, eventually spreading through much of Southeast Asia. Initially the haze was thought to be caused by forest fires in Sumatra and Kalimantan, though there seem to have been several factors. Some airports were closed for weeks, ships were lost at sea, and at times in the worst areas visibility was down to a few hundred meters. Although there was no natural cloud, I did not see the sun at all for three weeks in Kuala Lumpur and Singapore. Because the amount of sunlight reaching the ground was only about one third of normal, growth of rice and vegetable crops was reduced by about twenty percent. Some Malaysian hospitals registered marked increases of bronchial asthma and pneumonia in children and conjunctivitis in adults. Yet this disaster was almost completely ignored by the overseas media.

The Malaysian Prime Minister, Dr. Mahathir, stated at the opening of the ASEAN Economic Ministers' meeting in Kuala Lumpur on 7 October 1991 that the fires were causing much more damage than the controlled extraction of timber, while providing no benefits whatsoever; yet, "There has not been one squeak of protest from environmentalists of the North." Why, he asked, if the governments of rich countries are so concerned to save rain forests, did they not send teams of experts and heavy equipment that the poor countries of the region cannot afford? Where were the experts who had put out the oil fires in Kuwait with such skill and at such profit? "Perhaps it is because the haze does not spread to their countries. Perhaps it is because they cannot sound noble as they do when they champion the Penan." In response to what a television reporter called "a cynical question," whether Malaysia had asked for aid, Mahathir replied:

> We did not ask the people of the North to come to Malaysia to champion the cause of the Penan, but they came anyway. If they want to save the forests from burning, they won't wait to be asked.[4]

Cynical or not, it got results. Almost immediately, the Indonesian government appealed for help in fighting the fires and eventually received offers from several countries.

Whose Version of the Facts?

A necessary condition of having standing to make moral judgments on any issue is a knowledge of the situation being judged. It is unfortunate that ignorance is often no barrier to people having opinions, in environmental as in other matters. The proper opinion for the ignorant is no opinion. However, the facts about rain forests are not easy to come by: a major problem in evaluating the arguments over logging is evaluating the conflicting factual claims on which they are founded. For instance, even Malaysian environmentalists[5] acknowledge that national parks are properly protected and that on paper both Federal and Sarawak State government forest management plans are some of the best in the world. Officially, the policy is rigidly enforced, and one independent study of an area in Malaysia showed that 91 percent of mammal and 77 percent of bird species persisted after selective logging had occurred.[6] However, critics charge that, in practice, logging is highly destructive because little attempt is made to enforce the policy, and that if present rates of cut are continued, "we can expect the primary forests in Sarawak to disappear in five to ten years time."[7] Whom should we believe? How should we *decide* whom to believe?

Apart from selectively logged production forests, considerable areas are also clear-cut for conversion to palm oil and rubber plantations, orchards, and agriculture. The Malaysian government maintains that these uses are sustainable, just like the rice, softwoods, or dairy products grown in Bali, Canada, or New Zealand. In contrast, critics say that the plantations, especially, are environmentally damaging and not sustainable.

How are we to evaluate these claims? Despite the huge and growing literature on sustainability, there is no agreement on what precisely it means.[8] In fact, it ought to be possible for reasonable people to agree on a general account of sustainability in terms of types of land uses that can be continued without reducing the overall options for future generations. On this basis, experience has shown beyond all possible doubt that cattle ranching is unsustainable in former rain forest in Amazonia: by 1986, *all* the ranches established between 1965 and 1978 had been abandoned.[9] It is equally clear that rice and vegetable growing can continue indefinitely in deforested areas of Java and Bali. Presumably by now there is experience relevant to resolving the conflicting claims made about plantations. Whatever the conclusion, it is important to point out that removing forest to make land suitable for agriculture has happened everywhere. All over the world, both sustainable *and* non-sustainable agriculture have just one thing in common: the land which it uses has been massively altered from its original state, whether forest, scrub, prairie, swamp, or estuary.

Meanwhile, people have to live, and if logging were stopped immediately, the economies of Kalimantan and Sulawesi in Indonesia and Sarawak and Sabah in Malaysia would collapse. In Sarawak, which has a population of only 1.8 million, the state economy earned about $1 billion[10] in 1990 from log exports alone. However, critics charge that the economy in its present form provides relatively few benefits to ordinary people, and that native people such as the Penan, Kayan, Iban, and others suffer severe deprivation due to the destruction of their traditional resource base.

The Malaysian government's Federal Land Development Agency has settled hundreds of thousands of unskilled landless people on plantations. Again, the foreigner is at a loss to evaluate the success of this policy. For many years, the projects have provided an income comparable to that of middle income urban workers, as well as the opportunity for settlers to own their land and homes. Critics claim that housing and pay are poor and conditions dangerous. Certainly a great deal of pesticide is used, mostly paraquat. But even finding out who is at risk is difficult. According to one source,[11] 80 percent of the sprayers are women, but an expatriate Malaysian (who asked not to be named) told me that he doubted this figure: because of an acute labor shortage, many plantation workers are illegal immigrants, nearly all of whom are male.

Who is the expert, and when experts disagree, who can decide between them? Governments of the South are constantly receiving advice from overseas. But what works here may not work there, and projects funded by outside agencies have often turned into expensive economic, environmental, and social disasters, such as the Aswan High Dam. There are lots of less dramatic examples.

Environmental experts from rich countries frequently generalize about life in the South in ways that they would not dream of generalizing about life in their own countries. For instance, a recent book from the highly respected Worldwatch

Institute refers to "the wattle-and-daub villages and urban shantytowns where most of the Third World lives,"[12] a clause that cries out for deconstruction.

More generally, and perhaps on the basis of such confident stereotypic generalizations, we often set ourselves up as expert diagnosticians of the South's problems. The Worldwatch Institute in 1992 listed six "unmet human needs." Presumably calling them "*human* needs" is meant to indicate that they are universal. In addition to child malnutrition, lack of safe drinking water, preventable fatal childhood disease, and preventable fatal reproductive health problems in women, the list includes the facts that about 1 billion adults cannot read or write and that more than 100 million children of primary school age are not in school.[13] These are "needs" only within certain cultural contexts; yet we are meant to rate them on the same scale of urgency as conditions that are life-threatening, not just to humans, but to any animal.

I believe that we should generally accept that on social and environmental questions, local people know best; nevertheless, the selection of local experts is not a straightforward matter. Local people sometimes disagree with the development policies of their own governments. For instance, the New Zealand-assisted reafforestation project in Bukidnon, in the Philippines, is bitterly opposed by some locals; among other reasons, they fear that many small farmers will have their land confiscated, leaving them no option but to work as laborers in industrial forestry plantations.[14] Clearly there are different views about the appropriate direction of development. Critics argue that plantations destroy traditional life, force people into a cash economy, and lead to greater dependence on foreign markets. The Malaysian economy, they argue, should be self-sufficient, which means replanting forests, not industrial plantations.

The Sarawak State Government claims that the Penan are better off living a "modern" lifestyle in settled communities than as nomads. The Dayak leader, Dr. Palau Rubis, who is himself a minister in the State Government, believes that moving into the cash economy has saved his people from "self-destructive habits," and the headman of a Penan village, Kampung Jambatan Suai, is quoted as saying that his people "have been seeking a better life ever since they moved out of the jungle 70 years ago."[15] However, an article in *Singapore Business* takes the view that outside the "showplace" settlement near Mulu National Park, "the Penan live in poor conditions."[16] An exhibit on the lifestyles of traditional people in the Kuching Museum, Sarawak, paints a similar picture. Next to an exhibit of traditional longhouse and nomadic life, a notice states without comment that it is now government policy to persuade the Penan to abandon their nomadic lifestyle and become settled: "The series of photographs shows life in a typical Penan village." The forty pictures show a spartan, decrepit village inhabited by listless, sullen, and generally deprived-looking Penan glowering at the camera. Obviously, someone at the museum doesn't rate the policy very highly. Again, as a foreigner whose experience is limited to a brief tourist visit, I have no way of evaluating these views.

The role of shifting cultivators is itself disputed. The Sarawak government claims that slash and burn is practiced over 3.3 million hectares and is the main cause of forest degradation. Evelyne Hong, formerly of West Malaysia's Institute Masyarakat, says that the farmers clear less than 18,000 hectares of primary forest per year and that the practice must be sustainable because it has been going on for centuries.[17] Their role in the fires of 1991, and even the nature of the fires, is unclear. According to Indonesian government figures, 50,000 hectares of forest were destroyed. However, the Australian Rapid Assessment team found no evidence of any extensive fires in the area of East Kalimantan that it was able to visit in October, but rather a series of small fires, all intentionally lit, in forests, plantations, and farmland.[18] There is certainly evidence that major fires have begun as small fires deliberately lit by shifting cultivators, but logging and the 1991 drought (the worst in thirty years) help to create conditions that are conducive to fires getting out of hand, in which a cigarette butt or broken bottle can cause a major fire.[19]

Tropical Forests: Whose Heritage?

The Dutch did their best to suck Indonesia dry. Yet, Java's huge population is largely a legacy of the Dutch who by their extortionate tax system forced the rapid conversion of the island to farmland, thus enabling the population to increase from six million in 1825 to over 100 million today.[20] A friend said wryly that all the Dutch have bequeathed to his 90-percent Muslim country are some Heineken breweries. Now the Netherlands has banned the import of tropical timber. Such moves are often angrily rejected by local people, who claim that we have no business criticizing their policies.

Some people in the South take seriously the view that foreign environmental organizations are secretly funded by producers of softwood and non-timber building materials who hope to remove the cheap competition of rain forest timber. Many more consider anti-development arguments that begin with the premise that tropical forests are part of a global heritage to be attempts to promote Northern economic interests under the pretence of concern about the global environment:

> All sorts of campaigns are mounted by the richer countries against [destroying forests in order to generate] hydroelectricity, which have already developed their hydroelectric potential. Now of course the World Bank will be used to deprive poor countries of cheap hydroelectric power. Can we be blamed if we think that this is a ploy to keep us poor?[21]

They object very strongly to demands that their development strategies should be determined by our concerns about global warming and ozone damage. Most of the damage has been caused, and continues to be caused, by people in rich countries; brown people do not run much of a risk of skin cancer; in the tropics, climate change will be both less marked and less damaging to agriculture than in temperate countries. Yet, we expect people in the South to make sacrifices that are largely for our benefit.

The rhetoric of global heritage is, in fact, highly inappropriate. For there to be a shared heritage, there must be a *community*. A global heritage needs a global community in which resources are shared and decisions made by consensus. This is not how our world operates. If we are to have some say over tropical forests located in other sovereign states, we have to stop expropriating most of the world's resources and effectively excluding the South from decisions about Antarctica and the oceans. In other words, much more must be done than offering to cancel a few hundred million dollars of debt in return for not logging a few hundred square kilometers of forest.

Even when our criticisms are genuinely inspired by a desire to reduce human misery, we may be seen as lacking standing to make such criticisms. For instance, it is true that the economic benefits of development have not been shared equally by the people of the South, and I share the view of critics, that it is wrong for a few to have immense riches while others are malnourished, homeless, or deprived of basic health care. Nevertheless, it is the colonialist countries themselves, and neo-colonialist agencies such as the International Monetary Fund and the World Bank that are largely responsible for both establishing and enriching the new elites in the South. In any case, many countries of the North present the same contrast *or even greater* contrasts between the enormously rich and the desperately poor. The income of the richest twenty percent of Indonesians is eight times that of the poorest twenty percent—but in the United States the rich earn twelve times more than the poor, and in France, thirteen times.[22] The United States Medicaid program—intended to provide free medical treatment for the poor—covers only 40 percent of the residents of the state of Georgia living below the federal poverty line; yet at any one time, 40 percent of the hospital beds in Georgia are empty.[23] In cities of grace, style, high culture, and enormous wealth—London, Sydney, Seattle, and Washington D.C.—homeless people wander the streets; there are so many beggars in Seattle that the city passed an "Aggressive Panhandling Law" proscribing them from excessively importuning passersby. I was solicited by far fewer beggars in Southeast Asian cities than in the cities just mentioned. Meanwhile, the six largest corporations in the United States, which have a combined market value of $376.41 billion, use up vast quantities of nonrenewable resources, and grossly pollute the Earth and its human inhabitants.[24]

Rich countries, then are seen as hypocritical. We appeal to countries of the South to reduce their population growth; yet, over his or her lifetime each baby born in the North will consume many times the resources of a baby born in a poor country.[25] Even our environmentalists live fabulously rich lifestyles compared with the 30 million Indonesians who are below the official poverty line of $90 per year, and the millions more who earn less than $1 per day.[26] When Indonesians see foreigners, they see people like me whose idea of economy is a $15 hotel room. Even Malaysians, whose per capita income will be up to the lower end of the OECD nations within twenty years, consider foreigners to be generally rich, and are mostly happy to go along with forest exploitation because they see it as the basis for their increasing prosperity. Malaysia's *Vision 2020* plan calls for the country to become "fully developed" by 2020. Singapore's Strategic Economic Plan calls for a per capita GNP equal to that of the United States by

2030, and even on the lowest growth scenario, the achievement of the level of the Netherlands by 2017. Given our prosperity, any views we might have of their development programs should take account of these aims.

However, it is generally considered in the North that wide-scale industrialized development is impossible. Alan Miller's is a representative view:

> We may ethically [desire] the extension of our own standard of living to less privileged people in the developing world, but the very fact that the 5 percent of the Earth's population now living in North America consumes almost one third of total global resources demonstrates the limitations of such a fantasy of global consumption. If only 15 percent of the total world population consumed food, fiber, energy, and raw materials at the rate that the U.S. Americans and Canadians do, . . . virtually nothing would be left for the remaining 85 percent of the world's people. Furthermore, . . . capital stocks and mineral resources could never supply the developing world with the same materials infrastructure needed for its rapid development to advanced industrialization. There is simply not enough steel, lumber, mineral resources, and energy available to enable developing nations to replicate the Western systems.[27]

While there is some truth in this view, it takes no account of the reality of the conditions of life in the South: the issue is much more complex than such a simplistic comparison suggests. One cannot just extrapolate without qualification to the South from what is used (and wasted) to provide a given level of material culture for people living in areas with wide seasonal variations in climate, many of whom reside in sprawling, widely separated cities, where the infrastructure and industries are already in place, and are therefore inefficient. True, it seems very likely that our way of life cannot be sustained even within the North for much longer, let alone exported globally. I was born in 1945; perhaps my children and I will be the only generations in the history of the Earth to spend our whole lives in a society where the meaning of life is shopping. But what are we to conclude from this state of affairs? Surely not that we are licensed to tell everyone else what material standard of living they should aim at.

Nations of the South view outside attempts to influence their policies as attempts at recolonization,[28] for example, our restrictive trade practices and our unwillingness to share innovations in science and technology. Former colonial nations are seen as especially lacking in standing because they have caused massive damage both to their own environments and to those of their former colonies in order to enrich themselves. As early as 1839, Darwin noted that "Wherever the European had trod, death seems to pursue the aboriginal."[29] It was the British who introduced the plantation system into Asia, and indeed in some cases the crops: tea in India and rubber in Malaysia. As Dr. Mahathir notes:

> When the British ruled Malaysia they burnt millions of acres of Malaysian forests so that they could plant rubber. The timber was totally wasted because in those days nobody wanted Malaysian timber. . . . Malaysians got nothing from the felling of the timber. In addition, when the rubber was sold practically all the profit was taken to England.[30]

Yet, Britain now condemns Malaysia for continuing the British policy of converting forest to plantation.

Our reply is often an admission that we made mistakes, that we sacrificed irreplaceable natural wonders in order to pursue economic goals, but that we have now seen the error of our ways. However, this sort of argument tends to be seen as patronizing: "When you're older and wiser like me, you'll be glad I warned you against making the same mistakes as I did." In any case, it is what we do that counts, not what we profess to believe. As Aristotle noted, "Mere wishes or intentions remain in the vague, and even the unjust *pretend* that they wish to act justly."[31] For instance, the United States, where most people eat more protein than their bodies can absorb, annually imports well over $500 million worth of beef from Brazil alone.

It is not as if we rich countries are even now setting a very good example of looking after our own environments. Politicians in the South are struck by our wasteful habits, such as the amount of food thrown away in restaurants, the air conditioning left on when no one is at home, and the depletion of underground water resources in Southern California to maintain golf courses and artificial lakes around luxury hotels. In August 1991, Bruno Manser, who by then was heading an international rain forest organization, chained himself to the top of a utility pole in London. The Malaysian Deputy Prime Minister Ghafar Baba commented:

> Why did he chain himself to an electric pole? It is obvious that there are not many trees in London, nor are there many trees in other similar big cities. So, why should we listen to these Western people who do not themselves practice conservation?[32]

This criticism, to be sure, is not true of serious environmentalists, and many of the same people in the North who want to end logging in tropical forests also want their own forests to be preserved. Still, few developed countries have more than 20 percent of their original forest cover. Japan, alone, far exceeds this percentage, but its citizens' affluent lifestyle is maintained only by massive imports of timber and food, and there is a major worry about the future of its plywood industry as Sarawak phases out exports of raw logs. Millions of tons of topsoil in North America, Europe, and Australia are blown or washed away every year, or poisoned with chemicals. New Zealand is often considered to be a "clean and green" country, and so it is in many ways; yet almost all its sustainably farmed land is former forest or wetland. In Britain, attempts at commercial afforestation are strenuously resisted by the same environmentalists who oppose tropical logging because they do not wish to lose the parklike English countryside. Meanwhile, to meet the ever increasing demand for paper products, Britain imports ever increasing amounts of rain forest timber.[33]

Should Northern Greens Have Standing?

When people in the South reject our criticisms of their policies—usually politely, but sometimes angrily—they are not just saying that they will do as they like and that we should shut up. Their position has a sound moral base, which is that we have no business criticizing their policies.

Questions about the standing of a person to judge morally the actions of others arise in at least three ways. First, I may claim that the behavior in question is not your business because it is part of my private life—for instance, in Mill's sense of behavior that affects no one except the agent, or which has only trivial effects on others. Or it may be the legitimate concern of some other persons, but not of yours: my family or my church have an interest in my religious activities that others do not. That is, it is simply not your place to comment on my religious beliefs or church attendance, unless, for example, they cause nontrivial harm to others. Mill's harm principle is not the only way to draw the line, of course, but only the most authoritarian moralist insists that *every* aspect of the lives of others is his or her business.

If we wish to influence environmental management in the South, we will have to establish standing in this sense. For instance, we could try to show that their policies are unjustifiably causing nontrivial harm to other people. Or we could try to demonstrate to them that they are unjustifiably destroying things of considerable intrinsic value. Or perhaps we could try to convince them that we know better than they do what is in their best interests, even though given the history of colonial exploitation, this approach is unlikely to succeed. Whatever moral basis we may think we have, we need to do more than merely assert that they ought to change their ways.

Second, it may be claimed that our dirty hands disqualify us as moral critics. The vice of hypocrisy consists in advocating or professing a commitment to behavior but not following one's own precepts. A "successful" hypocrite maintains credibility by concealing the discrepancy between what he or she says and does. The story is, of course, more complicated. For instance, an adequate account of hypocrisy needs to distinguish it from socially understood and accepted forms of saying one thing while intending to do something else, as in games of bluff or in societies where it is polite to make extravagant offers of assistance that no one would dream of accepting. However, I firmly resist the temptation to begin with such an account, noting merely that central to hypocrisy is the attempt to deceive. Hypocrisy is wrong for the same sorts of reasons that lying and deception are wrong: it is coercive; it undermines trust; if widely practiced, it is damaging to the social fabric, and so on.[34] Not only are we not morally justified in preaching to the South principles that we do not practice and have no intention of practicing ourselves, but we have a moral obligation to refrain from doing so.

It may be said that I am being too harsh in making such sweeping charges of hypocrisy. Someone may fail to notice the discrepancy between the principles that they preach to others and their own behavior. They may not be aware that the rich countries cause most of the environmental damage, such as ozone depletion and global warning. Perhaps they do not realize that in driving a car to the grocery store to buy a liter of milk they are wasting over 99.9 percent of the energy used. Or they may know these facts, but fail to connect them with their strictures on the behavior of others. Ignorance may be justifiable, or it may be a form of self-deception, as when someone resolutely avoids finding out facts which would be unwelcome. For instance, suppose that a university requires all instructors to have their students complete course evaluations, but allows instructors to design their own evaluation forms. A sexist or racist professor may profess to be very fair in his dealings with students, but it may not seem necessary to him to include a question on fairness, and thus he will not discover that he discriminates. If this omission is genuinely due to ignorance, is it immoral? Whatever we make of this case, I suggest that there is a continuum from honest and excusable ignorance through deliberate hypocrisy, and that in educated societies the environmental ignorance mentioned above is near or at the highly culpable end of the scale.

It is true that the acceptability of an ethical view ought to be independent of the sincerity of the person who is presenting it. However, we do not trust people whom we discover to be hypocrites, just as people in the South do not trust rich outsiders. "Unsuccessful" hypocrites lack moral credibility—for instance, parents who drink or smoke while preaching the evils of drugs to their children—because they exhibit precisely the behavior of which they profess to disapprove. Certain high profile American religious leaders who had sought to project a general air of moral righteousness and publicly condemned the sins of others were completely discredited when the details of their sexual behavior were revealed. Such persons, because they have demonstrated a complete contempt for morality, simply have no right to tell others how they should live. They have ruled themselves out of the ethical debate because they refuse to play by its rules. Perhaps their credibility can be rehabilitated if they show clear evidence of moral improvement, though not if it seems to be just a result of getting caught (or getting AIDS).

A third respect in which one might lack standing is a matter of motives. If I advise or exhort you to behave in certain ways because (I claim) it is your duty or in your best interest to do so, you are entitled to assume that my choice of advice or exhortation is based on a sincere conviction about what *really* is your duty or in your interests. If you come to believe that my real motive is to benefit myself, I lose standing. For instance, if I set up business as Acme Computer Consultants, you will expect me to give you independent purchase advice. But if you discover that I am really a local distributor for a line of products, you will no longer trust me to be impartial.

Romancing the Stone Age

Any ways of life that turn out to be sustainable for people in developed countries will presumably be very much less wasteful than our present ones. However, I know of very few environmental philosophers who make any attempt to live a much simpler life than the average suburbanite, let alone follow Callicott's advice on "the ethical question on what to eat":

> Purists like Leopold prefer . . . to hunt and consume wildlife and to gather wild plant foods, and thus to live within the parameters of the aboriginal human ecological niche. Second best is eating from one's own orchard, garden, henhouse, pigpen, and barnyard. Third best is buying or bartering organic foods from one's neighbors and friends.[35]

To the extent that traditional societies such as hunter-gatherers lived sustainably—and presumably those that have existed for thousands of years must have been sustainable—they have much to teach us. However, it is not their specific skills that we need—who can live off the land in a city?—so much as the principles behind the ways of life in such societies. Whether or not these are articulated and consciously pursued, they will include an appreciation of limits, an acknowledgment of ecological necessities, a willingness to share resources, and a recognition that we are not the owners of the Earth to do with it as we like.

However, environmentalists from rich countries who write in praise of the simple life often come across as deluded Romantics. Consider this nonsense, written by a teacher who decided to build himself a woodland cabin in Massachusetts:

> My house was decidedly primitive—had no running water or electricity—but there was a satisfying, elemental comfort about life there. . . . I was living a sort of "Third World" existence. . . . I came to relish the elemental experience of hauling water, trimming the wicks of oil lamps, and cooking over a camp stove or open fire.[36]

Yet, another European "discovery"! Of course, he continued to work as a teacher, and when he got bored with playing at being poor, he drove off to have showers and meals at friends' houses, or to go to concerts and the ballet in Boston. The family seems to have had an in-built capacity for self-delusion: his brother Hugh "once won an award for a series of poems he wrote about the experiences of a survivor of Hiroshima."[37]

No doubt this example is extreme; still, as one author notes, admiration of a greatly simplified low entropy lifestyle seems to be typically a "post-affluent" phenomenon: "Only those who have been reared in affluent suburbs can rebel against overconsumption and the banality of materialism." In contrast, he quotes a welfare rights organizer, speaking to a group of young ecology radicals: "We will have some difficulty understanding one another, for our welfare mothers want what you are rejecting."[38]

Rather tentatively, I suggest that the type of oversimplistic value distinctions criticized above reflect distinctions at an epistemological level, for instance, the common division of cultures into "Western" and "Eastern." This division is unfortunate, for a number of reasons: I briefly discuss two here.

First, in the writings of many deep ecologists, as Ramachandra Guha notes, "Complex and internally differentiated religious traditions—Hinduism, Buddhism, and Taoism—are lumped together as holding a view of nature as believed to be quintessentially biocentric."[39] This view of Eastern philosophies (and day-to-day cultural practices) is, as he points out, romantic, monistic, and simplistic. I might add that to talk of a monolithic "Western culture" is equally crass. The Basque independence organization ETA has put out a "culture map" of Europe which indicates the many and diverse cultures of the region, the many different languages, artistic traditions, and so on.

It is both intellectually obtuse and insulting to lump all "non-Western" cultures together. The following story, for whose truth I cannot vouch since I was not present, illustrates this point well. A philosopher was presenting a conference paper about the relevance of Native American culture to environmental ethics. At the end of his presentation he was severely criticized by a young woman of Native American extraction. As any introductory anthropology text will make clear, the many groups who had lived in North America before the Europeans came were very diverse, she said, and he had no business talking about "Indian culture." There were hundreds of different languages, religions, and ways of life. There were societies that were patriarchal and some that were matriarchal; some that were peaceful and egalitarian and others that were warlike, who tortured their prisoners of war and took slaves. There were nomads, cliff dwellers, city builders, hunter-gatherers, artisan and trading cultures, and others that depended largely on banditry. They did *not* all spend their time praying to the buffalo and the bear prior to killing and eating them.

More prosaically, what happens outside the North tends to get lumped together by aggregating statistics from dozens of different countries, each regionally diverse, as if they were all collected by identical means and as if all measured the same phenomena by the same measuring stick. According to the Worldwatch Institute, more than one billion people survive on less than $1 a day.[40] Nevertheless, most people—especially poor people—in the South live their lives only partly in the cash economy, and in any case foreign exchange rates are largely irrelevant to poor people, because they are determined by factors such as the price of internationally traded goods and a country's level of indebtedness. The official exchange rates which are used to calculate per capita income are, therefore, not an indication of purchasing power. There is no comparison between the food that a person with "$1" ends up with in Los Angeles and rural Java. Thus, dollar comparisons overemphasize the gap between people on different parts of the Earth.

A second reason for rejecting the simplistic distinction is that it doesn't fit the facts. As one writer notes:

> When people involved in the management of protected areas use these terms [*traditional, native, indigenous*], the distinction they seem to want to make is between those who live harmoniously (in a strictly ecological sense) with their immediate environment and those who don't. In the current management literature it is too often presumed that traditional societies, native peoples, and indigenous peoples do; modern societies, colonists, and other non-indigenous peoples don't.[41]

As I noted earlier, we have much to learn about sustainability from traditional peoples. Nonetheless, the myth of "the primitive conservationist" cannot be sustained in the light of such evidence as wasteful and environmentally destructive Stone Age megafauna hunting practices in North America, Madagascar,

New Zealand, and other areas, and the damage caused by tribal peoples' pigs, goats, and sheep.[42]

Is There a Role for Foreigners?

Local people's objections to logging are often said to be a result of improper foreign interference. This claim was a major theme in the Sarawak elections in late September 1991—in which all anti-logging candidates were decisively defeated—with one cabinet minister advising voters to "ignore Westerners instigating natives to set up timber blockades."[43]

Again, it is hard to know how to evaluate these claims—everyone in all countries, given half a chance, tries to blame their problems on foreigners. Penan people themselves certainly reject the claim that their protests are instigated by foreigners: "It is only the foreigners who bother to come and spend time with us."[44]

One thing is quite clear, though: almost no one, least of all Malaysian environmental groups, welcomes public criticism or anti-logging protests by foreign individuals or non-governmental organizations. Such activities are useless, partly because, as a Singapore journalist puts it, "Pressure tactics and western style straight talking and logic do not work in Asia, much less in former colonies."[45] No doubt they are often counterproductive, even in terms of their own objectives. Eight foreign environmentalists entered Sarawak in July 1991, posing as tourists, and then proceeded to stage protests, including chaining themselves to cranes in a log pond. They were all arrested, charged with trespass, and sentenced to short prison terms.

> In the end, what became the issue? The foreigners were seen as being persecuted by the repressive Malaysian government that violates human rights to protest. The real issues—the rights of native people, the merits or otherwise of government policy—were pushed into the background.[46]

This case is reminiscent of the moral outrage in Australia and New Zealand at the hanging in Malaysia of two Australians and at the death sentences (later commuted) originally passed on New Zealanders, Lorraine and Aaron Cohen, in all cases for importing heroin. Now it is really impossible for anyone who can read to enter Malaysia (or Singapore) without knowing the penalties for drug offences—even Australians who may not have noticed the warnings printed in their passports—since they are emblazoned on entry forms, presumably to allow the truly ignorant one last chance to flush their stash down the toilet. However, the fact that Australians and New Zealanders do not regularly protest about Malaysian drug laws presumably suggests to Malaysians that we do not care whether Malaysians are executed or flogged for drug offences, but that we want foreigners who freely choose to import narcotics into Malaysia to be treated differently.

Local Greens apparently do not share the widespread admiration for Bruno Manser. The Malaysian environmentalist quoted earlier (note 11) said, "He *probably* has good intentions, but he has really become like a white adventurer, a

Tarzan."The media emphasis on his role has made it look as if the Penan are the only people affected by logging—in fact, now only about 300 to 400 of the 9,000 Penan live as hunter-gatherers—and has also "made them look stupid" because of the exaggerated leadership role attributed to Manser. The video, *Blowpipes and Bulldozers,* presents him as a romantic outlaw, dividing his time between playing his panpipes, recording his observations of the picturesque local flora and fauna (including the local people), and organizing the Penan, before being "smuggled out." Presumably the government was prepared to tolerate him up to a point, for it is hard to believe that he was really a fugitive for several years in a state which waged a successful campaign against armed and highly organized Communist guerrillas not so long ago and where Swiss environmentalists—and Australian film crews—stand out.

The major reason why foreign individuals and governments should avoid public ranting, though, is that most people in the South think that we have no business telling them how to run their affairs. National sovereignty is taken very seriously in states that have won their independence only comparatively recently. This distaste for foreign interference may be difficult for people in the rich countries of Europe and the British Commonwealth to understand, since, unlike the United States (where even many Americans seem to have forgotten their revolutionary origins), we did not "win" our independence. Although all the countries of Southeast Asia except Thailand were European colonies and also suffered under Japanese occupation, none were colonial settler states. We should try to imagine how we would feel if we had gained independence only a few decades ago, perhaps after a bloody war, as in Indonesia. We should imagine too that, like Malaysia or Singapore, only thirty years ago our country was desperately poor, but has since made great progress economically; unlike almost everywhere else in the world, people in Southeast Asia enjoy constantly rising living standards as measured by improvements in health, life span, infant mortality, literacy, and per capita income. Finally, we should imagine that about 99 percent of our population is of non-European origin. Then we should consider how we would react to our former rulers telling us how to run our affairs.

There is a lesson here for those Countries that are no longer colonies, but whose populations are overwhelmingly of European extraction. There are direct parallels between the rights of the people of a former colony and those of a colonized minority. When Native Americans, Australian Aborigines, or New Zealand Maori demand control over their own resources, they are asserting an identity. The remark by the Chief Minister of Sarawak that "the forests are our heritage and we will use [them] to benefit our people and not because some Western countries are short of oxygen due to their own environmental pollution"[47] will strike a chord with colonized indigenous people everywhere.

Conclusion

As I hope I have made clear, I do not believe that people in the North have any right to dictate environmental policy to people in the South. However, if they do wish to have an influence over the environmental policies of the South, they should certainly not go ego-tripping off to protest in Sarawak. Rather, they should look at their own responsibilities. International trade is largely dominated by the North. For instance, almost everyone agrees that the world would be a better place if narcotic drugs were not produced. But how much effort do the governments of the North make on the demand side? South American and Asian farmers who produce the raw material for cocaine and heroin would not be growing it if no one were buying. Indeed, cash cropping is itself largely a product of colonialism. If we want the growers—who in any case receive only a tiny fraction of the street price—to stop producing for our markets, we should obviously offer them economically acceptable alternatives, rather than simply assuming that we have a right to dump paraquat all over them from helicopters.

We in the North should also stop preaching a simple life to the South, or if we can't stop preaching, at least live at the level we prescribe for them, and acknowledge that *of course* people in the South have a right to a standard of living equal to whatever level we believe we are entitled to claim for ourselves. We should demonstrate a commitment to global justice by responding to requests from poorer countries for technological and other assistance in ways that they choose, for projects that they see as necessary. We do have a great deal of expertise to offer in areas such as environmental engineering, agriculture, and public health. Although, as I noted above, it is not always easy for foreigners to identify local needs, it is not impossible. The needs of local people can be identified basically by asking them what they want. A good example is the work of the engineers' organization, Water for Survival, which donates technical expertise, materials, and skilled labor for water projects in many areas of the South. These usually involve the provision of safe, conveniently located drinking water supplies at village level and are undertaken in response to requests from local people.

Finally, we should stop making self-serving claims about "global heritage" that amount to a demand that the South stop developing.

---

*Department of Philosophy, University of Waikato, Private Bag 3105, Hamilton, New Zealand. Gunn is a well-known environmental ethicist who has frequently contributed to this journal on such subjects as endangered species, environmental engineering, and ecological restoration. He is a member of the journal's Editorial Advisory Board. Much of the work on this paper was conducted during a short period of sabbatical leave spent at the National University of Singapore. Grateful thanks are due to the Department of Philosophy for hospitality and helpful comments on the paper, especially by Mr. Gob Swee Teh, Dr. Arun Balasubramaniam, and Dr. Padmasiri de Silva.

Endnotes

1. Produced by the Australian organization Gaia Films. Manser lived with the Penan for several years and took part in their protests against logging of the forest areas that had traditionally been their homes. The movie, which is undoubtedly sincerely intended, is almost laughably romantic in places: maidens in sarongs frolic under waterfalls and the sun seems always (in a rain forest!) to be casting dappled patterns of light and shade on the forest floor. It has been criticized by Malaysian environmentalists as giving far too much attention to Manser's role: see below.
2. Information in this section is drawn from newspapers in Singapore (*Straits Times*), Malaysia (*New Straits Times, Star*), and Indonesia (*Jakarta Post*) September and October 1991, and Malaysian television news, 7 October 1991.

3. Throughout this paper, "we" means "people in the North," unless otherwise indicated.

4. Certainly, the Penan are a sensitive issue. When *Time* ran a cover story on the demise of tribal peoples, a section on the Penan was blacked out in copies sold in Malaysia. Eugene Linden, "Lost Tribes, Lost Knowledge," *Time,* 23 September 1991, pp. 38–46.

5. The views of Malaysian environmentalist organizations are available via books such as Evelyne Hong, *Natives of Sarawak* (Penang: Institute Masyarakat, 1987); S. M. Mohd Idris, ed., *The Battle for Sarawak's Forests* (Penang: Sahabat Alam Malaysia, 1989); Philip Hurst, *Rainforest Politics* (London: Zed Books, 1990); and from newsletters of organizations such as Greenpeace and Friends of the Earth.

6. A. P. Johns. "Tropical Forest Primates and Logging: Can They Coexist?" *Onyx* 17 (1983): 114–18.

7. "Sarawak: Who Are You Fooling?" *Utusan Konsumer* 236 (August 1991): 6. This criticism is echoed by Nirmal Ghosh, "Sarawak and the Environmentalist Bogey," *Business Times* (Singapore), 5–6 October 1991.

8. One of the referees of this article for *Environmental Ethics* saw a concern with definition as "a cultural tradition beginning with the shift to literacy." However, an obsession with definition and the supposed difficulties of getting it right seems to me to be an unfortunate legacy of 1950s analytical philosophy, and a worrying sign of the slide of environmental ethics into what John Passmore (personal communication) refers to as scholasticism. Evidently, a balance is needed.

9. R. J. A. Goodland, "Environmental Aspects of Amazonian Development Projects in Brazil," *Interciencia* 11 (1986): 16–24.

10. All figures are in U.S. dollars. Information in this and the following paragraph from A. Balasubramaniam, ed., *Issues in Environmental Education in ASEAN* (Singapore: Friedrich-Naumann-Stiftung, 1986) and Hurst, *Rainforest Politics.*

11. Anonymous Malaysian environmentalist quoted in *ECO Newsletter,* December 1991.

12. Lester R. Brown et al., *Saving the Planet* (New York: W.W. Norton, 1991). The six ASEAN nations—Brunei, Indonesia, Malaysia, Philippines, Singapore, Thailand—which have a population of over 320 million and contain a huge proportion of the Earth's biodiversity, get a total of four references in the index of this book, versus fifteen references to the United States. Indeed, Brunei, Malaysia, and Singapore are not indexed at all, unlike Cleveland, Davis (California), San Jose, and Seattle, each of which gets one mention.

13. Lester R. Brown, *State of the World* 1992 (New York: W.W. Norton, 1992), pp. 4–5.

14. David Robie, "Cloud Over Bukidnon Forest," *New Zealand Listener,* 22 April 1989, pp. 22–23, 35.

15. Kalimullah Hassan, "Not His Uncle's Man," *Straits Times,* 6 October 1991; James Ritchie, "A Better Future for the Penan," *New Straits Times,* 3 October 1991.

16. Nirmal Ghosh, "Jungle Fever in Sarawak," *Singapore Business,* October 1991, p. 36.

17. Ghosh, "Jungle Fever," p. 28.

18. Reported in *Jakarta Post,* 29 October 1991.

19. Dr. S. Martopo, Director, Research Centre for Environmental Studies, Gadjah Mada University, Indonesia, personal communication.

20. For detailed discussion of the points touched on in this section, see *Our Common Future.*

21. Mahathir, quoted in "Boycott of Tropical Timber."

22. Brown, *State of the World,* p. 138.

23. Reported in *Atlanta Journal-Constitution,* 7 April 1992.

24. According to *Financial World,* April 1992, the top six are in the following order: Exxon, Philip Morris, General Electric, Wal-Mart, Merck, and Coca-Cola. The entire external debt of all Asian countries in 1989 was $369.02 billion. In the same year, India's GNP was $287.38 billion, while the GNP of Indonesia, Malaysia, the Philippines, Singapore, and Thailand

together was $251.86 billion. (Figures from World Resources Institute, *World Resources 1992–93* [NewYork and Oxford: Oxford University Press, 1992]). It is hard to believe that the economies of these Asian countries were less sustainable, more environmentally damaging, and less focused on meeting basic human needs, than Philip Morris, Exxon, and the rest.

25. North Americans are not, of course, the only over consumers. For instance, "Per capita, Australians and New Zealanders consume about seventeen times as much of the world's resources as people in the poorest nations." Ann Graeme, "Environmental Choice: The Green Dilemma," *Forest and Bird* [New Zealand] 266 (November 1992), p. 19.
26. As I note later, the resources actually available to poor people in the South are not accurately measured by adding up their cash income and translating it into an internationally convertible currency.
27. Alan S. Miller, *Gaia Connections* (Savage, Md.: Rowman & Littlefield, 199 1), p. 83.
28. The following quotation from Mahathir is typical: "We hope people do not latch on to causes and crusades in order to justify political and economic re-colonization." "Malaysia's Growth Model a Success," *New Straits Times*, 27 September 1991.
29. *The Voyage of the Beagle*, quoted in Alfred W Crosby, *Ecological Imperialism* (New York: Cambridge University Press, 1986), p. vii.
30. Letter from Mahathir to D. Abercrombie, August 15, 1987, quoted in INSAN, *Logging Against the Native of Sarawak* (Petaling Jaya: Institute of Social Analysis, 1990).
31. Aristotle, *Ethics* X 8.
32. Examples quoted in *New Straits Times*: "States Urged to Draw Up Labour Plan," 25 September 1991; "Boycott of Tropical Timber 'Not the Solution,"' 26 September 1991, "Rich Nations Told to Curb their Consumption," 27 September 1991.
33. Norman Myers, *The Primal Source* (Boulder: Westview Press, 1983).
34. Sissela Bok, *Lying* (NewYork: Pantheon Books, 1978).
35. J. Baird Callicott, "Animal Liberation: A Triangular Affair" *Environmental Ethics* 2 (1980): 336.
36. John Hanson Mitchell, *Living at the End of Time* (Boston: Houghton Mifflin, 1990), pp. 6–18. As I understand it, the rural women who in many countries spend several hours each day hauling water do not find this "elemental experience" especially rewarding.
37. Ibid., p. 56.
38. Norman J. Faramelli, "Environmental Quality and Social Justice," in Ian G. Barbour, ed., *Western Man and Environmental Ethics* (Reading, Mass.: Addison-Wesley, 1970), p. 191.
39. Ramachandra Guha. "Radical American Environmentalism and Wilderness Preservation: A Third World Critique," *Environmental Ethics* 11 (1989): 76.
40. Brown, *State of the World*, p. 4. Most of the points that follow are taken from Ziauddin Sardor, *Science, Technology and Development in the Muslim World* (London: Croom Helm, 1977).
41. David Harmon, "Cultural Diversity, Human Subsistence, and the National Park Ideal," *Environmental Ethics* 9 (1987):156.
42. See for instance George Uetz and Donald Lee Johnson, "Breaking the Web," *Environment* 16, no. 10 (December 1974): 31–39.
43. "20pc Ruling for Sarawak Loggers," *New Straits Times*, 25 September 1991.
44. Ghosh, "Jungle Fever," p. 38.
45. Ghosh, "Sarawak."
46. See note 11.
47. "20 pc Ruling."

LAND-USE STRATEGIES IN THE AMAZON FRONTIER:

Farm-Level Evidence from Ecuador

FRANCISCO J. PICHÓN

This article uses farm-level data to assess the role of ecological factors, household characteristics, and policy factors in shaping overall land-use strategies among settler farmers in the Ecuadorian Amazon. The survey data which form the basis for the analysis were collected by the author in 1990 from a probability sample of 450 colonist households. A descriptive cluster analysis is used to highlight the difference across and within the observed land-use strategies regarding the underlying resource base available to farmers, the socioeconomic characteristics of farm households, and the policy environment that affect them. The findings question the inevitability of a generalized pattern of forest clearing over time constrained by a "straitjacket" of natural resources and suggest that the range of land-use options open to farmers is narrowed or widened under different socioeconomic circumstances and policy factors. The results given are exploratory and intended to stimulate further discussion.

Key words: farms, frontier, settlement, land-use strategies; Amazon, Ecuador

The expansion of Latin American agricultural frontiers into Amazonia underscores the role of frontier settlements as an evolutionary process which offers a changing social and natural environmental landscape for migrants. These migrant settlers have in common a production system characterized by intensive use of family labor and simple agricultural technologies, a strong drive for cattle ownership. and over-exploitation of land through continuous incorporation of new areas, with little regard for the long-term preservation of the natural resource base. Government policies have also played a significant role in the frontier via subsidized credit, regional fiscal incentives, differential taxation, and a series of other policies that have encouraged inefficient and destructive forms of land use and forest exploitation.

Systems of land use, crop selection, and forms of human occupation that have been introduced or practiced in frontier regions have been often considered incompatible with the rainforest ecosystem. Settlers who have recently settled in the Amazon are said to bring to the region agricultural practices they are familiar with in places of origin, but which may be poorly adapted to the intricacies of rainforest ecology. Also, settlers are viewed as having an expansionist attitude towards new lands which leads to extensive deforestation, exhaustive use of farm land, and failure to consider the long-term effects of cultivation practices (World Bank 1992; Oberai 1988; Gillis and Repotto 1987; Nelson 1973).

The growing concern about the negative social impacts of frontier expansion into the Amazon basin has also raised important questions about the viability of small-farm colonization as a policy option. Plagued by institutional failures and market bottlenecks, settlement schemes have been unable to absorb a significant portion of the agricultural migrant population seeking to settle in the Amazon region. Colonist farmers have often failed in frontier regions, losing their lands to larger interests or abandoning them in the face of increasing natural resource degradation and land conflicts. There is also growing concern for the socioeconomic and political stratification of frontier society (often referred as the "Andeanization" of Amazonia) which appears to be leading to unsustainable farm sizes, second generation out-migration, and rising unemployment in frontier towns, as well as violence and reliance on illegal crops such as coca. Confirmation of this trend toward replication of origin-area inequalities would severely question the efficacy of frontier settlement as a natural resource and social development policy option, and belief the idealistic goal of frontier settlement as bringing together a "land without people to people without land" (Fearnside 1992).

Since relatively uninhabited land still exists in many Latin American countries, governments in the region are increasingly struggling to develop policies to rationalize the settlement and development process of their Amazon frontiers. However, as evidence has accumulated in the past two decades, contrary views have emerged on the question of how small farmers in the frontier can advance beyond the level of subsistence without destroying their natural resource base. For example, it is widely believed that farmers move through a similar progression of land-use patterns over time, constrained by a "straitjacket" of natural resources. This process has been described by the World Bank and others as the "peasant pioneer cycle" (World Bank 1992; Oberai 1988). The occupation of forest lands necessarily requires some initial deforestation to establish ownership and produce food crops to meet immediate food needs. Colonists later clear additional lands for perennial crops as they become settled. As soil nutrient levels fall, farmers clear new lands again for both perennial and annual crops, leaving the initially deforested areas for pasture and/or fallow. In this way, driven by environmental constraints and survival needs, and the absence of affordable and/or available alternative forms of agricultural technology that better sustain soil fertility, farmers have little choice but to encroach more and more on forested land.

Many analysts have interpreted this instability as evidence that declining crop yields and increasing poverty forces settlers to abandon their farms and seek new lands. However, recent studies in the Brazilian Amazon appear to suggest that rapid farm turnover and land abandonment can be better explained by changes in property rights regimes than by fertility decline (see Schneider 1994 for an excellent discussion of this work). According to these studies, frontier areas have special circumstances determined by abundance of land and relative scarcity of human labor and capital which make development and adoption of natural resource-saving systems unattractive. Moreover, income levels are usually too low to permit (financing) technological changes or investments to preserve the natural resource base.

Whether settler instability is evidence of loss of agricultural productivity or can be explained by rapid changes in property rights, the available evidence points to the need for more accurate diagnosis, including distinguishing the relative importance of socioeconomic and agronomic factors in determining the speed and manner of the settlement process in the frontier. Indeed, one of the most dominant impressions one has traveling in the Amazon region is the extreme variability in land-use strategies from one farmer to the next.

Based upon farm-level survey data collected in the Ecuadorian Amazon, this article highlights the differences across the land-use patterns observed in this region regarding the underlying natural resource base available to farmers, the socioeconomic characteristics of farm households, and the policy environment. To examine what combinations of land-use activities arose most frequently in this frontier area, a cluster analysis was performed. Four types of land-use patterns emerged from the data based upon a mix of land shares allocated to particular production and fallow activities: forest, pasture, annual food crops, perennial crops, and food/perennial intercropped crops. The analysis questions the inevitability of a generalized pattern of forest clearing over time constrained by the straitjacket of natural resources and explores the role of household-level and policy factors influencing land-use strategies and forest clearing patterns. In focusing on Ecuador, this article also introduces detailed consideration of another Amazon country within a topic that has been previously dominated by research mainly in Brazil.

Survey Site, Sample Design, and Data

Unstructured agricultural colonization is an extremely important factor in the redistribution of population across Ecuador's three ecological zones: coast (*Costa*), highlands (*Sierra*), and Amazon region. Historically, the development of new export commodities has paved the way for migration to agricultural frontier areas. Since 1970, a new export commodity (oil), combined with land scarcity and rising unemployment in settled rural areas of the Sierra and Costa regions, has induced a major population movement to the Amazon. Following the discovery of important oil reserves in the northern Amazon in the late 1960s, agricultural migrants from the heavily populated Costa and Sierra regions settled along roads built to support oil development, resulting in major encroachments on the rainforest environment of northeastern Ecuador.

Initially, approximately 10,000 people came to the labor-scarce Amazon to work on a variety of tasks related to the oil boom; 5% to 10% of them remained to establish family farms along the new roads or businesses in the frontier boom towns (Hiraoka and Yamamoto 1980; Bromley 1981). The current population of the Ecuadorian Amazon is estimated at close to 400,000 (unadjusted for the probable significant undercount), half of which lives in the oil-producing provinces of Napo and Sucumbios and over two-thirds of which is rural (INEC 1991). The population which settled near the petroleum zones grew at 8% per annum over the 1974–82 period, the highest rates in the country (the national average was 3% per year). Several of their cantons or counties (including Francisco de Orellana and Lago Agrio, where the survey was undertaken) recorded double-digit annual growth rates. Today the population of the Amazon region of Ecuador is growing faster than that of any other country's Amazon population. According to the last 1990 census, the Amazon region as a whole experienced a rate of population growth of over 5% per year, or double that of the country as a whole, primarily from in-migration. To this day, newly built roads have continued to serve as locational determinants for further influxes of settlers.

Two Amazonian provinces, Napo and Sucumbios, served as study sites for a survey of a random sample of settler households which I conducted in 1990 (Pichón 1993). The heart of the oil producing region, embracing the river basins of the Napo and Aguarico rivers, represents a laboratory for the study of smallholder agricultural settlement since it is the part of the Amazon which has received the largest number of in-migrants. Most land colonization in the region has been spontaneous: no significant settlements have been coordinated or directed by government agencies, though steps have been taken by the Instituto Ecuatoriano de Reforma Agraria y Colonización in the Amazon (or IERAC, the land titling and reform agency) to legalize, *post facto*, the land claims of colonists once they have established themselves.

Because of the availability of fairly up-to-date maps (with settler plots indicated and numbered) in the Amazon offices of IERAC, it was possible to construct a sample frame that permitted the selection of a scientifically representative sample of plots, which has apparently not been possible in earlier, small-area and nonprobability-sample surveys in Ecuador and other Amazon countries. The original sample of 450 households represented about 6% of the rural settler population of this part of the Amazon. This, in turn, makes the generalization of survey findings to the entire settler population of the Napo and Sucumbios region. The sample design was a two-stage cluster sample, with the first stage units comprised of sectors, or areas of settlement, and the second stage consisting of a cluster of households in each sector.

On the basis of two pre-tests of the questionnaire, I revised the questions and questionnaire format and decided, for both substantive and practical reasons, that separate instruments would be administered to the head of the household and spouse. In addition, two-person teams of one male and one female interviewer were determined to be safer for the interviewers in the jungle as well as non-threatening to the isolated and sometimes suspicious colonists. Administering separate questionnaires also allowed for simultaneous interviews with couples to gather more detailed information.

The final questionnaire administered to the head gathered information on migration history and background; land acquisition and land tenure; land clearing; land use and farm production and inputs; off-farm employment participation; etc. The spouse questionnaire covered a household roster; migration background; out-migrants from the Amazon household; fertility, mortality, health problems. etc. In addition, perceptions and attitudes of settlers towards the local climate, quality of soils, and forests were assessed for both the household head and spouse. Each questionnaire took on average less than one hour to implement and presented no significant problems in its field administration.

The fieldwork was extremely rigorous—taking place in a hot, humid area where it rains almost daily throughout the year—involving long hikes through the jungle to settler dwellings. Despite the difficult conditions, only one of the original 10 interviewers dropped out during the four months of fieldwork. Once the initial suspicions were overcome, most colonists were willing to be interviewed by the enumerators. Of the 450 settler households in the original sample, virtually complete questionnaires were obtained for 419 households. The 31 missing cases comprised temporarily unoccupied farms, without residents and/or agricultural activity. Only four households refused to cooperate—resulting in a refusal rate of less than 1%. Some households (66) owned a plot(s) outside the sample area in addition to their sample plot. This analysis, however, focuses on those 353 households which owned plots exclusively in the sample area and did not hold any additional outside plots.

Description of the Sample

Population and Farm Characteristics

Most sample settlers in the Amazon (90%) came from other rural areas of Ecuador, nearly 70% from the Sierra region, and 28% from the western coast. Prior to arriving in the Amazon, over half of the households had periods of residence in other areas away from their places of birth. For example, 8% in-migrated to the Amazon from various provinces in the Sierra while over 42% came from coastal environments, primarily from the Cantón of Santo Domingo (an area of heavy earlier colonization on the western plain) before finally settling in the Amazon. The vast majority of settlers arrived in the 1970s–1980s following the oil boom, with few arriving before 1972, and the peak years being 1979–1981. The average duration of residence in the Amazon was nine years, but respondents had arrived anywhere from the month previous to the survey to as far back as 30 years ago, before the oil boom in the region.

Over two-thirds of the sample had no land of their own before migrating while 14% had labored on small rented parcels of less than 8 ha on average. There were, however, striking exceptions: over 35% of the nearly one-third which had previous ownership of land in their areas of origin (or 11% of the sample) had at least 20 ha, with 20% of these (or 5% of all respondents) controlling 50 ha or more. Thus, although most sample migrants brought little or no assets with them to the Amazon, colonists with significant personal financial resources were not completely absent. Nearly 11% of sample households still maintained ownership and management of assets in their areas of previous residence at the time of interview.

Education levels were lower than elsewhere in Ecuador. About a quarter of household members had completed primary school, and only 10% had formal education beyond primary school. The mean level of education for household heads was 3.9 years, but over 52%, and one-third of all spouses, had either no formal schooling or had attended but did *not* complete primary school. Only 5% of the heads had formal schooling beyond primary school. Household size was well above the national average (6.6 in the total sample of 419 households versus 5.4 nationally), and varied widely from 1 to 26, comprising in total 2,761 persons. The frequency of small households with only one to three persons (17%) was as high as that of ten or more members. Most households were in the 4–6 and 7–9 size categories. Demographic analysis of the data indicates very high fertility and natural population growth and high retention of migrant settlers in the study area (Pichón 1993). Even though nearly 40% of the households already had at least one member out-migrate from the household *subsequent* to their coming to live in the Amazon region, most of the children of settlers are remaining in the region when they grow up. In addition, over 69% of the heads of household intended to remain on their Amazon farms in the future. But those intending to migrate within one year (3%) or even "after one year" (9%) were so few (all the other responses were "not sure") as to indicate only a vague desire to leave someday, which may never come to pass. Thus the "second generation" population pressures are likely to be substantial, adding to the population pressures from continuing in-migration from outside the region.

Landholdings were similar in size, the mean being 44.3 ha, but the standard deviation was high. Landholdings ranged from 1.15 to 400 ha, with over 90% smaller than 60 ha, and 80% between 25 ha and 60 ha. Over 92% of sample households had acquired at least a provisional title and nearly half had a full legal claim to their land (involving a time-consuming and expensive process). The remaining 8% with no title primarily comprised very small plots sold (illegally, strictly speaking) by settlers to later arrivers. Over 61% of the sample settlers obtained their land from previous colonists. Nearly 85% of the households had only one plot in the region, relatively few individuals had inherited the land, and renting and sharecropping arrangements were even rarer (only 2%). About 16%

of the sample colonists had managed to claim or purchase additional plots, the highest number in the survey being seven.

The distance from sample farms to the nearest market town averaged 27.5 km (21.9 km by unpaved primary and secondary roads plus 5.6 km by walking and canoeing), but standard deviations were quite high. The difficult field conditions in the Amazon required that over 50% of the sample plots were accessible only by long walks over muddy jungle paths and/or by canoe along the many rivers and streams in the region (up to 77 km, in the case of one group of sample farms). Still, despite half of sample households lacking immediate access to roads, over 85% engaged in market sales. Although most farms had trees of significant commercial value, only 12% (mainly those next to a road) engaged in timber sales. Forest exploitation at existing levels of extractive technology has labor requirements so high that costs almost equal returns and probably exceed them at a short distance from the road.

Natural Resource Base and Cropping Patterns

In many ways, farmers in northeastern Ecuador inherit a similar situation: the area experiences uniform rainfall (averaging over 2,800 mm) and temperature (an average 24°C) throughout the calendar year, with little geographical differentiation (Estrada *et al.* 1988). However, land quality varies widely throughout the region, including across farm households in the sample. Table 1, describing sample farms' soil quality and topography, shows that good soils and flat lands are common in the study area: 50% of farms had fertile or moderately fertile soils and over 40% were on level land. Still, notable differences in the natural resource base—the "raw materials" available to the farmer upon arrival—do exist within the study area, and are likely to influence agricultural production decisions both initially and over time, making the pattern of land use more complicated than the above *cycle* scenario suggests. Such differences alter the possible land-use "starting points" for any such cycle, as well as the rate of soil deterioration, and accompanying land-use change. For example, some 30% of the moderately fertile or fertile plots occurred on land characterized as hilly or steep, making some land uses such as annual crops or pasture more difficult than others (e.g., perennial crops) from the beginning. For nearly half of the sample farms located on moderate or steep hills, forest clearing can also precipitate severe soil erosion.

Modern inputs, a way of relaxing the natural resource straitjacket, were used on 59% of farms (see Table 1), although actual expenditures on purchased inputs were strikingly low. More importantly, users were uniformly distributed across soil and topography types, showing that farmers were just as likely (or unlikely) to seek out ways of enhancing the productivity of their natural resource base, regardless of the initial productive quality of the land. Low expenditure on land-improving inputs is also consistent with frontier perceptions that it is simply cheaper to exploit indiscriminately the agricultural resource base and move on to new lands than to invest in the sustainability of the productive base. However, few settlers complained about soil quality, supporting the belief that superior soils (Inceptisols) exist in this part of the Amazon. Yet two-thirds of the settlers

Table 1. Soil Types, Topography, and Modern Input Use by Farm

Soil Type	Topography: Level	Hilly	Steep	Combination	N (#)	N (%)	Using Modern Inputs[*] (#)	Using Modern Inputs[*] (%)
Organic, fertile (black)	106	36	12	6	160	45	88	55
Moderately Fertile	10	5	2	0	17	5	10	59
Acidic, low fertility	3	43	39	2	87	25	47	54
Poorly-drained	3	1	0	0	4	1	2	50
Combination of organic and acidic	16	23	10	15	64	18	41	64
Other combinations	9	2	5	5	21	6	17	61
N	147	110	68	28	353		205	
(%)	42	31	19	8		100		58
Using Modern Inputs[a]	89	65	33	18	205			
(%)	61	59	49	64		58		

[a]Number of farms using *one* (or more) of the following modern inputs: chemical inputs, lime, modern seeds, pesticides, livestock medicine, mineral salt, and others.

reported observing declines in crop yields during their time of settlement in the Amazon.

Almost all new settlers, nearly 90% in the survey area, adopt a polyculture system incorporating both cash and subsistence crops. The polyculture is based upon a slash-and-*mulch* variant of the more widely known slash-and-*burn* cultivation system. In the slash-and-*mulch* system, the felled vegetation is not burned due to uniformity of rainfall and temperature throughout the year. The system has four steps. The *socola,* or the clearing of undergrowth, may be started at any time, followed by the *plantio,* during which seeds and cuttings are planted under the shade of larger trees. When sprouts appear, the remaining vegetation, except economically useful trees, is cut down, a process called *tumba de montaña.* The felled vegetation crushes or covers some seedlings (but sufficient plantings survive to provide harvests), and lies unburned where it has fallen, serving as mulch for the cultivated plants. Depending on the intensity of weed growth, there may be one or two *chapeos,* or weedings, before the harvest of annuals.

Under the polyculture system practiced in this area of the Amazon, farmers intercultivate annuals such as corn and rice, semiperennials such as plantains, bananas, and manioc, and perennials, especially tree crops such as coffee, cacao, *guaba, achiote,* and citrus. Table 2 gives an idea of these crops' prevalence in the study area. The varied shares of gross planted area (cleared land minus land devoted to pasture) dedicated to each crop across farmers' plots suggest the broad outlines of the polyculture system in northeastern Ecuador. For most colonists in the study area coffee and then cacao are the main cash crops occupying the larger amounts of cultivated land on the farm, but settlers also allocated considerable areas to grow plantains, manioc, maize and rice. Coffee and cacao were cultivated by 96% and 34% of the surveyed households, respectively, while plantains (cultivated by 84% of the sample), manioc (58%), corn (52%), fruit trees

Table 2. Share of Gross Cropped Area (GCA)[a]

	Farms (%)	Farms (#)	Mean (#ha)	sd (#ha)	Range (#ha)
Coffee	96	339	0.69	0.21	0.14–1
Plantains	84	298	0.13	0.12	0–0.74
Manioc	58	206	0.06	0.07	0–0.50
Corn	52	183	0.21	0.16	0–1
Fruit	47	165	0.03	0.06	0–0.6
Cacao	34	120	0.08	0.11	0–0.83
Rice	16	56	0.10	0.09	0.01–0.45
Af. Palm	1	5	0.16	0.25	0.01–0.45
Vegetable	1	3	0.01	0.01	0–0.02
Others	4	15	0.06	0.03	0.02–0.11
GCA	100	353	7.79	5.92	0–42.1

[a]excluding pasture

(47%), and rice (16%) were the main subsistence crops. Mean land-use values for sample farms were as follows: 6 ha (or 16% of the farm area) were planted in perennial crops, 2 ha (5%) in food crops, 10 ha (22%) in pasture, and 28 ha (57%) remained in undisturbed forest.

Cattle ranching on some scale is also the rule in the study area, with its favorable environmental conditions for pasture. Livestock activity is analogous to the system found in other tropical humid regions of the Amazon and Central America (Jones 1990). Cattle ranching fulfills multiple objectives within the farmer's food security, asset accumulation, and cash flow strategies (Vosti and Locker 1989). Cattle's high productivity to labor ratio is also critically important to the labor-scarce household economy of the frontier.

At the time of the survey, nearly two-thirds of the households owned some cattle, varying from 1 to 422 head, but only 27 households had over 20 head, and only seven over 40 head. Herd size averaged 12 head for cattle owners and seven for all sample farms. The norm was 1 to 4 cattle, used mainly to provide milk for the settler's family and as a source of savings in cases of emergency. Chickens were also raised on 93% of the farms, and pigs on nearly two-thirds; these animals provide protein, cash income, and insurance against crop failures, low crop prices, or"lumpy"medical and other household expenses. Over 45% of the respondents said that, given the chance, they would undoubtedly switch from agriculture to some kind of specialized cattle raising to improve their welfare.

A great number of farmers in the study area get along without agricultural extension. A little less than one-third of the sample households had received agricultural assistance from government, religious, or other organizations since coming to the Amazon region, though over 70% had been visited by the Malaria Control Service, which sprays settlers' dwellings with DDT. During the year prior to the survey, 60% employed hired labor. Only 18% had ever received credit—invariably those who could post sufficient collateral in the form of land titles or fixed assets, with over two-thirds of these having significant debts at the time of the survey. Credit was used mainly for cattle and pasture expansion. Nearly 35% of the households in the survey had at least one family member employed off the farm in the previous year.

Perceived Quality of Life

About two-thirds of the household heads felt that life was better in the Amazon, 20% thought life was the same, and 12% felt that their expectations of the Amazon had not been realized, evaluating their current status as worse than in the former place of residence. These perceptions differed somewhat from those of the spouses (almost always female), who tended to be more negative in evaluating family adjustment to the challenges imposed by the new Amazonian environment. Only 53% of the spouses felt that life was better for them in the Amazon, and over 25% thought it was worse compared to their previous place of residence. Settlers' most keenly felt needs included access to services, such as education, health, water, and participation in community life.

Serious health problems appeared common in the study area, as 75% of the respondents reported that an average of two persons per household (or one out of four family members in the sample) experienced a significant illness during the three months prior to the interview. Over two-thirds felt that health problems were worse in the Amazon than in their area of origin, and over 45% considered them the most fundamental constraint to their ability to produce. Nearly 30% of the farm households had at least one death (the average being 1.5) in the family since migrating to the Amazon.

Land-Use Patterns in the Ecuadorian Amazon: A Descriptive Cluster Analysis

In this section I explore how the natural, social, and institutional environments together shape the land-use systems practiced by farmers. Using a clustering technique, the analysis examines different scenarios of land-use combinations based upon the percentages of total farm area allocated to undisturbed forest, annual and semiannual food crops (including plantains, corn, manioc, rice, vegetables and others), perennial and tree crops (including mainly coffee, but also cacao, African oil palm, and fruit trees), food/perennial intercropped crops, and pasture (including fallow land or, as it is known in the region, *rastrojo*) most frequently observed in the study area. To determine cluster membership, homogeneous land-use subgroups or clusters were generated using a nearest centroid sorting algorithm (see Nerlove *et al.* 1993).

After several initial explorations, it was decided that four clusters, each representing farms with "similar" mixes of the above land-use categories, best reflected observed differences in land-use patterns in the study area. The percentages presented for each cluster represent land-use shares on the cluster's *representative* farm. Figure 1 presents the results of the cluster analysis based on land-use shares.

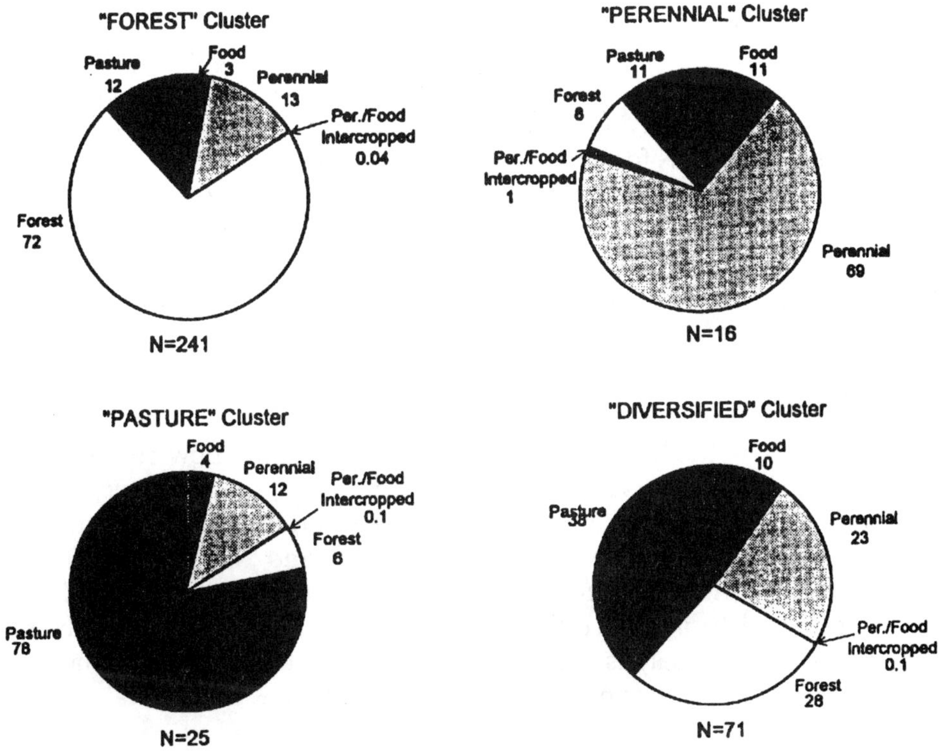

Figure 1. Land-Use Shares (%) on "Representative Farm." by Cluster

Four land-use typologies emerged: *predominantly forest*, with 72% of the representative farm under undisturbed forest; *primarily perennial*, with 69% of the representative farm under perennial crops; *predominantly pasture*, with 78% of the representative farm allocated to pasture; and *diversified land use*, with no predominance of any particular land-use type. It is important not to reduce each cluster to its primary component; rather, it is precisely the *mix* of land-use shares that cluster farms exhibit that caused them to be classified similarly. Therefore, the four typologies capture the *combinations* of land use, including the *types* of land use present on individual sample farms and their relative *importance* to the farmer.

The bulk of the sample (68%) fell into the *predominantly forest* cluster, which also included farms with substantial portions (about 12% of the representative farm) in pasture and perennials. Some sample households (20%) exhibiting a more *diversified land-use* had substantial shares of area in pasture (38%), forest (28%), and perennials (23%). Much smaller portions of the sample (5% and 7%, respectively) fell into the remaining two clusters: the *primarily perennial* cluster, with some diversification (close to 10% on the representative farm) into each of pasture, food, and forest shares; and a less diversified *predominantly pasture* cluster, which involved some land (12% on the representative farm) to perennial crops.

In no cluster did intercropped food/perennial crops cover more than 1% of total farm area. Hence, most intercropping implied intensification of a single land use (as defined above), rather than a combination of uses. Note that the primarily *perennial* and *diversified* cluster farms had a higher share of farm area under food crops than did the predominantly *forest* or *pasture* clusters, indicating potentially important concerns for household food security.

While almost no overlap across clusters existed in share to the predominant land use for each cluster, substantial overlap did exist for other land uses. In other words, two farms could devote similar land shares to, say, perennial crops, and still fall in different clusters because it is the *combination* of land uses, not any one in isolation, that determines cluster classification.

Cluster Results

The factors shaping the land-use pattern in each cluster can now be considered. The *natural resource* base (soil quality and topography), *socioeconomic and farm characteristics* (farm size, education, farming background, duration of settlement, household size, sex and dependency ratios, etc.) and the *policy/institutional environment* (road infrastructure, land tenure, etc.) faced by each settler household are expected to influence the cluster or combinations of land uses they decide to adopt. The analysis below also explores the extent to which land-use decisions may occur in an identifiable cycle driven by natural resource constraints.

The initial quality of the natural resource base (the"straitjacket"within which farmers have to work) "sets the stage" for land-use strategies under given technology in important ways. Differences in the quality of the natural resource base across the four clusters were evident. Examination of environmental factors such as soil quality and topography across clusters shows that *diversified* and *pasture* cluster farms were found primarily on land that allowed greater flexibility in land-use decisions; 63.4% of all *diversified* farms and 64.0% of *pasture* farms tilled fertile black soils, while 62% of *diversified* farms and 68% of *pasture* farms were located in flat lands. Farm households in the *forest* cluster appeared to be more constrained on land-use strategies as they were far more likely to farm infertile soils and steeper slopes. This finding suggests that a resource base ill-suited for agricultural production may, in fact, retard the process of forest intervention.

Even though the initial quality of the agricultural resource base may affect forest clearing and subsequent land-use choices, ecological factors do not tell the whole story. Socioeconomic characteristics and policy environment also influence land use in the study area. In some ways, the results both substantiate and challenge the idea that farmers follow a uniform cycle of land use in the Amazon frontier. Table 3 compares descriptive statistics on farm household characteristics across clusters and for the sample overall.

If farmers moved through a common cycle of land use, plots established earlier could be expected to be "further along" in the process of forest clearing than would plots more recently settled. On average, sample plots had been

Table 3. Infrastructure/Policy Factors and Household Characteristics by Cluster

	Forest Mean (SD)	Perennial Mean (SD)	Pasture Mean (SD)	Diversified Mean (SD)	Sig*	All
Time						
Years Plot Opened	12.3(5.0)	15.9(3.1)	16.0(3.5)	15.0(3.2)	F-PE, PA, D	13.3(4.8)
Years Settled	8.4(5.0)	8.7(5.6)	11.9(5.5)	9.1(5.2)	F-PA	8.8(5.2)
Infrastructure/Policy						
Distance to market (km)	31.0(18.9)	18.0(17.1)	19.0(16.2)	20.6(10.8)	F-PE, PA, D	27.5(18.0)
% primary	73.0(31.2)	76.0(25.2)	75.7(31.0)	72.5(32.1)	–	73.2(31.0)
% secondary	8.0(17.4)	13.3(24.0)	12.4(24.2)	18.8(28.5)	F-D	10.7(21.2)
% walked	10.7(13.0)	10.7(13.5)	11.9(16.3)	8.7(12.7)	–	10.4(13.2)
% boat	8.3(26.0)					
5.7(21.8)						
Tenure						
No title	9(3.7%)	8(50.0%)	1(4.0%)	10(14.1%)	28(7.9%)	
Provisional title	123(51.0%)	3(18.8%)	7(28.0%)	17(23.9%)	150(42.5%)	
Secure title	109(45.2%)	5(31.3%)	17(68.0%)	44(62.0%)	175(49.6%)	
Farm Size (has)	47.4(29.3)	18.9(18.8)	40.4(13.5)	40.7(19.6)	PE-F, D	44.3(27.0)
Household Characteristics						
Origin						
Amazon	11(4.6%)	1(6.3%)	1(4.0%)	1(1.4%)	14(4%)	
Sierra	167(69.3%)	11(68.8%)	18(72.0%)	44(62.0%)	240(68%)	
Costa	63(26.1%)	4(25.0%)	6(24.0%)	26(36.6%)	99(28%)	
Rural	218(90.5%)	13(81.3%)	23(92.0%)	63(88.7%)	317(90%)	
Urban	23(9.5%)	3(18.8%)	2(8.0%)	8(11.3%)	36(10%)	
Na	241	16	25	71	353	
Education						
None	15(6.6%)	5(31.3%)	1(4.4%)	4(5.6%)	25(7.4%)	
Primary	206(90.8%)	9(56.2%)	20(87.0%)	60(84.5%)	295(87.5%)	
Secondary above	6(2.6%)	2(12.5%)	2(8.7%)	7((9.9%)	17(5.1%)	
Na	227	16	23	71	337	
Household Composition						
Household Size	6.2(3.5)	8.4(5.4)	6.6(3.6)	7.4(3.8)	–	6.7(3.6)
Sex Ratio (males/hhsize)	0.57(0.18)	0.52(0.19)	0.55(0.17)	0.55(0.15)	–	0.6(0.2)
Dep Ratio (<14, >55)	0.47(0.24)	0.46(0.17)	0.48(0.25)	0.49(0.22)	–	0.5(0.2)
Na	230	16	23	71	340	

*Clusters with significantly different means at the 5% level in pair-wise testing are noted (F for forest, PE for perennial, etc.), with all clusters appearing to the right of the hyphen significantly different from that appearing to the left. A dash indicates the test showed no significant differences.

[a]Sample size differs for some variables due to missing cases for that variable.

opened for just over 13 years, with the *forest* cluster farms opened for significantly less time (averaging 12.3 years, based upon pair-wise tests of means) than farms in all other clusters. Still, these included not only new arrivals facing wholly forested plots, but also older settlers maintaining significant areas of forest cover on their farms. Thus less than 10% of these 241 *forest* cluster farms had been opened for less than five years while over 30% for fifteen years or more. Moreover, the data provide no evidence of a clear progression through other land-use patterns involving more land clearing over time. This is the case whether farmers leave a substantial portion of land forested and diversify into perennials, food crops, and pasture (such as in the case of the *diversified* cluster farms); or deforest more extensively, using more of the land for pastures (such as in the *pasture* cluster farms), or for perennial crops (as in the *perennial* cluster).

Duration of settlement of current farm residents (as opposed to time since the plot was first settled) was significantly shorter for the *forest*, *perennial* and *diversified* clusters (8.4, 8.7, and 9.1 years, respectively) than for the *pasture* cluster (nearly 12 years). These results suggest that land cleared over time was allocated primarily to pasture, but there is no indication that land use changed the same way for all farmers over time. This finding thus does not vindicate the idea of all farmers following a common cycle of land use. Nevertheless, duration of settlement does play an important role in farm land use, although how much of this effect is due to soil deterioration itself, and how much to other factors also changing over time and not controlled for in the analysis is difficult to discern.

Regarding policy/institutional factors, the results suggest that accessibility to markets and quality of infrastructure significantly affected land-use patterns. Farms in the *forest* cluster were the most remote, averaging 31 km from the nearest market town compared to farms in the *perennial*, *pasture*, and *diversified* clusters which averaged 18 km, 19 km, and 21 km, respectively. In addition, while farmers in all clusters traveled roughly 75% of the total distance to market on primary (unpaved) roads, and about 10% on foot, those from *forest* farms had significantly less of their journey on secondary (not all-weather) roads than did those from the *diversified* cluster. Plot inaccessibility is associated with greater diversification of land use (including greater dependence on food crops), since with less access to markets, farmers are likely to be increasingly reliant on their own food crops for subsistence. Yet diversification of land use grew only when the journey involved more reliance on secondary, but *not* primary roads, vis-a-vis no road at all.

Security of land tenure also differed across clusters. Secure title was most prevalent among the *pasture* and *diversified* clusters (68% and 62% of cluster farms, respectively, compared to 50% for the sample overall). *Forest* cluster farms, on the other hand, were more likely to hold provisional titles (over 50%) than the sample overall (43%). In general, farmers with some title to land (definite or provisional) converted less forest to agricultural land, and had smaller shares of farm area cultivated in perennial and food crops and pasture than households without formal tenure.[1]

Perennial cluster landholdings were significantly smaller than either the *forest* or the *diversified* cluster farms. This result suggests that having a smaller plot could lead to more intensive use of the soil, viz., more forest clearing for perennial (cash) crops. On the other hand, farmers with larger plots were less likely to have cleared large proportions of forest, although the importance of pasture on large rather than small plots was also observed. This is shown in separate (unreported) cross-tabulations of the data which indicate that the average area in pasture increased with farm size (3 ha on 10–24 ha farms, 7 ha on 25–39 ha farms, 10 ha on 40–59 ha farms, 18 ha on 60–99 ha plots, and 44 ha on 100 ha farms and larger).

Regarding household and farm characteristics, the agroecological zone of origin of the head of the settler household did not appear to influence overall land-use patterns. However, nearly 80% of farmers coming from the southern Amazon were settled on *forest* cluster farms compared to 69% and 63% for colonists migrating from the Sierra and Costa regions, respectively. Presumably, the former appeared more knowledgeable about Amazon soils, and therefore, tended to rely on traditional farming systems based on subsistence food crops better adapted to the rainforest environment. Highland and coastal settlers grew more cash perennials but fewer food crops than their Amazonian counterparts. Although no large differences in the land-use pattern of highland and coastal settlers were observed, colonists from the coastal region had been expected to have more experience with cattle ranching in their areas of origin than their highland counterparts, making it more likely that they would engage in livestock activities in the Amazon as well. However, settler's farm background did not appear to affect the share of farm area under pasture. Finally, most households in the sample (90%) were of rural origin and also fell into the predominantly *forest* and *diversified* clusters.

No large differences in the educational achievement of the head of the household were observed across the clusters. However, it is remarkable that over 40% of the most educated farmers in the sample (heads of household with formal schooling beyond primary school) occupied *diversified* cluster farms compared to 20% and 16% for heads of household with primary and no formal schooling, respectively. This result appears to suggest that more education may result in greater awareness by an individual of the value of diversifying land use on the farm, including preserving some of the land in forest. However, more education may have also a positive effect on forest clearing to the extent that it can stimulate desires to increase household consumption (and production) to improve living standards. Indeed, a dominant impression one had interviewing these more educated farmers was the higher entrepreneurial skills and knowledge of farming and business management that exist among such a group. Such interests are likely to lead farmers to clear more land to increase agricultural production in order to raise their quality of life in the frontier.

Demographic factors also appeared to shape land-use strategies on settler plots. Although demographic indicators include age-sex composition, fertility, and mortality, the focus here is on the effects of household size and demographic composition. Household size was largest among farms in the *perennial* cluster (averaging 8.4 compared to the *diversified, pasture,* and *forest* clusters which averaged 7.4, 6.6. and 6.2 respectively), confirming the labor-intensive nature of coffee cultivation in this area of the Amazon. Households with fewer members were more likely to keep larger areas in forest, while greater availability of family labor was positively associated with more farm land planted in perennial crops. Smaller household sizes were also associated with greater shares of farm areas allocated to pasture, which is not surprising given the low labor intensity of cattle ranching. For example, sample plots over 100 ha, which were mostly cattle farms, had an average person-to-land ratio of 0.06, compared with 0.14 for the more common 40–59 ha farms. Cattle ranching is thus a labor-saving activity, demanding considerably less labor per unit of area than raising crops.

Although household size may capture both the supply- and demand-side effects on land use, the analysis here cannot distinguish empirically whether a *production* or *consumption* effect is more significant in shaping land-use patterns.[2] Nor can it ascertain here differences in the impacts by sex. However, it is expected that the effect of males are bound to be greater than those of females, at least from the supply side, to the extent that males are more directly involved in land clearing. Still, for both demand-side and supply-side reasons, a larger household size is expected to be associated *a priori* with a larger proportion of the plot cleared.

In addition to the effects of demographic factors on land clearing on *existing* plots, it is clear that there is wide scope for further deforestation throughout the region. Demographic factors are likely to continue to play an important role in forest conversion because of high retention of migrant settlers in the region, high natural population growth due to the settlers' high fertility, and substantial second-generation effects as many of the settlers' children who leave their parental households remain in the Amazon region, quite possibly creating new settler plots and clearing new areas of forest.

Conclusion and Implications

Farm-level data from the Ecuadorian Amazon were examined to assess land-use patterns among migrant farmers, and to identify certain ecological, socioeconomic, and other household-specific factors associated with the observed land-use patterns. This broad approach questions the extent to which a common cycle of land use (i.e., moving from deforestation to annual cropping, to perennial cropping and pasture), dictated by soil nutrient depletion, occurs. Several conclusions can be drawn from the analysis:

1. It is not enough to look at any one land use in isolation from the rest; clusters of land-use shares (suggested by the data) characterize the patterns emerging in the Ecuadorian Amazon frontier: highly forested farms typically devote substantial tracts to pasture and perennials; farms consisting mostly of pasture specialize more than others, and are also somewhat involved in perennial cropping; farms specializing in perennials, on the other hand, tend to diversify across other land uses; and a substantial number of farms diversify across all land uses examined.
2. Differences do exist across land-use clusters in natural resource base and other factors related to policy and the socioeconomic characteristics of the sample population. Although settlers are vulnerable to constraints imposed by the natural resource base, they inherit different production potentials in their land, as they control different *quantities* and *qualities* of natural resources on their farms. Moreover, government policies affect farmers with varying socioeconomic characteristics differently, even where the natural resource base is similar. The diverse socioeconomic character-

istics of households and the institutional or policy environment may also loosen or tighten the natural resource base "straitjacket" a farmer inherits in his farm, making it likely that land use practices deviate from a generalized pattern of forest clearing over time, dictated by natural resource constraints alone.

3. The findings presented above call attention to the variability in settler land-use strategies that can result from settlers' differences in security of land tenure, labor availability, farm infrastructure, road access, soil quality, farm size, education, and other farm and household socioeconomic and demographic characteristics. But the *separate* effects of these and other factors associated with the natural resource base, socioeconomic situation, and policy environment facing settlers are impossible to tease out in a descriptive analysis and need to be investigated in a multivariate context. For example, time since a plot was opened to agricultural production captures much more than just soil deterioration: access to and quality of infrastructure, family size, or having a secure title may be also functions of duration of settlement in a particular area. A much improved, empirically-based understanding of factors influencing land-use decisions among small farmers is thus essential for identifying policy-manipulable entry points to alter farmers' production patterns in desirable ways.
4. Given the interrelatedness of land-allocation decisions, preconceived notions about land use that focus on only one or two dimensions (e.g., area deforested or land used as pasture) may lead to an inadequate understanding of the amalgam of land-use choices available to farmers, and subsequent policy inferences. Each part of the amalgam of land-use decisions a farmer makes has implications for the other parts; a full analysis must take the interrelatedness of these decisions into account. More land may be put into pasture, for example, by converting land from its use in perennial crops, annual crops, or forest. These conversions have different ramifications for the settler family's food security, production, and income, as well as the long-term sustainability of its agricultural resource base.

Notes

1. Even though, theoretically, plots without title are expected to be farmed more extensively, presumably as a result of less interest in the long-run sustainability of the plot, the effect of land tenure in this sample must still be viewed with caution due to sample selection bias associated with this variable. The original sample frame did not include any farm households with the usual (40–50 ha) size which lacked any form of title because they would not have registered their land claims with the local office of IERAC (the source of data for the sample frame). Invariably, this is the case for households which had settled within the six-month period previous to the 1990 survey. Hence, their landholdings were not yet indicated on IERAC's official survey maps (and therefore could not have been known at the time the sample frame was prepared). The few sample households in the final sample with no title (28 cases) were mostly small subdivisions of the original farms "sold" to later settlers or newcomers. These untitled plots are thus primarily small farms with high person-land ratios which logically has undergone aggressive deforestation to support their occupants.

2. It would be instructive to examine the possible effects on the extent of forest intervention considering both the *demand* side (more mouths to feed stimulates more land clearing, and, possibly also, more use of land for food production) and the *supply* side (more adults on the farm means more *brazos* to clear the land). From the supply side it is desirable to know which age-sex groups customarily participate in land clearing operations and in each type of land use. Separate (unreported) tabulations with both a demand-side variable (number of equivalent adult consumption units), and a supply-side variable (number of adult equivalent-persons that worked on the farm) produced rather similar-looking tables, suggesting the importance of both effects (Pichón, 1993). Evidently, future multivariate analysis will be necessary to help distinguish empirically whether supply or demand-side factors are more significant in the forest conversion process.

References Cited

Bromley, Ray. 1981. The Colonization of Humid Tropical Areas in Ecuador. *Singapore Journal of Tropical Geography 2*(1):15–25.

Estrada, R. Seré, C. and H. Luzuriaga. 1988. Sistemas de Producción Agrosilvopastoriles en la Selva Baja e la Provincia del Napo. Ecuador, Quito: AID, CIAT, CIID, IICA, INIAP and MAG.

Fearnside, Philip. 1992. Human Carrying Capacity Estimation in Brazil's Amazonian Settlements as a Guide to Development Policy. Paper presented at the IUSSP-ABEP Conference on Population and Deforestation in the Humid Tropics. Campinas, Brazil.

Gillis, M. and R. Repetto. 1987. *Public Policies and the Misuse of Forest Resources.* Cambridge: Cambridge University Press.

Hiraoka, M. and S. Yamamoto. 1980. Agricultural Development in the Upper Amazon and Ecuador. *Geographic Review* 70:423–445.

INEC (Instituto Nacional de Estadística y Censos). 1991. V Censo de Población y IV de Vivienda. Quito: INEC.

Jones, J. 1900. *Colonization and Environment: Land Settlement Projects in Central America.* Tokyo: United Nations University Press.

Nelson, Michael. 1973. *The Development of Tropical Lands: Policy Issues in Latin America.* Baltimore: Johns Hopkins University Press.

Nerlove, M., S. Vosti, and W. Basel. 1993. Role of Product Mix and Farm-level Diversification on the Adoption of Modern Technology: The Case of the Zona de Mata de Minas Gerais, Brazil. Washington, DC: International Food Policy Research Institute (IFPRI) Research Report.

Oberai, A., ed. 1988. Land Settlement Policies and Population Redistribution in Developing Countries: Achievements, Problems and Prospects. New York: Praeger.

Pichón, Francisco. 1993. Agricultural Settlement, Land Use and Deforestation in the Ecuadorian Amazon Frontier: A Micro-Level Analysis of Colonist Land-Allocation Behavior. Ph.D. Dissertation, Department of City and Regional Planning, University of North Carolina at Chapel Hill.

Schneider, Robert. 1994. Government and the Economy of the Amazon Frontier. Washington, DC: The World Bank, Report No. 34.

Vosti, Steven and William Loker. 1989. Some Environmental Health Aspects of Agricultural Settlement in the Western Amazon Basin. Paper Presented at an IFPRI/World Bank Symposium on Sustainable Agriculture. Washington, DC.

World Bank. 1992. Brazil: An Analysis of Environmental Problems in the Amazon. Washington, DC: The World Bank, Internal Discussion Paper, Latin America and Caribbean Region, Report No. 9104-BR.

The author is a social scientist (Young Professional) at the World Bank, currently working with the Agricultural and Environment Division of the Southern Africa Department in the World Bank's resident mission in Zimbabwe. The work presented in this article was, however, completed while he was a visiting scholar at the International Food Policy Research Institute (IFPRI), Washington, DC. The author would like to thank Steven Vosti and July Witcover from IFPRI, and Catherine Marquette, Richard Bilsborrow, and Laura Murphy from the University of North Carolina at Chapel Hill for their guidance and many helpful discussions. Also, the Carolina Population Center, the World Wildlife Fund, the National Science Foundation, the Compton Foundation, the Pew Charitable Trust, the Charles A. and Anne Morrow Lindbergh Foundation, the Center of Latin American Studies at the University of Pittsburgh, and the Howard Heinz Endowment all provided valuable support. A number of institutions in Ecuador, including the National Planning Organization (CONADE), the Ministry of Agriculture and Livestock, the National Land Titling and Reform Agency (IERAC), the Instituto de Estrategias Agropecuarias, and the Centro de Documentación e Información de los Movimientos Sociales del Ecuador (CEDIME), provided important assistance. The author also thanks Gustavo Rodríguez, Gloria Rivadeneira, Rosie Obando, Luis López, Germania Estévez, Cecilia Estévez, María Inés Palacios, Norman Castillo, and Jaime Ayala for their valuable assistance during fieldwork in the Ecuadorian Amazon.

GRASSLAND ENCLOSURES:

Catalyst of Land Degradation in Inner Mongolia

DEE MACK WILLIAMS

Since decollectivization, Chinese government policies have promoted household enclosures as the best solution to maximize pastoral productivity and control desert expansion in grassland areas. Recent fieldwork from Inner Mongolia contradicts this optimism. Data and participant-observation reveal the enclosures, as implemented through village level social context, actually compound grazing problems for most residents and the wider ecosystem. Expanding household enclosures function to intensify hyper-critical stocking ratios on highly vulnerable rangeland, exacerbating wind and soil erosion processes across vast territories only to protect small isolated fields dedicated to poorly financed fodder cultivation. This case study indicated that privatization of land in modernizing pastoral societies can be less meaningful for good resource management than other factors, such as secure tenure, equitable access to community resources, and meaningful institutional supports in the form of credit, production services, and legal protection.

Key words: environmental degradation, pastoralism, common property resources, land tenure; China.

Many influential people still insist that common property resources need only be parcelized and privatized to promote conservation in an increasingly crowded world. For pastoral societies, this facile doctrine has frequently instigated government policies that dramatically narrow access to forage and water across national rangelands. Typically, large tracts of land that once accommodated multiple uses and users become partitioned and distributed among single or small cluster households, always with highly significant consequences for land use and local social relations. New restrictions of access to land resources are widely conceded to introduce troubling social inequities, but policy makers usually justify this result as the necessary cost of good stewardship. The disposal of common herders and common lands, however, does not always enhance prospects for sustainable land use. As recent government initiatives come of age, evidence mounts that privatization is no tidy solution to land management problems.

Anthropologists have labored at length to demonstrate the many oversimplifications inherent in the popular interpretive framework described above that Hardin (1968) once (mis)coined the "tragedy of the commons." (For a comprehensive critique, see McKay and Acheson 1987; for a more narrow focus, see Feeny *et al.* 1990; Dove and Rao 1992). During the last decade, numerous studies have stressed the general inapplicability of the model to traditional societies, demonstrating from ethnographic research that indigenous forms of land utilization do not inevitably lead to range degradation (see Goldstein *et al.* 1990; McCabe 1990; Buzdar 1992). Other field studies argue conversely that privatization does not exactly guarantee conservative rangeland management (Little and Brokensha 1987).

This article documents yet another case study—an ongoing enclosure policy in the Inner Mongolia Autonomous Region (IMAR) of Northern China—and offers detailed field data as evidence to support a growing literature that privatization can easily complicate problems of grazing management among modernizing pastoral societies. In this example, privatization is a losing proposition for both the majority of residents and the wider ecosystem over a time frame of sufficient duration to merit serious concern. Furthermore, the article brings evidence to bear from a relatively uninvestigated corner of the world that land use is a socially embedded phenomenon of widely underestimated complexity. Whether a management regime be private or collective, human-land proprietary relations are not necessarily as straightforward as once imagined. Data from Chinese grasslands indicate that privatization avails no simple solution to sustainable land use, and requires definite institutional supports (not merely coercion) to prevent broad systemic distortions.

A village-level case study of land use from pastoral China is long overdue, and deserves attention for three important reasons. First, China has a huge pastoral industry of geopolitical significance, though scholars do not often realize it. Dwarfed by the numbers and political centrality of ethnic Han cultivators, minority pastoralists in China remain obscure, marginalized on the fringes of Chinese geography, scholarship, and national economic priorities. Yet China has 260 pastoral counties accounting for 39 million people. It has some 400 million hectares of grassland comprising 42% of its land mass. Northern China contains the world's third largest grassland, which supports the world's largest population of sheep and goats and the fourth largest concentration of cattle (Zhang Xinshi 1992:47). Inner Mongolia itself constitutes fully one fourth of China's total rangeland area, and over recent years has produced more wool, cashmere and camel hair than any other province (Zhang Cungen 1990:57).

Second, little published ethnographic data are available about contemporary life and land use in the grasslands. The limited information available in English can be explained by problems of long-term access to sensitive border regions that have only recently relaxed (see Sneath 1991; Pasternak and Salaff 1993; Longworth and Williamson 1993). The need for information from this area only grows stronger with China's own reassessment of its importance for continued national development. Within a domestic agenda, China hopes to increase the dietary intake of protein among its citizens by eating more red meat. It also hopes to secure more fully the strategic border regions that are populated by minority ethnicities by raising the low standards of living. Within an international agenda, China hopes to expand exports not only in meat and leather products but also in light industry, which relies heavily on raw materials provided by sheep and goat husbandry.

Third, a critical situation prevails across most of China's northern grasslands. Severe and widespread land degradation has seriously limited prospects for sustained agro-pastoral production. Official reports from China routinely assert some figures that have become familiar in mainstream media: formerly fertile grassland has been lost to moving sand at a rate of 2,100 square kilometers per year over the last decade, critically affecting eight percent of total land mass and compromising the livelihood of about 170 million people (China State Council 1994:180–181). This article testifies to the direct relevance of anthropogenic factors on wind and soil erosion processes that contribute to desert expansion.

Case Study

Not unlike the critical juncture of the Dust Bowl era in American history, accelerating land degradation in the Inner Mongolia Autonomous Region especially challenges Chinese policy makers to reconsider development strategies for endangered rangelands. The sandy lands of Inner Mongolia are widely considered to be man-made disasters, prime examples of an environment "sabotaged" for centuries by excessive grazing, inappropriate cultivation, and deforestation (see Xu 1993; Zhu 1990:70; Hu 1990:204–207). In recent years, central and provincial level governments have formulated explicit grassland policies to address human dimensions of resource management. Ironically, interventions imposed from outside have in many respects only made matters worse for much of the indigenous population. Whatever combination of physical, biological, and social forces may have originally turned grass into sand, over the last fifteen years it has been the policy-initiated grazing management system itself that has magnified wind and soil erosion processes.

The favored policy since decollectivization has been an enclosure movement that, once implemented at the village level and mediated by local social realities, actually accelerates the very chain of ecosystem decline which the policy intended to bring under control. That assertion may well sound counter-intuitive to anyone who has toured the grasslands, where even casual observation confirms a consistent structure to the degraded landscape: enclosed territories appear green from light grass cover while land outside the fences appear patchy at best, marked by semi-fixed dunes. Given the stark contrast, it is easy to misperceive the reality of the situation and imagine enclosures to be a promising solution to the problems of household resource management. Perhaps it is just such superficial grassland tourism that accounts for the longevity of this erroneous but popular view among so many China policy makers. Despite appearances, fences are not saving the grasslands. If one stays long enough to observe local behaviors, a much different perspective emerges.

My research indicates that as privately enclosed land area has steadily increased since 1980, local elites have advantageously manipulated stocking ratios in such a way as to intensify grazing pressures on precisely those tracts of land most immediately vulnerable to erosion processes, and at the worst possible period in the vegetation growth cycle. In essence, the greenery of private enclosures is purchased at the expense of the larger regional ecosystem. A presentation and discussion of data to support this finding will follow a brief review of the post-reform policy initiatives intended to facilitate discussion.

Policy Initiatives

Grassland degradation processes in large areas of Inner Mongolia were already advanced at the time when decollectivization began in the late 1970s. Chinese leaders and specialists, apparently hoping to protect against further ravages expected to follow from minority pastoralists herding private livestock on public range, instituted the new "double contract" household production responsibility system of management whereby local production brigades distributed land use rights (in 1984),[1] as well as animals (in 1981), among independent herding families. Privatization of rangeland resources was intended to be an initial step in a long series of adjustments intended to "rationalize" the animal husbandry sector. Li Yutang, Director of the Grassland Division of the Ministry of Agriculture in Beijing, has outlined the basic reform strategy: first, distribute animals to private households; second, distribute grazing lands; third, assign carrying capacities for each plot of land; and finally, implement incentives and sanctions to enforce a sustainable balance between animals and vegetation at the household level (NRC 1992:33; Li 1992). Prominent scientific publications have likewise endorsed a similar formula to attain the desired "rational management system" (see NRC 1992:49; Huang 1989; Jiang 1989; Zhang 1989).

Chinese policy makers have assumed that private enclosures would force independent households to confront the contradictions between forage demand and forage availability among their separate herds. The logic of that policy requires animals to be contained within a bounded territory. Furthermore, household managers were expected to gradually substitute labor and capital for increasingly scarce fertile land, and thus a traditional, "rely upon heaven," extensive grazing regime would eventually yield to a conservationist, "scientific," intensive regime made possible through technical inputs and motivation born of private ownership. Despite the visionary objectives, neither assumption has proven valid. One simple reason is because the vast majority of households have been left to their own devices to cover the high costs of capital investment. However, more complicated social factors are also involved, with the result that most animals still do not graze within private enclosures, and more sustainable land use practices have not evolved.

To investigate and better understand the situation from a household perspective, I conducted fieldwork over a period of twelve months during 1993–94 in the remote Nasihan Township of Wengniute Banner, Chifeng City Prefecture. This area is located about 500 kilometers northeast of Beijing in the triangle of land between the Xilamulun and Laoha rivers. I engaged in participant-observation and collected data throughout the entire township, but the bulk of my contact occurred with residents in my host village of Wulanaodu, where I enjoyed relatively unimpeded access to government documents, household registries, and personal interviews. Wulanaodu is a community where the proliferation of household enclosures is relatively advanced. It is likely that their experiences thus anticipate somewhat the transformations that are just now beginning, or yet to occur, in other pastoral areas of Inner Mongolia.

In 1993, Wulanaodu had a population of 740 people divided among 174 households. The population is 98% Mongol ethnicity, and stock herding of cows, sheep, and goats still accounts for 87% of community income. Wulanaodu is situated in the western portion of the Keerqin Sandy Lands, with only 51% of total land area officially classified as "usable pasture." Local vegetation is characterized as transitional (from forest to steppe system) and includes weedy plants such as *Alnus pumula* and willows, *Betula microphylla* in dune areas, and *Ephedra distachya* and *Caragana microphylla* as predominant shrub species. The predominant grass species is *Pennisetum flaccidum*. This remote window onto pastoral China revealed some interesting and previously unreported facts about contemporary resource management in the northern grasslands.

Policy Impact

Distribution of privately managed real estate and the explicit promotion of enclosures resulted in the creation of distinctly different categories of land throughout pastoral IMAR. Within residential districts where herders have lived a relatively settled existence since at least the 1950s, there are four categories of land: 1) privately enclosed land, utilized either for supplemental hay production, gardening, or tree nurseries; 2) private unenclosed land and 3) public unenclosed land, both of which are made up of narrow sections of compacted turf functioning primarily as thoroughfares; and 4) enclosed collective property, held in reserve from June to mid-October as fields for hay and fodder production to carry the community through lean months of winter and spring. User rights to the hayfield were originally distributed in 1982. Mowing strips were assigned to each family by lottery, with the size determined according to a formula based 40% upon family livestock and 60% upon male population.

Out on the range, removed quite a distance from residential districts, only two principal varieties of land exist: enclosed and unenclosed. Theoretically, each hectare of rangeland has been equitably distributed among all households in allotments roughly ranging from 47 to 67 hectares in size, depending upon quality of pasture conditions and remoteness from the village residential center. Private contracts issued at the time of decollectivization did not create clear property boundaries; they refer only to boundary outlines roughly demarcated by indistinct landmarks such as sand dunes, hills, or treestands. Over the years, barbed wire of various quality has come to define vast stretches of privately enclosed land, creating disconnected pockets that etch out territorial claims among semi-fixed dunes.

Within the rangeland enclosures, some areas may be more intensively nurtured for fodder production than others. Such areas are designated "improved pastures," in distinction to the term "semi-improved pasture," which applies to any fenced land. Territory that remains outside the fences constitute what is still the largest category of land in all of Inner Mongolia: unenclosed land that is utilized as public range.

The impact of enclosures on local land use and grazing management strategies has been dramatic. Contrary to the intended logic of privatization and parcelization, residents graze their animals as sparingly as possible on their own enclosed land. Since decollectivization, those households who could actually afford enough costly wire to protect their pasture allotments have faced no pressure whatsoever to alter their traditional grazing habits, and so quite rationally keep livestock outside enclosures so long as forage is available on the wide unenclosed range. This strategy permits them to maintain larger herds. They essentially pick clean the grass of those too poor to fence, saving their own for hay production or emergency grazing during winter and spring. Indeed, it is not uncommon for households with large enclosures to graze greedily all year long outside their own fences. Generally, cows and valuable transport animals (horses, donkeys, camels) are stall-fed throughout winter and spring, while sheep and goats are enclosed anywhere from one to the full six months of cold weather. But from the beginning of spring growth until late autumn, only select transport animals ever see the inside of a private enclosure.

Those residents who managed to acquire fencing early, either through direct purchase or as part of the decollectivization process, or by gaining access to limited bank loans through social connections, have enjoyed a tremendous advantage in local competition for present and future grassland resources. The most immediately productive tracts of land were coveted and enclosed first, regardless of proprietary contracts held by others. Furthermore, those with financial leverage have enclosed far more than their allotted share of rangeland, essentially squatting until the day that someone dares to push them back on rightful boundaries. To the majority of Wulanaodu households, the single most limiting factor of production has not been land, labor, or livestock, but barbed wire. Over time, more and more households have gained the means to claim some portion of rangeland, but they have in turn adopted the same exploitative grazing strategies.

Both the ambiguity of boundaries and the exercise of power through nepotism on the part of local officials have been instrumental in allowing this situation to arise. But more entrenched social factors are also at work. Limited access to rural credit is one problem. Since the enclosure policy was launched, only eight different families in Wulanaodu have managed to secure loans significant enough (in excess of 2000 yuan, or about 250 USD) to finance production investments. These loans went to village elites—members of clout belonging to either the dominant kin network, the Communist Party, or both.

Furthermore, exploitative land use practices are actually encouraged, rather than curtailed, by grassland regulations. After the distribution of land use rights to households in 1984, policy statements circulating among township and village level officials stipulated that land must be well managed as a condition of tenure. Otherwise, pasture resources might be confiscated and entrusted to families more capable of using it productively. Good management is defined as a two-step process involving first "protection," which means surrounding property with fencewire, and second "construction," which means developing productive capacity by planting grass and trees (PGWC 1984). At village level implementation, this condition has been construed as license for illicit possession of property; that is, whoever has the leverage to enclose land, by definition becomes the rightful caretaker. Policy thereby becomes an unintended source of economic exploitation and chaotic grazing practices, despite official rhetoric to the contrary. This is the context in which influential scholars and policy makers in Beijing, apparently oblivious to local conditions, repeat their calls for "rational" land use that extol coercive measures (China State Council 1994:169). Under prevailing conditions, their sanctions would not penalize exploitative elites, but instead would punish a large number of poor households who are losers in the race to control community resources.

Beyond the problematic implications of such land use for social justice and community harmony, the enclosure movement has both perpetuated and aggravated preexisting land tenure insecurities that have long exerted a detrimental influence over grassland management strategies. In some areas of Inner Mongolia, over the span of just one generation personal property rights have changed some nine times since 1949. Variations in the exercise of livestock ownership have especially fluctuated through several stages moving toward both collectivization and decollectivization. For a review of these various changes, see Longworth and Williamson (1993:42–47). Substantive changes in land tenure relations have also occurred at least four times, not including all the instances since decollectivization when the government announced alterations in the contractual duration of private land use rights. Property rights have changed far too often for these settled herders to see the sense of investing much energy or capital into pasture reconstruction. Residents of Wulanaodu simply laugh when asked how long they expect to enjoy utilization rights to the land over which they now tentatively assume control. When interviewed, at least 52% of Wulanaodu household managers openly doubt the long term stability of current tenure relationships. The free-for-all atmosphere created by unregulated enclosures has only fed this insecurity for all.

In addition to the problem of illicit enclosures on the range, land tenure insecurities are aggravated by changes in policy regarding access to the collective hayfields. For example, after 1982, mowing rights were redistributed in 1987 to accommodate the population growth and provide for newly created households. In 1993, however, in some villages of Nasihan Township, local authorities decided to raise revenues by charging residents substantial new fees for their mowing rights, which must be paid in advance. People are free to refuse the new tax, but they simultaneously forfeit their rights to harvest precious winter fodder. Further, these changes occur without warning or any guarantees about future rates.

Land Degradation

As enclosures expand, grazing pressure and erosion both intensify and the poorest residents bear the brunt of ecosystem decline. Seeking to prevent a "tragedy of the commons"scenario, enclosure policies in Nasihan since decollectivization have actually precipitated it. I have arranged the relevant summary data into three related tables.

Table 1 reports the increase of privately enclosed land in Wulanaodu since 1980. Table 2 reports yearly sheep equivalent units (SEU) for summer/fall months from 1984–1993 in the village. It also indicates yearly stocking ratios (SEU/hectare) on unenclosed and enclosed public range during the critical vegetation growth season from late spring to late autumn. Table 3 likewise reports yearly sheep equivalent units for winter/spring months from 1988–1994 (the only years Wulanaodu officials collected such data).[2] It also tracks stocking ratios on household enclosures. I collected enclosure data and herding strategy for each household through private interviews conducted in the home, while livestock data came from annual village registries. Combining these data sets, I then computed how many total cows, sheep, goats, horses, donkeys, and camels for the entire village grazed outside all private enclosures during each season of the year.

Figures for Tables 2 and 3 are based on a total land area defined as 100,000 mu, or 6.670 hectares (1 mu = 0.067 hectares). The most recent land survey estimates total land area of Wulanaodu to be 132,914 mu, but roughly 33,000 mu must be subtracted to account for moving dunes (22,000 mu), forested area not included among household enclosure data (10,200 mu), land reserved for cultivation of fodder crops (600 mu), and unenclosed pond area (200 mu). Following the Chinese convention as outlined by the Chifeng Grassland Regulatory Office (CGRO 1990:217), one SEU is defined in Tables 2 and 3 as any combination of animals with a total forage demand roughly equal to that of one adult ewe, or 2.4 kg of dry matter per day. Each cow counts as 5 SEUs based on an assumed forage demand substitution ratio of 5, or 12.0 kg of dry matter per day. Each goat counts as 0.9, each horse as 6, each donkey as 3, and each camel as 7 SEUs.

The data reveal several important points. First, stocking ratios on the unenclosed range have climbed well beyond recommended levels year after year. Officially sanctioned stocking ratios in this degraded area have an upper threshold of 2.15 SEU/hectare. County level officials estimate that the range in its

Table 1. Yearly and Cumulative Wulanaodu Land Enclosure Totals (Hectares)

Year	Total (ha)	Cumulative (ha)	Rangeland (%)
1980	471.47	471.57	.03
1981	113.72	585.29	.09
1982	667.93	1253.22	.19
1983	183.76	1436.98	.22
1984	270.67	1707.65	.26
1985	223.11	1930.76	.29
1986	407.87	2338.64	.35
1987	394.00	2732.63	.41
1988	368.52	3101.15	.46
1989	244.46	3345.61	.50
1990	318.03	3663.63	.55
1991	82.51	3746.14	.56
1992	50.89	3797.03	.57
1993	29.41	3826.45	.57

Table 2. Summer/Fall Sheep Units and Rangeland Stocking Ratios (SEU/ha)

Year	Sheep Unite	Unenclosed Ratio	Enclosed Ratio
1984	16531.8	3.307	0.093
1985	15111.6	3.151	0.114
1986	16252.0	3.700	0.116
1987	15890.6	3.969	0.112
1988	19290.3	5.292	0.148
1989	20272.2	5.961	0.154
1990	14865.1	4.769	0.161
1991	14682.1	4.827	0.169
1992	17203.7	5.773	0.181
1993	16204.8	5.483	0.178

Table 3. Winter/Spring Sheep Units and Rangeland Stocking Ratios (SEU/ha)

		Enclosed		Unenclosed	
Year	Sheep Units	(winter)	(spring)	(winter)	(spring)
1988	17361.8	2.876	2.327	2.415	2.793
1989	15577.7	2.606	2.140	2.361	2.492
1990	11814.1	1.855	1.422	1.690	2.110
1991	13171.3	2.092	1.633	1.855	2.322
1992	14243.3	2.381	1.846	1.877	2.439
1993	13281.5	2.116	1.604	1.857	2.405

current condition can produce on average only about 925.19 kg of forage per hectare (or 61.71 kg/mu) over the entire growing season. At such low output levels, the nutritional requirements of each adult cow is said to require 35 mu of rangeland, with each sheep and goat requiring 7 mu (NTOD 1988). A look at Table 2, however, reveals that in Wulanaodu. summer/fall stocking ratios on unenclosed land have not fallen below 3.0 since decollectivization, even climbing as high as 5.96 in 1989. This figure is more than 2.7 times the estimated critical stocking rate.

Second seasonal stocking ratios (and thus grazing pressure) can increase significantly over time without any remarkable change in livestock numbers. In recent years at least, relentless stocking pressure derives not so much from ever larger numbers of animals, but from the effects of a shrinking land base during summer/fall. As the total land area of private household enclosures increase yearly, grazing pressures from late spring to late autumn intensify on the unenclosed public range even though total SEUs in the village may remain steady, increase somewhat, or even decline. For example, since 1990, the stocking ratio on unenclosed land during summer/fall surpasses that of 1984 despite the fact that sheep equivalent units are generally down by an average five percent.

Third, late spring and early summer grazing seasons on unenclosed range can no longer be assumed to provide ample nutrition for livestock trying to recover from bitter winter months. With only 130–150 annual frost free days, the traditional animal husbandry production bottleneck of winter/spring feed has now, in a sense, been stretched into the summer months as well. Livestock in Wulanaodu are increasingly more likely to experience nutritional distress for longer periods. This phenomenon manifests itself in recent years through decisions to extend by up to ten days the usual June 1st cutoff for open grazing on collectively enclosed hayfields.

Fourth, the data testifies to what the majority of grassland residents already know to be true. That is, in Wulanaodu, any pre-reform imbalance between livestock forage demand and available forage has only been exacerbated in post-reform years by an unregulated enclosure policy that drops all animals on a shrinking land base for the entire growing season year after year. Fully 100% of interviewed household managers assert that the productive capacity of unenclosed rangeland has declined significantly since privatization. At least 39% of those same people dare to assert openly (in front of a foreigner!) that enclosures, as they function in their community, are not a good policy initiative from the collective point of view. This accords with the experiences of herders all across IMAR grasslands, as reported by Hinton (1990:107): "no matter where we went the people all said the grazing was worse than it had been ten years ago and much worse than in their childhood." If ground cover is crucial to impede effects of wind and soil erosion so integral to land degradation processes, then increased stocking ratios on large tracts of sparsely vegetated scrubland amounts to a grazing regime that deliberately puts the majority of land at risk in order to protect an occasional oasis of enclosed greenery.

Enclosures as implemented in Wulanaodu doubly jeopardize the wider ecosystem by training grazing pressure on the same vulnerable land throughout the entire growing season. Any potential relief that enclosures might provide the system through rotational grazing with public access land is largely diluted because of uncoordinated herding routines among households. Nor does rotational grazing occur within enclosed land on any large scale. When households manage to acquire fencewire, they typically seek to expand their net holdings, rather than parcel up what they have by sub-fencing.

There are serious ecological implications from this uncoordinated management system. Contrary to what seems to be a common perception among grassland officials, intensive grazing during the growing season is just as harmful to the range as overgrazing during early spring or late autumn. Soil studies conducted in the region stress the need to allow grass to regrow at least 15 cm between periodic grazing, as well as to prohibit grazing during the first 12–18 days of spring growth, and for 30 days at season's end to maximize vegetation performance and protection (Liu 1990). Soil carbon levels may decline by as much as 15% when 50% of plant production is removed during summer grazing, with losses estimated to be even higher on sandy Chifeng soils (Chuluun and Ojima 1993). Furthermore, above-ground biomass is believed to decrease "heavily" when grazing intensity during growing periods approaches 40% of plant growth, with additional detrimental consequences for soil salinity, pH, nutrient cycling, and organic matter (Kou *et al.* 1993).

Loss of soil cover from intensive grazing is particularly detrimental in eastern Wengniute County where yearly average wind speeds are 4.0 m/sec and gale force winds greater than level seven occur at least 70 days per year (Xu Lan 1990:235). Sheet and gully erosion from water constitutes a significant hazard as well, despite average precipitation levels of only 368.8 mm for the past thirty years. Rainfall tends to be concentrated within two or three months, causing frequent summer floods.

The ratios listed in Tables 2 and 3 should not be construed as absolute values, given the unavoidable potential for informant error in the estimation of either livestock or enclosure data. Nevertheless, they do constitute the best available record, and accurately reflect upon a phenomenon that hundreds of household interviews across the township confirm to be true. The figures are certainly reliable enough to demonstrate the essential trend of a grazing management system that is contributing to ecosystem decline.

Policy Adjustment

It was in tacit recognition of some of these problems that a new direction in grassland management policy emerged in 1988. By then, some Chinese grassland specialists began to acknowledge that grassland reconstruction could not keep pace with rangeland degradation (Ba 1993:19). Subsequently, officials have enthusiastically promoted a different policy, known as the "small grassland enclosure" (*xiao cao ku lun*). The policy hopes to encourage independent households to devote their resources to more modest goals on a restricted scale, like intensive production of hay and fodder crops on enclosed plots of land no bigger than 20–30 mu. The small enclosure would serve as an artificial pasture, bringing together for maximum productivity five key elements: water, grass, grain feed, trees, equipment.

The plan first calls for sinking private wells and constructing a system of water conservancy. Next, leguminous grasses should be planted on 40% of the land area, with another 30% dedicated to silage grain (usually corn), and 20% reserved for other grain feed (usually soybeans). A shelter belt of poplar, willow, or elm trees should occupy the remaining ten percent of land area, both as a means to manage enclosure micro-climate and to diversify family economy with tree farming. Finally, electric water pumps operating 4–6 hours a day should complete the set, but additional recommended equipment includes: sprinklers, tractors, harrows, hay balers, and other processing tools.

The program is believed to be an efficient way to quickly raise technological levels and managerial skills where they are sorely lacking among "backward" communities (Ba 1993). Implementation of these goals, however, remains subject to the initiative and financial capabilities of each household. According to optimistic officials, after an initial downpayment of 2,500–3,000 yuan (roughly 300–375 USD) toward the purchase of fencewire, seeds, tree saplings, and water pumps, 60% of the investment is recouped within the first year of production, and profits accrue by the second year.

Even if productivity could meet these high expectations, not many families can afford such expensive start-up costs. The average net income for residents of Wulanaodu is only around 400 yuan per year (about 50 USD). And despite policy plans to make available benevolent loan options, credit remains hard to come by in rural China. As a result of peer financing, the artificial pastures rarely work according to plan, often missing two or three elements from the intended set of five. Even the crucial component of grass production seldom meets specifications, as households typically forego managed alfalfa cultivation and make do with what grows naturally. In Wulanaodu, the best managed artificial pasture yields about 370 kg. of haystraw per mu, an admirable harvest by local standards but far short of the projected 541.9 kg/mu for heavily fertilized fields (Hu 1990:216).

Epilogue

The Chinese have an expression that translates: "going out empty-handed to capture wolves." Rural people say this usually in the context of inadequate or improper preparation to tackle a difficult task. In Wulanaodu, some residents use it to mock grassland policies that call upon common herders to fix moving sand dunes and check advanced wind and soil erosion processes without the benefit of investment capital or even meaningful institutional support. For the majority of anxious herders who cannot exercise their land use rights, the smug enjoinder to flourish through the means of small enclosures amounts to a government invitation to lie vulnerable before the twin wolves of blowing sand and plundering neighbors.

Whatever the original intentions of the enclosure policy, the way it has been implemented at the village level does not promote sustainable pastoral production over short or medium-term considerations. The policy may yet produce a conservationist effect over much longer term considerations as the entire rangeland becomes parceled into separate enclosures and some guidance emerges to coordinate rotational grazing. As the situation now stands, however, chaotic grazing practices may substantially redefine the range and the community before any presumed long-term benefits have a chance to evolve.

Notes

1. Technically speaking, the 1984 distribution of rangeland resources allocated land use rights to *"lian hu"*, or small household clusters (usually ranging from 3–5 households), who were supposed to manage and develop the land together. In reality, many of these small households functioned independently from the start. By 1988, discord was so prevalent that local authorities conceded to a second distribution which formally divided the range among independent households. My fieldwork investigations found that some land was claimed and enclosed for private use as early as 1980 in anticipation of rural decollectivization (see Table 1).
2. The reason for the large difference between summer and winter livestock figures is that large numbers of animals are sold or slaughtered in the later months of the year, after they are fattened somewhat on autumn harvest but before the lean cold season begins.

References Cited

Ba Gen, and the Chifeng Grassland Work Station. 1993. "Overview of Householder Herd and Grassland Construction Policy," Nei Menggu CaoYie 2:19–22. [in Chinese]

Buzdar, Nek. 1992. The Role of Institutions in the Management of Commonly-Owned Rangelands in Baluchistan. *In* Sociology of Natural Resources. Michael Dove and Carol Carpenter, eds. Pp. 218–238. Karachi, Pakistan: Vanguard Press.

Chifeng City Grassland Regulatory Office (CGRO). 1990. "Collection of Grassland Management Rules and Regulations." Chifeng, Inner Mongolia Autonomous Region. [in Chinese]

China State Council. 1994. China's Agenda 21: White article on China's Population, Environment, and Development in the 21st Century. Beijing: China Environmental Science Press.

Chuluun, Togtohym, and Dennis Ojima. 1993. Simulation Studies of Grazing in Mongolian Grassland Ecosystems. Paper prepared for the Grassland Ecosystem of the Mongolian Steppe Research Conference, Racine, WI, Nov. 4–7.

Dove, Michael and Abdul Latif Rao. 1992. Common Resource Management in Pakistan: Garrett Hardin in the Junglatt. *In* Sociology of Natural Resources, Michael Dove and Carol Carpenter, eds. Pp. 239–252. Karachi, Pakistan: Vanguard Press.

Feeny, David, Fikret Berkes, Bonnie McCay, and James Acheson. 1990. The tragedy of the Commons: Twenty-two Years Later. Human Ecology 18(1):1–19.

Goldstein, Melvyn, Cynthia Beall, and R. P. Cincotta. 1990. Traditional Conservation on Tibet's Northern Plateau. National Geographic Research 6(2):139–56.

Hardin, Garrett. 1968. The tragedy of the commons. Science 162:1243–48.

Hinton, William. 1990. The Great Reversal: The Privatization of China, 1978–1989. New York: Monthly Review Press.

Hu Ming Ge, ed. 1990."Chifeng Grasslands."Beijing: Agricultural Press. [in Chinese]

Huang Wenxiu. 1989."The Prospect and Pathway for Pastoral Animal Husbandry Development in China."*In* Zhongguo Caodi Kexue Yu Caoye Fazhan, Beijing, pp. 42–66. [in Chinese]

Jiang, Su. 1989."A Recommendation on Reasonable Utilization of Grassland Resources and Development of Animal Husbandry in China." *In* Zhongguo Caodi Kexue Yu Caoye Fazhan, Beijing, pp. 15–18. [in Chinese]

Kou Zhenuw, Song Bo, and Dennis Ojima. 1993. Model Analysis of Grassland Dynamics to Different Management Measures in Wulanaodu. Paper prepared for the Grassland Ecosystem of the Mongolian Steppe Research Conference, Racine, WI, Nov. 4–7.

Li Yutang. 1992. (Executive Vice President of China Pratacultural Association). Interviewed by Author at the Ministry of Agriculture, Grassland Division, Beijing, June 5.

Little, Peter and David Brokensha. 1987. Local Institutions, Tenure and Resource Management in East Africa. *In* Conservation in Africa: People, Policies, and Practice. David Anderson and Richard Groves, eds. Pp. 193–209. New York: Cambridge University Press.

Liu, Yuchen. 1990."Sandification Process of Keerqin Sandland and its Comprehensive Treatment." *In* Guoju Caodi Zhibei Xueshu Huiyi Lunwenji, pp. 621–626. [in Chinese]

Longworth, John and Gregory Williamson. 1993. China's Pastoral Region: Sheep and Wool, Minority Nationalities, Rangeland Degradation and Sustainable Development. Wallingford, UK: CAB International (with Australian Centre for International Agricultural Research).

McCabe, J. Terrence. 1990. Turkana Pastoralism: A Case against the Tragedy of the Commons. Human Ecology 18:81–104.

McCay, B. J. and Acheson, J. M., eds. 1987. The Question of the Commons: The Culture and Ecology of Communal Resources. Tucson: University of Arizona Press.

National Research Council (NRC). 1992. Grasslands and Grassland Sciences in Northern China. National Academy Press: Washington, D.C.

Nasihan Township Official Document (NTOD). 1988. Government document circulated to village officials (in Mongolian), September 23.

Pasternak, Burton and Janet Salaff. 1993. Cowboys and Cultivators: The Chinese Inner Mongolia. Boulder: Westview Press.

People's Government of Wengniute County (PGWC). 1984. Wengniute Renmin Zhengfu Wenjian. Document #36, Article 100. [in Chinese]

Simon, Davie. 1993. The Communal Lands Question Revisited. Third World Planning Review 15(1):R3–R7.

Sneath, David. 1991. Post-revolutionary Social Economic Change in the Pastoral Population of Inner Mongolia. Draft Ph.D. thesis, Department of Social Anthropology, Cambridge University.

Xu Lan and the Shenyang Institute of Applied Ecology. 1990. "Landscape Characteristics of the Western Keerqin Sandy Lands."Jingguan Shengtai Xue. Pp. 235–238. [in Chinese]

Xu Youfang. 1993. Speech at the National Desertification Conference in Chifeng, Inner Mongolia, as quoted in Xinhua news agency (English report), Beijing, September 24.

Zhang Cungen, 1990. An Overview of Wool Production and Processing in the Inner Mongolia Autonomous Region. *In* The Wool Industry in China. John Longworth, ed. Pp. 57–73. Victoria, Australia: Inkata Press.

Zhang Xinshi. 1989."A Research Project on Optimized Ecological Model for Main Types of China's Northern Grasslands."*In* Zhongguo Guojia Ziran Kexue Jijin Zhongda Xiangmu Lixiangshu. Beijing. [in Chinese]

Zhang Xinshi. 1992. Northern China. *In* Grasslands and Grassland Sciences in Northern China. Pp. 39–54. National Academy Press: Washington, D.C.

Zhu Zhenda. 1990. Desertification in the Northern Territory of China: Present Status and Trend of Development. Journal of Chinese Geography 1(1):61–70.

The author is a Visiting Fellow at East-West Center, Program on Environment, Honolulu, Hawaii. Grants from the National Science Foundation and the Committee for Scholarly Communication with China made this research possible. Columbia University and the East-West Center funded the analysis and writing. A word of thanks is due to James Reardon-Anderson for assistance with field placement and to Michael Dove for constructive comments on an early draft of the paper.

AGROFORESTRY

SARA J. SCHERR

In the tropical world, rapid growth in rural populations and consequent demand for food and fiber coincide with the closing of the agricultural frontier. Most newly cultivated lands are "marginal" for conventional intensive annual crop production, and easily degraded. Intensification has been associated with loss of forests and woodlands which are important for tree products, rural livelihoods, and environmental services.

Agroforestry—the deliberate growing of woody perennials together with crops, pastures, and livestock—offers a new paradigm for intensifying tropical land use and integrating production, livelihood, and ecological concerns. This case study of agroforestry intensification in western Kenya illustrates the potential of this approach, and suggests some implications for planning and policy.

New Roles for Agroforestry in Tropical Land Use

Agroforestry has been widely used for millennia. Indigenous practices range from extensive systems, such as valuable fruit and timber trees scattered in cropland or woody fallows in swidden cultivation, to intensive systems such as the multi-species agroforests and home gardens of Southeast Asia.[4] While industrial and scientific agriculture over the past century have emphasized monocultures in both trees and crops, recent research and experience have focused on the potential technical and socioeconomic contributions of multi-species agroforestry.[8]

Globally, a gradual transformation appears to be under way, from reliance on gathering of naturally growing forest products to domesticated and cultivated products: a parallel with the agricultural revolution of 10,000 years ago. In land-abundant countries this is resulting in more intensive natural forest management (increasingly regulated to achieve multiple use for biological, aesthetic, or other reasons) and plantation forestry. In land-scarce countries (especially those with limited capacity to import), farm tree production—agroforestry—is becoming a dominant source of "forest" products.

At the farm level, pressures for intensification and improved resource husbandry have in the past led to land-use innovation, through both endogenous and induced processes.[5] Agroforestry can play a key role in such intensification. Crops may yield more (or cost less to produce) by their association with trees or shrubs that fix nitrogen, modify microclimate, or protect plants from wind or hail. Other effects of trees—such as nutrient cycling, soil stabilization, water drainage, pest control—may contribute to the sustainability or stability of crop yields.

Smallholders, who cannot afford to take land out of food production, often prefer intercropping. If the value of tree products is high, economic returns from joint production may be higher than those from crops alone. Where there are no reliable low-cost markets for tree products important for household use (such as fuelwood, medicines, building materials) and for farming (such as stakes, fodder, mulch), farmers may want to produce them. Many tree products can substitute for annual food and fodder crops.[1] Tree husbandry can stabilize rural livelihoods, reducing production and consumption risks and improving the use of required farm labor.

From an environmental perspective, agroforestry promotes a wider range of biodiversity and other forest services in agricultural landscapes. On steep slopes, well-managed agroforestry systems can protect watersheds and control soil erosion.

Land-use Challenges in the East African Highlands

The East African highlands (Figure 1, inset), as defined here, include those areas of >1000-m altitude, with a bimodal rainfall regime (>1000 mm annually in most areas other than the dry highlands of Ethiopia). The region presents major challenges for food production, sustainable land use, and rural livelihoods. The highlands are densely populated with as many as 1000 persons/km^2 in some parts of rural Kenya and Rwanda. Population growth rates in the region are high (4.2%/y in Kenya) and, despite high rates of rural emigration, average farm size is declining.

Highlands agriculture is the most important economic sector and source of exports for all of the East African countries. Most cultivation is permanent, be it annual crops, fodder crops for dairy production, or perennial export crops such as coffee and tea. While the volcanic soils of the region are inherently fertile and resilient, much of the population resides in areas of poorer, erodible soils suffering widespread nutrient depletion. Deforestation of naturally forested land has been dramatic: <5%, of an original 20%, of Kenya's land is now in closed forest. Deforestation rates are similar in the other countries.[2]

Intensification of land use in the highlands has been associated with the widespread use of agroforestry systems, both indigenous and introduced. Contrary to common belief, but confirmed by aerial photography and farm

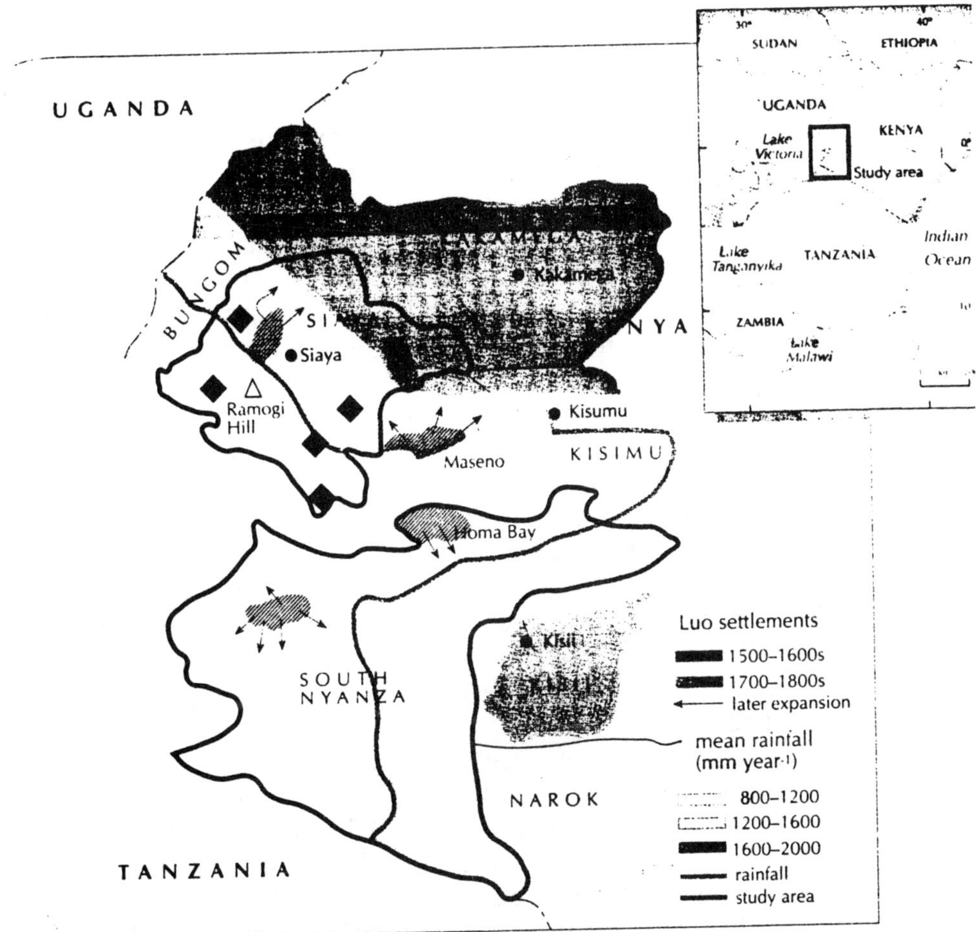

Figure 1. Historical settlement and rainfall patterns in Siaya and South Nyanza districts, Kenya, and a general map of the East African highlands (inset).

surveys, tree density on farms has risen as population density has risen in many parts of the highlands.[9] This association is present also in the history of land use in the Luo areas of western Kenya.

Past Patterns of Land-use Intensification and Agroforestry, Siaya and South Nyanza Districts, Kenya

The Luo, a Nilo-Hamitic people, migrated into Kenya from the Sudan in the 1500s and 1600s as agropastoralists, using shifting cultivation methods, and fishermen. During the 1800s, permanent settlements under chiefships became common, and some groups moved into the more humid areas, borrowing more intensive agricultural practices from neighboring Bantu farmers. Their current area of settlement is in the lower-altitude highlands of Siaya, Kisimu, and South Nyanza districts (figure 1).

The transition from fallow-based to permanent agriculture began during the colonial period, with the introduction of maize and associated row-planting, and later, sugarcane and cotton. Average population density rose to ~100 persons/km^2 by 1948. During the 1950s and 1960s the districts were net food exporters, but intensive maize production with inadequate fertility supplement and without erosion control led to serious soil degradation and declining yields by the 1970s and 1980s. Average population density rose to 250 persons/km^2, reaching 400 in the high rainfall zone. As the agricultural frontier essentially closed, average farm size became smaller, migration of laborers increased, and poverty and malnutrition reached high levels. These farming system changes led to changes in farmers' demand for agroforestry and in patterns of farm tree supply.[6,7]

Under shifting cultivation, all subsistence tree products were gathered from the surrounding woodland. Trees were rarely planted, although certain ritually important or highly valued trees were protected. The development of fallow-based agriculture brought new uses for tree products in fish- and food-processing, boats, rope, and baskets. Live hedging became important to mark boundaries, and fencing became critical to protect cropland from livestock. Indigenous fruit trees were protected when fields were cleared, and some wildlings were transplanted. Mango (*Mangifera indica*) and guava (*Psidium guajava*) were introduced and protected. Woody fallows were important, but not deliberately enriched.

With massive clearing of the woodland tree cover for settlement, cultivation, fence and construction wood, and tsetse fly control, sources of gathered products began to disappear. Protecting useful tree species when clearing land, and transplanting wildlings became important in the 20th century. Tree tenure rights evolved so a tree was "owned" by the person who planted it or whose ancestors had (*yadha*). Boundary hedges came to occupy a significant area. The colonial government introduced exotic fruit-tree species, sisal, and some timber species. Smallholders began to experiment with the new species, but only largeholders adopted the woodlot model. Farmers in more humid areas attempted to modify the fallow broadcasting seed of *Sesbania sesban*, a short-cycle tree with prolific leaves. In cropped fields, farmers used the leaves from the indigenous timber tree *Markhamia lutea* as green manure.

After the 1950s, economic integration and population growth led to development of new local and regional markets for construction wood and fruits. Access to the subsistence products from naturally growing trees sharply declined. Farmers practiced more intercropping, both in traditional scattered arrangements and in lines. By 1989, 36% of farmers interviewed in a survey of Siaya and South Nyanza farmers had purchased fuelwood during the previous year. Only 20% used off-farm sources as their primary supply while 41% depended primarily on naturally growing farm trees. For building poles, <1/4 of interviewed farmers still relied upon supplies gathered from naturally growing trees on- and off-farm.

Sources of innovation for agroforestry (new tree species and tree-crop systems) have been both external—from traders, neighbors, government tree nurseries, and nongovernmental organization projects—and internal, through local development of new methods of propagating and managing indigenous tree species.

Current Patterns of Agroforestry Practice

In 1984, the Agroforestry Extension Project, sponsored by CARE International and the government of Kenya, initiated a project in Siaya and South Nyanza to promote agroforestry. The project provided community groups with technical assistance to identify agroforestry opportunities, obtain tree germ plasm, manage tree nurseries, and establish and manage trees on farms. In 1987, the Agroforestry Extension Project and the International Centre for Research in Agroforestry together initiated a study of farmers' agroforestry practices and project effects. Data sources included individual and group interviews, and farm surveys.

A 1989 survey of 336 group members included a complete census of trees on farms, distinguishing those established before the project (which reflect typical agroforestry practices in the region) from those established after the project. The project's strategy was to acquaint farmers with the range of possible tree species and agroforestry systems, and to encourage them to select those that would address priority functions on their farms. Thus "postproject" trees represent farmers' preferences resulting from recent changes in land-use patterns.

Study results show significant changes in agroforestry practices, even though farmers had been involved with the project only an average of 3.2 years. The average number of free-standing trees per farm rose by 125% to 513, and the average length of dense hedges (used to demarcate homesteads and some fields) increased 29% to 317 m. For farm plots with agroforestry, average tree density rose from 220 to 504 trees/ha, or 129%.

Sites farmers selected for tree-growing reveal pressures for land-use intensification. Of all freestanding trees censused on farms, 25% were found in homesteads, 39% in or around cropland, and 17% in or around pastures. Cropland became the dominant site for new tree-planting (49% of all new free-standing trees). Density of trees in cropland tripled. Configurations for growing trees changed significantly. Of all free-standing trees in 1989, only 27% were in traditional scattered arrangements. Some 37% were planted in lines (typically around plot perimeters), 21% in small blocks, 9% in "alley-cropping" arrangements (with crops grown between lines of trees cut back regularly to control shading and provide green manure), and 6% in regularly spaced tree-crop mixtures.

The farmers also used more intensive methods to plant trees. Of freestanding trees established before the project, only 30% were grown from seedlings; another 9 to 10% each resulted from vegetative propagation, direct seedling, and transplanted wildlings, with a large share (42%) by natural establishment.

Table 1. Changes in Free-Standing Tree Species Grown by Farmers in Siaya and South Nyanza, Kenya

Species	Before Project	%	After Project	%	Total	%
(*N* =)	(330)	98	(333)	99	(335)	100
MAIN FORESTRY TREES						
+*Eucalyptus* spp	12698	17	20771	22	33469	19
+*Cassia siamea*	2282	3	5798	6	8707	5
Subtotal	14980	20	26569	28	42176	25
MAIN AGROFORESTRY TREES						
+*Markhamia lutea*	10747	14	12417	13	23164	13
•*Leucaena leucocephala*	348	0	16638	17	16986	10
Subtotal	11095	15	29055	30	40150	23
OTHER TREES FOR WOOD AND LEAF						
**Multiple naturally growing species*	6376	8	1600	2	7976	5
**Agave sisalensis*	6096	8	1760	2	7856	5
+*Cupressus lusitanica*	2178	3	5270	5	7448	4
**Lantana camara*	6065	7	1676	2	6741	4
•*Terminalia peruviana*	2915	4	2731	3	5844	3
+*Cassia spectabilis*	1141	2	3031	3	4558	3
+*Grevillea robusta*	230	<1	2867	3	3097	2
**Acacia* spp	2843	4	90	<1	2936	2
**Albizia coriaria*	1711	2	413	<1	2124	19
•*Melia azadirach*	531	1	1454	2	1985	1
•*Parkinsonia aculeata*	7	<1	1849	2	1856	1
**Sesbania sesban*	374	<1	1381	1	1755	1
**Callitris robusta*	335	<1	1133	1	1468	1
•*Calliandra calothyrus*	7	<1	1381	1	1755	1
**Euphorbia candelabrum*	646	1	452	<1	1098	1
**Sesbania bispinosa*	0	0	785	1	785	<1
**Bridelia macrantha*	424	1	349	<1	773	<1
**Terminalia brownii*	661	1	162	<1	773	<1
**Combretum molle*	731	1	38	<1	769	<1
•*Croton megacarpus*	373	<1	249	<1	622	<1
Subtotal	37896	50	29282	31	67715	39
FRUIT TREES						
**Psidium guajava*	2659	3	1502	2	4265	2
+*Citrus* spp	670	1	2461	3	3131	2
+*Carica papaya*	402	1	2001	2	2403	1
**Magnifera indica*	1043	1	690	1	1734	1
Subtotal	4774	6	6657	8	11533	6
ORNAMENTAL/SHADE TREES						
+*Jacaranda mimosifolia*	202	<1	887	1	1089	1
•*Terminala mentali*	50	<1	626	1	676	<1
Subtotal	252	<1	1513	2	1765	1
MINOR SPECIES	7067	9	2847	3	8749	5
ALL FREE-STANDING SPECIES	76064	100	95923	100	172088	100

* = species indigenous or introduced to Siaya and South Nyanza districts prior to 1900.
\+ = species introduced to these districts 1900–1950.
• = species introduced to these districts 1971–1989.

Table 2. Changes in Primary Use of Trees on Farms, Siaya and South Nyanza, Kenya

Species	Before Project	%	After Project	%	Total	%
(*N* =)	(330)	98	(333)	99	(335)	100
FREE-STANDING TREES						
ornamental	116	<1	433	<1	549	<1
boundary marker	3875	5	323	<1	4198	2
building poles	26637	35	36348	38	62985	37
charcoal	323	<1	32	<1	355	<1
animal fodder	681	1	1037	1	1718	1
live fencing*	7334	10	8193	9	15527	9
fruit	5091	7	7441	8	12635	7
fuelwood	18352	24	9948	10	28306	16
green manure	288	<1	13967	15	14255	8
rope	983	1	456	<1	1439	<1
soil conservation	254	<1	1590	<1	1844	<1
seed	1	<1	15	<1	16	<1
shade	3039	4	3733	4	6773	4
medicine	656	1	1030	1	1686	1
stakes or *fitos*	627	1	1027	1	1654	1
timber	6443	8	9011	9	15454	9
windbreaks	1264	2	1259	1	2523	1
TOTAL	75964	100	95843	100	171917	100
DENSE HEDGES						
ornamental	10	<1	10	<1	20	<1
boundary marker	14031	14	1640	5	15671	12
building poles	3754	4	2297	7	6051	5
charcoal	260	<1	260	<1	520	<1
animal fodder	192	<1	1205	4	1379	<1
live fencing*	50929	53	22951	72	73880	57
fruit	1316	1	466	1	1782	1
fuelwood	17796	18	1229	4	19025	15
green manure	92	<1	687	2	779	1
rope	3536	4	30	<1	3566	3
soil conservation	75	<1	60	<1	135	<1
shade	140	<1	410	1	550	<1
medicine	947	1	181	1	1128	1
stakes or *fitos*	2325	2	330	1	2655	2
timber	52	<1	93	<1	145	<1
windbreaks	1384	1	120	1	1674	1
TOTAL	96839	100	31969	100	128978	100

*Most live fences are established as dense hedges. Their length in meters was measured, rather than counting individual trees. Figures derived from a 14% sample.

During this project, the proportion of trees established by seedlings shot up to 83%, although this is more labor-intensive.

Tree species composition of farm-grown trees was diverse (Table 1); >170 species were recorded in the survey. While 4 planted species—*Eucalyptus* spp, *Cassia siamea*, *Markhamia lutea*, and *Leucaena leucocephala*—accounted for almost ½ of free-standing trees, 26 accounted for another 46%, and 137 others accounted for the remaining 5%.

About 39% of free-standing trees were traditional species, indigenous to the region or brought there during precolonial times. Another 38% of trees were species introduced during the colonial period, and 18% were species introduced after Kenya's independence in the 1970s, or more recently, with the CARE project.

While there is a clear shift toward nonindigenous species (most of which have undergone germ plasm selection to improve productivity), farmers seem to have chosen a mix reflecting their preferences. If we cluster species by their predominant use, we find that the biggest change has been an increase in use of *Markhamia* and *Leucaena*, species particularly suitable for intercropping.

Farmers' Rationale for Growing Trees On-farm

Why, when land is becoming scarce, did Siaya and South Nyanza farmers resort to planting more trees on more land? The findings of this study suggest that the predominant reason was to obtain critical consumption goods which would otherwise have to be obtained by paying scarce cash. Other reasons were to diversify their sources of cash income and to ensure that they would have enough food, especially since annual crop yields were declining.[7]

Before the project, most free-standing farm trees were used primarily for building poles and fuelwood (Table 2). Live fencing and boundary markers were also important uses. Nearly 1/4 of the trees were planted or protected for a wide range of other uses: timber, fruit, shade, windbreaks, stakes, or medicines. Hedges were used mainly as live fencing, boundary markers, or fuelwood.

In recent years, fuelwood declined as a principle use. This reflects a shift from harvesting fuelwood from naturally growing trees, which had no other significant use, to harvesting it as a secondary product from trees planted for another use. Building poles were the dominant use of trees established in recent years; they, together with timber, accounted for nearly 1/2 of all trees. Using planted trees for building poles reflects the scarcity of off-farm sources of forest products. That scarcity created a marked increase in cash costs for basic household construction, and also provided a market for surplus farm production.

Household response to scarcity may take many forms. Options other than intensified tree production are interregional trade, substitution by nontree products, more efficient use, and reduced consumption. In this region, scarce cash incomes and high transport costs combined to limit the potential for importing tree products or using nonwood substitutes. Some houses are now constructed in designs that conserve higher-value materials, and community standards for a minimum level of household fuelwood stores have declined.

But the most common response to tree-product scarcity was reduced consumption, through switching to lower-quality fuelwood and timber species, declining numbers of household animals that require fodder, and declining use of bush fallow and animal manure. These declines in living standards could only be countered, under prevailing economic conditions, by increasing local tree supplies.

Farmers also used trees to enhance food security. A high proportion (15%) of free-standing trees were grown for green manure (mainly in alley-cropping and mixtures with coffee), typically substituting for increasingly scarce manure. Some 44% of project-assisted households tested the innovative alley-cropping system, usually experimentally on small parcels of land suffering from nutrient depletion. Windbreaks were established in the zones where wind damage and desiccation significantly reduce crop yields.

There were also modest increases in tree-growing for fruit. The wide variety of indigenous and exotic fruits grown (27 species reported in the survey), in small numbers per farm, suggest their importance in the household diet. Fruits are particularly used as food for children, and during the preharvest period.

Finally, farmers grew trees to earn cash. Eucalyptus was widely grown for sale in urban construction markets, and citrus and papaya for local and regional fresh-fruit markets. Together, these primarily commercial species accounted for ¼ of all free-standing trees grown on farms in 1989. Interhousehold trade of building poles has also developed at the community level for locally used species, and, more recently, fuelwood markets have grown around towns and in more densely populated areas. The survey indicated that during the previous year, 51% of households had sold some building poles, 47% fruit, 32% fuelwood, 22% tree seedlings, 10% tree herbs, 4% tree seed, and 3% other products. With average monthly incomes in rural Siaya households estimated in 1990 to be KSH500 (at 30 Kenya shillings per US dollar), the periodic sale of 4-m building poles for KSH10 to 30 represented significant income.

Table 3. Primary Use of Free-standing Trees by Wealth

Use (*N* =)	Poor (98)	% 29	Average (224)	% 67	Wealthy (10)	% 3	Total (335)	%
fuelwood	9736	23	17 363	14	1203	19	28 302	16
fruit	2197	5	9829	8	573	9	12 599	7
fencing	3294	8	10 745	9	1148	18	15 187	9
building poles	16 313	38	44 920	37	1634	26	62 867	37
timber	1926	4	13 085	11	414	7	15 425	9
green manure	3856	9	9915	8	493	8	14 255	8
SUBTOTAL	37 322	86	105 858	87	5455	86	148 635	87
other	5900	14	16 114	13	905	14	22 919	13
TOTAL	43 222	100	121 972	100	6360	100	171 554	100

These figures do not include dense hedges. Totals are larger than the sum of the specific uses because some uses have been excluded. The total refers to all trees on farms. Figures are derived from 14% sample.

An important finding of the study was interhousehold variation in objective, scale, and type of agroforestry practices. Factors influencing the mix of practices included agroecological conditions, farm size, household wealth, gender and household status of the group member, household labor availability, nonagricultural income sources, and stage of the household demographic cycle. For example. on smaller farms proportionately more trees were grown in cropland, generating a demand for intercropping species.

Agroforestry use differed among poor, average income, and wealthy households (Table 3). Fuelwood is a much more common use among the poor; fruit and timber were more common among average income farmers; fencing was more common among wealthy farmers. Alley-cropping was more common on farms of average income households. Around 90% of trees on the farms of poor and middle income farmers were planted, while only 68% of farm trees were planted on wealthy farmers' land, a reflection of larger average farm size and woody fallows.

These patterns of interhousehold variation call for flexible extension approaches using a fairly large number of the potential tree species and practices for several different purposes. Tree-breeding specialists should also consider the fact that the same tree species may be used in different sites, under different management, and for different purposes, even in the same local area.

Policy Challenges for Agroforestry Development

The Kenya case study illustrates the potential of agroforestry to contribute significantly to land-use intensification in the tropics. To be fully unleashed, however, this potential requires institutional and policy support to facilitate innovation and diffusion.

There are clear technological challenges. With the exception of major food and timber commodities exported to temperate countries (e.g., coffee and teak), efforts to improve germ plasm of tropical trees have been modest, and these principally for forest plantations. Identification and selection of tree species and provenances suitable for intercropping with crops and grasses is at an early stage and requires public support. Scientific understanding of tree-crop-soil interactions is still weak. Strategic collaboration among farmers, research institutions, and local development agents will be needed to develop and adapt agroforestry technologies suitable for different agroecologies, farming systems, and market conditions.

Existing policies are rarely favorable to agroforestry.[3] Formal education, research, extension, and land-use planning institutions have only recently begun to consider agroforestry. Indeed, agroforestry is representative of a broader paradigm of integrated land-use systems which these institutions are currently struggling to address.

Secure rights to plant, manage, and harvest trees are fundamental to agroforestry. Farmers face strong disincentives to practice agroforestry in many countries where land-use regulations and land- and tree-tenure arrangements undermine that security. For example, Kenya has a history stemming from the colonial period of prohibiting growing trees in cropland, implemented through fines and restrictions on access to credit. In Ethiopia, government "ownership" of forest and tree rights, even on private lands, has discouraged tree planting.

Other disincentives result from market conditions and policies. Markets for many tree products have only recently emerged. Their effective functioning could be promoted by improving information systems among farmers, relaxing restrictions on transport and distribution, and investing in marketing infrastructure. Pricing policies for tree and crop products, their substitutes, and inputs for agroforestry production which result in artificially low relative prices for tree products, also discourage farmers' investments. A common example is below-cost sale of timber from public forests, which distorts incentives for farm-produced supplies.

The challenge of agroforestry development thus raises a broad research agenda in agroecology, landscape design, and geography; agricultural and forestry technology; socioeconomics and policy; and community development. That agenda calls for scientists who are willing to take on the challenge of innovative, interdisciplinary research, and policymakers who are willing to make the investments needed now to ensure sustainable land use in the 21st century.

References

1. Cannell, M. G. R. Food crop potential of tropical trees. *Experimental Agriculture* 25:313–326, 1989.
2. Getahun, A. Agricultural growth and sustainability: Conditions for their compatibility in the tropical East African highlands. Vosti, S., Reardon, T. & von Urff, W., editors: *Agricultural Sustainability, Growth and Poverty Alleviation: Issues and Policies,* pp. 451–468. Deutsche Stiftung für internationale Entwicklung and International Food Policy Research Institute, Feldafing, 1991.
3. Gregersen, H., Oram, P., & Spears, J., editors, *Forestry and Agroforestry Policy Research: Issues and Priorities.* International Food Policy Research Institute, Washington, D.C., 1992.
4. Nair, P. K., editor: *Agroforestry Systems in the Tropics.* Kluwer Academic Publishers and the International Centre for Research in Agroforestry, Dordrecht, the Netherlands, 1989.
5. Ruthenberg, H. *Farming Systems in the Tropics.* Clarendon Press, Oxford, 1980.
6. Scherr, S. J. The evaluation of agroforestry practices over time in the crop-livestock system of western Kenya. Dvorak, K., editor: *Social Science Research in Agricultural Development: Spatial and Temporal Dimensions,* pp. 118–143. Commonwealth Agricultural Bureau International, Oxford, 1993.
7. Scherr, S. J. Meeting household needs: Farmer tree-growing strategies in western Kenya. Arnold, J. E. M. & Dewwes, P. A., editors, *Trees in Farmers' Strategies: Responses to Agricultural Intensification.* Oxford University Press, Oxford, 1994.
8. Steppler, H. A. & Nair, P. K. R. *Agroforestry: A Decade of Development.* International Centre for Research on Agroforestry, Nairobi, Kenya, 1985.
9. Warner, K. *Patterns of Farmer Tree-growing in Eastern Africa: A Socioeconomic Analysis.* Tropical Forestry Papers 27. Oxford Forestry Institute and International Centre for Research in Agroforestry, Oxford, 1993.

Acknowledgements

I would like to acknowledge the farmers and staff of the CARE International in Kenya/Government of Kenya Agroforestry Extension Project, who collaborated in much of the research reported here. Field studies were undertaken while I was a principal scientist at the International Centre for Research in Agroforestry. Field research was funded by the Australian International Development Assistance Bureau, the Ford Foundation, and the Rockefeller Foundation. Japanese International Cooperation Agency supported my work at the International Food Policy Research Institute.

Sara J. Scherr, research fellow, International Food Policy Research Institute, Washington, DC 20036.

DEFORESTATION IN TANZANIA:

Beyond Simplistic Generalizations

SOLON L. BARRACLOUGH
AND KRISHNA B. GHIMIRE

Using national-level data, many analysts frequently blame deforestation in tropical countries on a country's population growth and poverty. Detailed, local-level case studies in Tanzania, however, reveal that the relationships between population dynamics and environmental degradation are much too complex to support reductionist generalizations about cause and effect. In one lowland rural area, large population increases led to more tree cover, while in a lightly-populated region, there has been rapid deforestation. In another degraded area, a conservation project rehabilitated deforested land by excluding local pastoralists and farmers—which resulted in social and ecological problems elsewhere. The principal impetus to deforestation in the country has come from world markets and state policies.

Nearly half of Tanzania, a country of some 94.5 million hectares, could be considered forested. Almost 98 per cent of this forest is *miombo* woodlands which cover most of the western and southern part of the country and range from rather dense closed forest to open woods in drier regions, blending into bushlands and grasslands.

Only about 1.4 million hectares of moist, closed tropical forest remain in Tanzania on the upper slopes of mountains, especially in the southern highlands and near the north-eastern coast; the East Usambara mountains support one of the richest biological communities in Africa. There are also 80,000 hectares of closed mangrove forests along the country's east coast bordering the Indian Ocean, 60 per cent of all remaining mangroves in East Africa.

There are no reliable data on recent deforestation in Tanzania; estimates that 300,000–400,000 hectares are being cleared each year have been repeated for over two decades. The UN Food and Agriculture Organization, however, puts overall deforestation in Tanzania at about 0.3 per cent annually and twice this rate in the humid tropical forest and mangrove areas.

The number of people in Tanzania, meanwhile, is estimated to have increased six-fold since 1913 when the population was about four million, reaching 25 million in 1990. Although urban population growth has ben rapid, four-fifths of Tanzania's population still lived in rural areas in 1990 and were for the most part agricultural.[1] Population increases and densities are unevenly distributed throughout the country.

Intense population pressure on available land resources is usually blamed for the country's deforestation. However, case studies of the interactions between local institutions, markets and state policies over several decades in three areas—a district in the Usamabara mountains, a delta near the capital, Dar es Salaam, and a semi-arid area in the centre of the country—reveal this to be a simplistic generalization.

Land Scarcity in the Western Usambaras

Several centuries before German colonization in the late nineteenth century (see Box, p. 000), settlers began to clear the Usambara mountain forests to expand and intensify agriculture.[2] A gradual population increase in the area was in part due to successive waves of migrants who fled there to avoid widely-feared warriors such as the Maasai; the mountainous terrain facilitated defence and had a moist and sub-tropical climate with several areas of good soils.

The agricultural systems the people used protected soils from erosion, improved water retention and protected biodiversity. For example, the most common *shambaa* farming system resulted in multi-storeyed vegetative cover similar to the forest it had replaced. Taller trees which yielded edible fruits and seeds dominated the canopy while the understorey consisted of bananas (a principal food crop) and vines yielding various useful foods and other products. Scattered patches were planted with yams, beans and other garden crops.

During German and then British colonialism, about one-third of agricultural land in the Lushoto district of the Usambaras was appropriated by European settlers to grow coffee, tea, sisal and other plantation crops. The colonial authorities gradually incorporated other areas into forest reserves so that by the time of independence in 1961, some 107,000 acres (13 per cent of Lushoto district) was demarcated. Colonial foresters' concern with deforestation had led as early as 1895 to the enactment of various forest protection laws, which prohibited forest clearance on hilltops, steep slopes and along valley waterways, and the establishment of a game reserve. By 1961, over two-thirds of Lushoto became unavailable for agriculturists and pastoralists who constituted the bulk of the population.

This land alienation occurred at a time when the local population was growing; the number of inhabitants in Lushoto district almost tripled between 1948 and 1988 from 128,000 people to 357,000, most of them living in the rural areas. Compared with the previous century, little of this increase was due to in-migration; in fact, many young men were leaving the area to find employment in the capital, Dar es Salaam, in the town of Arusha, and elsewhere. Improved health services contributed to a lower mortality rate, but there were also economic incentives for peasant families to have more children, such as increasingly labour-intensive farming practices and a need to deploy family members as migrant labourers to top up family finances. Population densities in the district ranged from less than 30 persons per square kilometre in some municipalities to over 300 in others, variations due partly to patterns of land tenure and partly to differences in productive capacity because of different soils, topography, rainfall and histories.

A tripling of the peasant population combined with a two-thirds reduction in traditional farm land and a lack of livelihood alternatives led to growing pauperization. Catastrophic deforestation, however, did not occur, largely because the peasants adapted their production and consumption systems to the scarcity of land.

Many peasants intensified their farming practices, using less land and more labour. Some began to produce fruits and vegetables for sale in the expanding urban markets of Dar es Salaam, Arusha and nearby Kenya. Traditional irrigation systems were repaired, while terracing and bunding became more widespread. Many peasants planted trees for their fuel, fodder and construction needs. Some peasant cultivators turned to livestock raising, previously the prerogative of only a few ethnic groups. They stall fed higher-quality animals rather than grazing them in open pasture, a practice which provided more manure for fertilizer.

Peasants adjusted their consumption patterns as well. Maize, potatoes, cassava, yams and beans became more important in local diets, gradually replacing traditional foods such as millet, bananas and animal products, which were not so high-yielding. Those who produced for the market began to consume more purchased foodstuffs. New house construction depended less on wooden poles and thatch and more on mud, grass and bricks. Most families curtailed their use of fuelwood.

Peasant adjustment to land scarcity was aided by government programmes.[3] The government's Lushoto Integrated Development Project, initiated in the mid-1960s, encouraged the marketing of vegetables, improved dairy production, and helped to spread better brick-making techniques and various other activities. A Soil Erosion Conservation Agroforestry Project followed in the 1980s, promoting soil conservation practices, intensifying agricultural production and providing seedlings and technical assistance. Although conflicts were generated when these programmes did not take local cultural norms (such as customary land tenure) sufficiently into account, the post-independence state was usually supportive of customary farming systems and land rights.

Forest Destruction in Rufiji

In Rufiji, a 1.3 million hectare delta area some 150 kilometres south of Dar es Salaam which is 90 per cent forested and includes half of Tanzania's mangroves, a very different picture emerged.[4]

Rufiji district is lightly populated with about 150,000 people and a population density of about 11 persons per square kilometre, only one tenth that of the peasant farming areas of Lusotho. Because of out-migration, population growth has been slow in Rufiji since 1948; during the 1980s, the total number of people actually decreased. Although much of the delta has good agricultural potential, less than four per cent of the district was cultivated in the early 1980s. Most of the agriculture was of rice, cassava, maize and other foods for self-provisioning, but cashew nuts and cotton were also grown as cash crops.

One of the most important sources of cash income is timber sold as logs, charcoal, construction poles and fuelwood. Wood and timber extraction and processing employ over one-fifth of the district's workforce. Over half the forest products are taken from the 40,000 hectares of rich coastal mangroves which have been heavily overcut since the 1970s.

The immediate cause of indiscriminate deforestation has been the lucrative markets for timber, wood and charcoal in Dar es Salaam, a city which increased in population from 70,000 in 1948 to 1.4 million in 1988. Whereas producer prices for cashew nuts in Rufiji district increased by 40 per cent during the 1970s and cotton prices by 100 per cent (only to fall again in real terms in the 1980s), the producer price of charcoal went up by over 400 per cent, as did pole and timber prices.

But attractive prices for timber, wood and charcoal do not fully explain why local people began to mine the mangroves, after having exploited them prudently for many generations. Also significant was the breakdown of customary land tenure and production systems as a result of "villagization", the post-independence government's programme of enforced resettlement of scattered rural people into government-planned villages (see Box, p. 213–215).

The first villagization programme in Tanzania began in Rufiji district in 1969 when flood victims were resettled in the delta. Customary tenure was in theory respected but in practice village boundaries were seldom clear. Long-established farming systems were disrupted and social differentiation increased. The government took legal ownership of the mangroves and included them in Rufiji's 143,000 hectares of state forest reserves, but it did not have the resources to protect or manage them. Many young people left to work in the capital, leaving communities short of able-bodied workers. Remaining villagers found mining forest resources, particularly those on government lands, the easiest way to earn their livelihoods. Customary rules governing forest exploitation were simply undermined by new lucrative markets, insecure forest tenure and social disruption. Merchants and some public officials who stood to gain did what they could to persuade local authorities that they too could benefit by destroying their forests.[5]

Deforestation Under Colonialism and "Villagization"

As in other parts of Sub-Saharan Africa, land in Tanzania belonged primarily to different ethnic groups or clans. It was allocated by a chief or elder responsible for land rights, to be worked or grazed by the community, family groups or individuals. Access to land was a birthright of all clan members.

Tree tenure was usually distinguished from land tenure, especially if the trees had been planted or tended for a particular purpose.

Clans dedicated principally to livestock raising could hold complementary tenure rights for grazing during certain periods on lands belonging to agriculturists. Clans might have been under the protection of a central authority or chief to whom they paid some tribute, but the principal property rights to land rested in the clan, which had the right to exclude outsiders or grant them access under certain conditions. Members of the clan usually had the duty to observe customary rules of land use and of distribution of its products.

European Colonization

Land tenure was disrupted when Germany colonized Tanganyika in the late nineteenth century (the country joined with the island of Zanzibar after independence to become Tanzania). Moist tropical forests were three or four times more extensive then than they are now. The colonial administrators promoted export crop plantations of coffee, tea and rubber which required soils and climates similar to those of the moist forests. White settlers were attracted to the highlands because of the pleasant, healthy climate. Other densely-forested areas were cleared for the production of sisal.

To facilitate crop exports, the colonial authorities constructed roads, railways and port facilities. The Germans built a central 1,600-kilometre railway through the country which caused much deforestation, as did the construction of the Tanzania-Zambia railway (TAZARA) half a century later.

In coastal areas and along the central railway, several hundreds of thousands of hectares of woodlands were cleared for sisal plantations. By 1954, sisal plantations covered about 300,000 hectares and employed about the same number of workers. The country became the world's largest sisal exporter.

Besides establishing their own plantations, the colonialists also attempted to induce and coerce peasants to produce for the market, especially for export. To do so, they had to bring about many profound alterations in land use, agricultural practices, work habits, land tenure and social relations more generally, both within and between existing communities.

The colonial administrators regarded any land that was not occupied or cultivated as "vacant", and disposed of these areas as they wished. Yet peasants frequently left much of this "vacant" land fallow for long periods to allow soils to renew their fertility under forest or other vegetative cover, while they used other "vacant" forest lands for hunting, grazing, and gathering useful plants and other products.

In addition to plantation crops in the moist forest region, extensive drier wooded areas were cleared for cotton and tobacco. Tobacco plantations had to be rotated frequently, while curing required vast amounts of fuelwood.

All these deforestation processes continued when the country became a British protectorate after the First World War. A failed attempt by the British to turn Tanganyika into a major groundnut exporter caused deforestation in some districts in south and central regions.

More successful was the promotion of cashew-nut cultivation in the south where local agriculturists used cheap migrant workers from Mozambique to plant large areas with cashew trees. This made a few peasants relatively wealthy as these trees became the individual property of the planter, increasing social differentiation in the area. Developments such as these meant that throughout the country, rules regulating land tenure, production and distribution began to break down.

Villagization

Following independence in 1961, land became state property, supposedly to be allocated and used for the benefit of all. In practice, large private plantations and farms owned by Europeans, as well as other foreign-owned properties and businesses, were nationalized (with a few exceptions). Traditional land tenure systems remained dominant in extensive rural areas except in parks and game reserves.

In the early 1970s, President Julius Nyerere launched a resettlement programme called Ujamaa, aimed at grouping dispersed rural people in villages. This had a negative ecological impact in many rural areas. "Villagization" was meant to provide rural people with amenities such as water, sanitation, schools, housing, social centres and the like, while facilitating the introduction of modern agricultural inputs, tools and practices, along with better marketing and cooperative organization. By 1975, some nine million rural people were living in nearly 7,000 villages compared with half a million people in 2,000 villages just five years earlier.

However, the programme was very uneven in its implementation; while some of this massive movement of people was voluntary, there were coercive pressures in many cases. The selection of sites selected for villages were frequently based on inadequate information about their natural resources and their potential uses. The result was serious environmental degradation in some instances. (Land use planning is now for the most part decentralized.)

Villagization often disrupted customary production and social systems. It was sometimes a direct cause of accelerated deforestation, as the new settlements required construction material and fuelwood—previously gathered and distributed over a wide area by dispersed groups—to be concentrated in a few small areas. Land-use assessments and planning for the villages were often patchy, leading to cultivation and heavy grazing of areas that were unsuitable for such intensive uses. Government attempts to promote agroforestry systems and community forestry in villages led to highly variable outcomes.

Although the villagization programme brought social and economic benefits to rural people in some cases, in others the social consequences were perceived by the relocated peasants as harmful. Villagization was a coercive violation of customary tenure rights, even though its aim was supposedly to strengthen and modernize traditional social and production systems. There was considerable opposition by villagers who could not follow the logic of moving to a village to be near services and markets that they used only occasionally instead of continuing to live near crops and animals requiring daily attendance. Today, while many relocated rural people have remained in the villages, others have returned to live near their fields, often leaving part of their families in the villages.

Deforestation and Desertification in Kondoa

Many of the complex interactions between land appropriation, demographic change and colonial and post-independence politics are illustrated by Kondoa in central Tanzania.[6] The district is frequently cited as a classic example of deforestation and desertification caused by population growth and supposedly careless pastoral and cultivation practices. Over half the district's 1.4 million hectares are badly eroded. Highly-visible land degradation—spectacular deep gullies, sand rivers and parched earth—has made Kondoa a target of conservationist interventions by successive colonial administrations, the independent state and foreign donors.

The moister highlands were once largely forested, but now less than one fifth of them support *miombo* woodlands. Rainfall averages only about 600 millimetres annually, is unreliable and is concentrated in a short rainy season. Before human settlement, the drier plains and lower hills were covered by grasses and bushes with scattered patches of woods.

Land degradation seems to have accelerated in the mid-eighteenth century, when an important trading route from the east coast to Lakes Victoria and Tanganyika crossed Kondoa to avoid the Maasai. Caravans often numbered over 100 members. Kondoa settlements became important provisioning depots for the traders and extensive forested areas were cleared to cultivate grain for them. Other forests were exploited to supply charcoal for iron smelting to furnish the caravans with tools and utensils. The traders' demand for *vyome* (bark containers for grain storage) contributed still further to deforestation.

Strategic settlements of Arabs and Waswahili arriving from the coast also grew. These immigrants converted many of the indigenous Rangi of Kondoa to Islam, used slaves to extend the area of cultivation and frequently appointed Waswahili as village headmen and tax collectors. All these processes weakened customary social institutions and values. The Rangi gradually abandoned many of their pastoral and agricultural practices which had been protective of soils and trees and lost control over their management of their natural resources.

In the late nineteenth century, a rinderpest epidemic decimated livestock in the district. The Rangi's cattle, which constituted their accumulated wealth, their insurance against bad years and their principal status symbol, were wiped out. This disaster coincided with the German colonial conquest, which was particularly harsh and ruthless in Kondoa. The district had few natural resources of commercial interest to the colonialists, but instead became an important source of forced labour for colonial projects elsewhere. During the First World War some two decades later, Kondoa became a battlefield between British and German armed forces. Both sides looted local grain supplies. The Germans' scorched-earth tactics during their retreat caused further environmental damage.

Around this time, many Rangi fled to the highlands and cleared the wooded areas for subsistence cultivation to escape fighting, conscription in the German army, or the sleeping sickness carried by the tsetse fly, which had begun to spread southwards into Kondoa following large-scale brush encroachment associated with the rinderpest epidemic, colonial labour exploitation and war.

After the war, containing sleeping sickness became a high priority for the new British colonial administrators of Kondoa. They undertook a major programme of brush clearance in infested areas of the hills and plains to facilitate resettlement from, and destocking in, the badly-eroded highlands. They also imposed labour-intensive soil conservation practices such as ridge cultivation, contour banking and reforestation. These measures did not take adequate account of the Rangi's customs and livelihood concerns, but the Rangi were too weakened to resist collectively. Some responded to harassment by the colonial officials with apathy, others by migrating to other regions. A net ecological result of the British administration's conservation interventions was to slow land degradation in the highlands but to accelerate it in the lowlands.

In the 1960s and early 1970s, several severe droughts forced many peasants of Kondoa to move once again to the highlands. Renewed deforestation and erosion as well as food shortages impelled the government to make Dodoma region (of which Kondoa district was a part) the second area of villagization in 1972. The Dodoma Land Rehabilitation Programme (HADO) was established in the district in 1973, with substantial financial and technical support from Sweden, to rehabilitate its soils and forest. The programme's approach was in many ways similar to that of the British land rehabilitation effort in the 1930s. It emphasized cattle destocking, soil conservation measures such as contour banking and tree planting for shelter-belts, agroforestry, and village woodlots. In severely eroded areas, cattle were excluded, effectively evicting their owners as well. Like the earlier efforts, HADO was a top-down and technocratic project with little real participation by the local people in setting goals or in designing and implementing the project.

The HADO programme did demonstrate that eviction of people and cattle could contribute to the restoration of vegetative cover on some degraded semi-arid lands within a relatively short time—but such a narrow technocratic approach only exported problems elsewhere.[7]

Challenges and Blame

These examples from Tanzania suggest that even in the absence of increasing numbers of people, forest degradation and deforestation will continue in developing countries as long as other conditions remain more or less the same. In Tanzania, demographic change (migration as well as birth and death rates) is only one aggravating factor among many. No intense population pressure on available land resources, which could have caused massive deforestation at national levels, existed. Deforestation has been primarily driven by public policies since colonial times that stimulate the expansion of agro-experts, commercial logging and massive land alienation.

Customary land tenure systems, while still influential in Tanzania, are vulnerable to disruption, depending on the extent of commercial activities, social differentiation and state land policies. How the land is used and cared for depends on the political alliances and conflicts between customary authorities and those of the national state.

A major challenge now in Tanzania is to ensure that rural communities retain relative equity in access to land, concern for environmental protection and the decentralized local management that customary land systems usually embodied. At the same time, policies are needed which enable local people to adapt and protect themselves in a rapidly-changing and usually hostile socio-economic environment. Meeting this challenge will not be facilitated as long as "experts" identify customary farming systems, poverty and population growth as the main causes of deforestation.

This article is an edited extract of *Forests and Livelihoods: The Social Dynamics of Deforestation in Developing Countries*, Barraclough, S. L. and Ghimire, K. B., MacMillan, London/St. Martin's Press, New York, 1995.

Protection and Plantation

The setting aside of land in what is now Tanzania for conservation and recreational purposes started during the German colonial period, when foresters became concerned about rapid deforestation. The expansion of protected areas continued under the British and after independence.

Today, nearly five per cent of Tanzania's 45 million hectares of forest land are classified as strictly protected. Another 29 per cent are in forest reserves while the remaining two-thirds are classified as public forest lands. Local people are usually excluded from farming, grazing, hunting or otherwise using protected areas in customary ways. This has generated many conflicts over the years between local people and park, game and forest reserve authorities.

Under Tanzania's present decentralized authorities, the central government's administrative structures for natural resource control and other functions are replicated in each region. While all land is legally state property, central government officials have to go through a regional director to contact a regional officer. Within each region, local governments (district councils) have considerable authority, including the power to levy taxes on the use of forests and other natural resources. Within the districts, villages have usufruct rights to their lands and considerable discretion in their use. Village-controlled forest boundaries, however, are seldom clear.

Given this four-tier administrative structure—centre, region, district, village—central government foresters have little scope for influencing management of the 29 million hectares of public forest lands or the 13 million hectares in forest reserves. In most cases, responsibility for their management reverts to regional governments or, in some instances, districts or villages. In addition, the national forestry service (the Forest and Beekeeping Division (FBD) of the Ministry of Tourism, Natural Resources and the Environment) has only some 500 technicians and professionals—one for every 90,000 hectares of forest land in Tanzania.

However, the FBD has achieved a degree of managerial control over a few of these reserved forests by declaring them special attention zones and obtaining funds from foreign aid agencies. In 1989, it was able to assert control over 80,000 hectares of mangrove and some reserves of closed forests in the Usambaras.

Tanzania also has some 150,000 hectares of plantations, over half of which are in national forest reserves. Softwoods (*Pinus* and *Cupressus)* predominate and current sustainable yields (if they were to be harvested) exceed effective industrial demand by about one third.

The German colonists established the first plantations of exotic species such as eucalyptus. The colonial and post-colonial state encouraged both broadleaf and coniferous plantations to meet anticipated demand for timber and fuelwood. They may also have been motivated by bureaucratic dynamics and a need to create rural jobs. Teak plantations in the north-east were apparently planted to meet a brisk demand.

In fact, where planting is undertaken in anticipation of future markets, demand has proved insufficient for much of the timber grown. Projections of national global supply and demand for timber may have little to do with the realities of markets for particular products in particular places.

Solon L. Barraclough and Krishna B. Ghimire are senior consultant and project leader respectively at the United Nations Research Institute for Social Development (UNRISD), Geneva.

Notes and References

1. This estimate is from national sources. The World Bank estimated in 1992 that 33 per cent of the population was urban. See World Bank, *World Development Report 1992*, Oxford University Press, New York, 1992.
2. See Mascarenhas, A. and Maganga, F. P., "Land Scarcity and Deforestation in the Western Usambaras", paper presented at UNRISD workshop on Social Dynamics of Deforestation in Developing Countries, Nanyuki, Kenya, 15–19 July 1991. Significant deforestation took place in the Usambara mountains about 2,000 years ago when several areas were partially

cleared to provide wood and charcoal for iron smelting. When iron production in the region declined, the forests recovered.

3. In retrospect, these programmes were sometimes rather ineffective and contradictory. The villagization programme had little disruptive impact in the district as most rural people were already living in villages.

4. Mascarenhas, A., "Deforestation by Proxy in Rufiji District", draft report prepared for UNRISD, July 1991.

5. Villagization also contributed to the breakdown of cashew nut production. Because owners had to travel long distances to tend and harvest their trees, many abandoned their cashew orchards and cut their trees to make charcoal.

6. Mascarenhas, A.,"Ecology and Deforestation in Semi-Arid Kondoa District, Tanzania", draft report preparted for UNRISD, 1992.

7. The HADO programme also suggested that genuine popular participation of the various local groups in the area could have helped make it more successful in meeting social as well as strictly environmental objectives. Indeed, the government and donors recognized this in the HADO 1986/87–1995/96 second phase master plan. The historical record from the eighteenth century onwards, however, suggests that achieving effective democratic participation for indigenous people on terms equal to those of aggressive coveting their land, trees and labour is a long way from being exclusively a local issue.

PART 3

Considering the Importance of Biodiversity

As the first article in this section demonstrates, biological diversity has recently become a central topic in environmental research. The author, Martin Holdgate, shows that the concern has originated in part from the recognized increased rate of extinction of plant and animal species. The surge of the extinction rate is demonstrably due to human alteration of habitats, particularly in tropical rain forests. He also goes to show that biodiversity is an indicator of the health and sustainability of habitats and ecosystems and, moreover, that the conservation of those systems is critical for major life-sustaining functions, such as cycling of carbon and oxygen and regulation of water flow. The second article, by Edward Grumbine, looks at the importance of biodiversity not in ecological terms but in environmental ones. He asks, specifically, how the value of biodiversity has been used as an argument for nature protection in the United States. He shows that it is a very powerful idea, since it taps into a value that many Americans share about the protection of species, especially endangered plant and animal species.

The third article presents a strong argument for the need to preserve biodiversity in the Amazonian rain forest. Likening Amazona to a library of information about plants and animals, Richard Schultes compares the deforestation and burning of Amazonia to the burning of a great library, one which contains the only information on topics crucial to humanity. He shows that the ethnobiological knowledge held by Native American residents of the rain forest can be considered encyclopedic and should be a primary goal of preservation. The loss of that knowledge and of the species on which it is based is an irreparable loss for humanity. The fourth article, by Joseph Dudley, presents a similar case for the importance of southern Africa as an area of high biodiversity and that critically needs conserving. Included in that region is the island of Madagascar, which has extraordinarily high levels of endemic plants and animals, i.e., found only there. As in Amazonia, the biological diversity is of prime global significance and deserves wider recognition as an international conservation priority.

Biodiversity is also highly important in domesticated plants and animals, given their fundamental place in the food systems of the world. Biodiversity not only provides for an array of different food items, but the number of genetic strains provides invaluable protection against disease and catastrophic loss. The reduction of genetic strains is a serious issue in that regard, since a new strain of disease can ravage huge areas of crops. The final three articles in this section look at issues associated with domestication and biodiversity. The first of those, by Grimaldo, Vásquez, focuses on plants and animals in the Andes region of South America and shows how genetic diversity is maintained through selective breeding and cultural knowledge about the desirability of the resultant biodiversity. The second article, by Alejandro Casas et al., describes the processes of plant management by Nahua and Mixtec populations in an area of Mexico. He describes an extraordinary depth of cultural information about the management of plants, wild, domesticated, and those in the process of being domesticated. Approximately 180 species of plants are edible, and of those, 83 are actually planted, whereas the remaining 97 or so are either gathered, tolerated, or enhanced. There are, in other words, gradations between wild and domesticated, and the culturally based management in those terms promotes and sustains biodiversity. The third article, by Vandana Shiva, looks at the question of international efforts at the preservation of seeds for food crops. He shows that there has been an erosion of variety of seed types, especially due to the impact of the "elite varieties" of the Green Revolution, and that as a consequence, the collection and preservation of as many varieties as possible should occur.

THE ECOLOGICAL SIGNIFICANCE OF BIOLOGICAL DIVERSITY

MARTIN HOLDGATE

Introduction

The term 'biological diversity', often contracted into one word, 'biodiversity', has burst into sudden prominence in recent years. This rise to prominence has been driven by increasing concern over the alleged surge in the worldwide rate of extinction of plant and animal species, as a consequence of human alteration of habitats, and especially the clearance of tropical rain forests. The total recorded loss since 1600 AD is 724 species, and that between 1900 and 1950 is 60 species. These figures may be compared with estimated natural background rates in both taxa of 1 species per 100 to 1000 years[1].

These figures are small compared with the estimates that have been widely (and alarmingly) quoted in the popular literature (Table 1)[1–8]. The discrepancy arises because like is not being compared with like. The larger figures are calculations based on three essential elements:

i. Studies of the canopies of tropical forests indicate that a very large number of species, especially of insects (mainly Coleoptera) and arachnids live there, and that only a fraction of these species has been described by science. Many appear to be confined to relatively restricted areas. This discovery[9–10] also led to an upward revision of the estimated total of species on Earth to between 10 mill. and 100 mill., of which less than 1–5 mill. have been described by taxonomists.
ii. Estimates of rates of forest clearance in the tropics suggest the conversion of 16.9 mill. ha. yr^{-1} in 1981–1990 (Table 2), mainly for cultivation.[11]

Table 1. Some estimated rates of species extinction (1)

Estimate	% Global loss per decade	Method of estimation and source
1 mill. between 1975 and 2000	4	Extrapolation of trend (2)
15–20% of all species between 1980 and 2000	8–11	Species/area curve and forest loss projection (3)
2000 plant species per year in tropics and subtropics	8	Loss of half species in area deforested by 2015 (4)
25% of species between 1985 and 2015	9	As above (5)
At least 7% of plant species	7	Loss of half species in 10 hot spots covering 3.5% of forest area in next decade (6)
0.2–0.3% all species per year	2–3	Half of species in forest assumed endemic and lost at clearance (7, 8)
2–13% of all species lost between 1990 and 2015	1–5	Species area curve and rates of deforestation (1)

iii. Assumptions of relationship between species and area (which, as Arrhenius pointed out in 1920[12] broadly follow a double logarithmic relationship) then permit the estimation of loss.[13]

The argument has been clearly stated by Raven[14] (as quoted by Heywood and Stuart,[15]):

> "The regions where tropical moist forests are expected to be destroyed in the next 30 years contain an estimated half of the world's plants. . . . If half these species are at risk when the forests have been reduced to less than 10% of their original extent, I estimate the extinction of perhaps 65,000 species, a quarter of the world's total, within the next several decades."

A very thorough estimate, not based on species/area curves, by May, Lawton and Stork[16] concludes that impending extinction rates today are likely to be at least 10,000 times greater than the background rate displayed in the fossil record.

This concern led to the international agreement of the United Nations Convention on Biological Diversity, signed by over 100 States at the United Nations Conference on Environment and Development in Rio de Janeiro, June 1992[17,18]. There were two kinds of motive behind the Convention. Many conservationists in northern developed countries sought international action to stem the destruction of tropical rain forests and other habitats, and the consequent loss of species. For them the priority was the identification and protection of the areas of highest value as centers of biological diversity. For developing countries, on the other hand, the attractions of the Convention were more utilitarian.

In recent years it has increasingly been appreciated that tropical rain forests and many other habitats can supply a diverse range of products and services of immense economic value. Some examples, from a variety of sources, are given in Box 1. The developing countries, within whose sovereign territories most of the

Table 2. Estimates of tropical deforestation for 87 tropical countries (11)

Continent	Number of countries	Forest area 1980	Forest area 1990	Annual deforestation
			(mill. ha)	
Africa	40	650	600	5.0
Latin America and Caribbean	32	923	840	8.3
Asia	15	321	275	3.6
Total	87	1894	1715	16.9

tropical forests lie, have become alerted to the potential economic returns that might be derived from such products, especially as pharmaceuticals. The Convention is seen as providing a legal mandate for the cooperative development of such resources, on an equitable basis, safeguarding the intellectual property rights of indigenous people and local communities. A model is provided by the cooperation in Costa Rica between national scientists and the United States pharmaceutical company Merck, which has funded a new joint exploration and development company, INBio.[19]

All this is relatively well-trodden ground. But there is another dimension. Ecological systems are clearly important as the regulators of major planetary processes. They are central to the cycling of carbon, nitrogen, oxygen, phosphorus, sulfur and other elements. They are important in the hydrological cycle and the maintenance of local climates. How far are these attributes related to biological diversity as such? How significant will losses of diversity be for the functioning of such systems and their response to stresses such as climate change?

Biological Diversity Defined

There has been a good deal of ambiguity in the use of the term 'biological diversity'. A neat, short, definition has been given by Solbrig[20]: "the property of living systems of being distinct, that is different, unlike." As he points out, diversity is a fundamental property of every living system, manifest at every level of the biological hierarchy from molecules to ecosystems.

It is important to discriminate between diversity as a property or attribute of a system, expressed in terms of how much variety it contains, and diversity as the aggregate of all genes, species, and ecosystems. In the aggregate sense, 'biological diversity' is simply a synonym for 'living nature.'[24] In the former sense it is a measurable parameter. What does measurable diversity in ecosystems mean for their functional integrity and adaptability?

Ecosystem Structure And Function

An ecosystem is an interactive system of plants and animals and the components of the physicochemical environment with which they interact.[25] Most ecosys-

Box 1. Some examples of values ascribed to wild habitats and species.[22,62]

The combined sustainable harvest of forest products such as rattan, oil, latex, tannin, dyes, medicinal products, lac, gums, resins, balsams, bamboos, genetic resources, wild meat, skins, hides and other products can bring in around USD 200 per hectare per annum whereas once-off logging yields only around USD 150 per hectare.[7] In some areas of Amazonia a figure as high as USD 700 per year has been estimated. Indigenous peoples in that region use from half to two-thirds of all forest tree species in one way or another.[23]

In Ghana and Zaire, 75% of the population depends largely on traditional protein from wild habitats: in Botswana over 50 species are used, yielding up to 90 kg per person per year.[20] In Zimbabwe the CAMPFIRE project has demonstrated that sustainable use of wild animals brings a higher economic return than ranching cattle.[24]

Genes from wild species are continually used in crop breeding: a "useless" wild wheat species from Turkey gave disease resistance valued at USD 50 mill. a year in the USA and one gene from an Ethiopian barley strain now protects California's USD 160 mill. a year crop from yellow dwarf virus. A Mexican wild relative of domesticated maize has yielded genes giving estimated savings of USD 4.4 billion a year worldwide.[22]

Traditional herbal medicine is the basis of primary care for over 2 billion people. The retail value of plant-derived drugs was estimated at USD 43 billion in 1985.[22]

Ecotourism based on wildlife, is the main foreign exchange earner in several African countries: a lion in the Amboseli National Park in Kenya has been estimated to be worth USD 27,000 per year in visitor attraction and a herd of elephants was valued at USD 610,000.[62]

Coastal wetlands in the United States have an estimated value of USD 4,938 per hectare per year as fish nurseries: in Boston Harbour a wetlands complex saved USD 17 mill. a year in flood protection.[62]

tems contain five main functional categories of living organisms. At the base, driven by the input of nutrients, solar energy and water and carbon dioxide are the green plants or primary producers, using the sun's energy to make carbohydrates. On them feed herbivores or secondary producers, which in turn support a tertiary level of carnivores. Dead material and faeces are used as energy sources by decomposers which in turn support microbivores, which may in turn support higher trophic levels. Energy and the various elements cycle through all these components and energy is released at each of the transformation processes, which is one reason why the total biomass of living matter in the higher trophic levels is less than that lower down the system.

Not all these components have to be present in every ecosystem (though it is rare to have any one missing entirely), and the relative proportions vary. On small and relatively newly-formed volcanic oceanic islands, for example, where species numbers are low because immigration is difficult, and where habitat diversity is small and the climate relatively uniform, there may be only ten to twenty vascular plants, no land vertebrates (the second and third levels in the hierarchy being dominated by insects and arachnids) and a restricted range of decomposers. Even on the continents, some major formations like boreal coniferous forests are dominated by only a small number of tree species and the ecosystem is, in all its functional components, less diverse by far than a tropical forest with several hundred tree species in the canopy.[26]

Very broadly, the pattern of diversity in land ecosystems can be related to gradients of temperature (or solar energy input), water availability and habitat age and structural variety.[27] Thus:

- the polar regions and high mountains are poorer in life than lower latitudes and altitudes;
- drylands are poorer in species than areas with abundant water;
- isolated, small, and relatively uniform patches of habitat (like some on oceanic islands) support fewer species than large, varied and accessible areas on continents;
- areas with long continuity of favorable habitat support more species than those which have experienced recent fluctuations.

The oceans are much less well known, and surprising discoveries continue to be made there, but it is clear that at the phylum level they are more diverse than land or fresh waters.[20] Shallow, saline waters are generally richer than deep or brackish ones, but the deep sea, until recently thought to be a biological desert, is now known to have an abundant biota with over 800 known species in more than 100 families, and the recently discovered deep ocean hydrothermal vents are known to support at least 16 families of invertebrates that were unknown to science 10 years ago.[21]

Diversity has increased steadily during the history of the Earth, but diversification has not been smoothly continuous: there have been particular periods with elevated rates of extinction.[28] However, these disruptions did not do more than halt and temporarily reverse the diversification process. Recent millennia have seen another extinction spasm which many have attributed to the dual impacts of the climatic and habitat oscillations during Pleistocene glaciations and the impact of human hunters, killing out large mammals including mammoths, mastodons, giant elk, north American camels, and many other taxa.[29]

The issue is whether this increasing diversity over time in any way enhanced the efficiency and resilience of ecosystems as entities, and how significant diversity is by comparison with other key ecological variables such as biomass, productivity, and resilience. As Solbrig[20] says, this is an old question in ecology. Clearly some diversity is necessary for ecosystem function, and where biogeographical barriers or extreme habitat conditions exclude major groups as they do on some oceanic islands and polar regions, ecosystems are atypical in structure and function. For example, the deep accumulation of well-aerated organic matter on the central South Atlantic Islands may be at least partly related to a lack of some major groups of invertebrate decomposers including earthworms, and the massive perennial tussock grasses that dominate the coasts of such southern islands certainly cannot survive when vertebrate herbivores are introduced to the ecosystem.[30]

But the key question is whether there are optimal levels of diversity for ecosystem function.[20,31,32] Can communities with high diversity maintain functional integrity better when a system is stressed because there are likely to be alternates to the current dominants, occupying subtly different ecological niches, and poised to gain prominence if those niches expand, for example as a result of climate change?

I would like to state three hypotheses: i) that biological diversity is only imperfectly correlated with ecosystem function; ii) that biological diversity is very closely related indeed to the structure and history of habitats and regions; and iii) that diverse systems are more likely to maintain their biomass and levels of productivity under stress, because they have built-in redundancy.

Diversity And Ecosystem Function

Solbrig[20] sets out 9 hypotheses concerning diversity at the ecosystem level, five being related to my primary hypothesis: that biological diversity is only imperfectly correlated with ecosystem function (Box 2).

Biomass (the total amount of carbon held in the living biota) is one indicator of ecosystem status. A global overview of the distribution of carbon in living ecosystems was given by Olson, Watts and Allison.[33] Their figures are summarized in Table 3. It is clear that tropical moist forests have the highest carbon densities of any world formation, followed by temperate and then by boreal forests, and that taken together forests are the principal reservoirs of living carbon (477 billion tonnes out of a global total of around 560 billion tonnes of carbon in living organisms). There is thus a broad positive relationship between biomass and diversity but it is not a strong one, in that there is a difference of several orders of magnitude in diversity between the most species-rich tropical forests and most species-poor boreal ones, whereas biomass carbon per unit area differs only by a factor of two.

Box 2. Hypotheses concerning diversity and ecosystem function.[20]

1. Primary productivity in an ecosystem depends on a minimum of diversity.
2. The relationship between primary productivity and diversity is not linear.
3. At large scales, diversity is not important in carbon, nutrient and water balance due to internal redundancies and compensations.
4. The more diverse the ecosystem, the more component species depend on diversity.
5. Ecosystems exhibit degrees of diversity several orders of magnitude higher than is required for efficient trophic function.

Source: Modification from Solbrig.[20]

Table 3. Biomass (carbon) in major ecosystems[33]

System	Area (mill. km²)	Estimated density (kg C m⁻²)	Global total (billion tonnes)
Tropical/Subtropical humid forest	12.7	13.5	169
Nonboreal mixed forest	12.0	10.5	128
Boreal forest	11.6	7.2	84
Dry woodlands and savannah	21.9	4.4	96
Wetlands (includes coastal)	2.8	2.7	7.8
Farms and marginal (incl. settlements)	15.9	1.4	21.3
Grass and shrubs	24.8	1.4	32.1
Tundra and fringes	13.0	0.97	12.6
Desert	18.2	0.32	5.9

In the tropics there is a proportionately much greater biomass of larger soil organisms: under cold moist conditions microorganisms predominate and rates of organic breakdown are less.[34] There are thus significant functional differences that relate to the proportionate biomass of different elements, but these are likely to derive in parallel from thermal and hydrological regimes.

The best single indicator of ecosystem function is net primary production—the fixation of carbon by green plants (Table 4). There is a positive relationship between productivity and diversity, and this extends also into marine and estuarine habitats, where some diverse formations are also highly productive. The faunal production (secondary and tertiary) also tends to be higher in more diverse formations. However, as with biomass the relative differences in diversity are much greater than those in productivity: the positive relationship is a weak one.

Table 4. Primary Production in major vegetation types (35).

Vegetation type	Net primary production ($g\ m^{-2}\ yr^{-1}$ dry matter) Mean	Range
Tropical rain forest	2200	1000–3500
Tropical seasonal forest	1600	1000–2500
Temperate evergreen forest	1300	600–2500
Temperate deciduous forest	1200	600–2500
Boreal forest	800	400–2000
Savanna	900	200–2000
Temperate grassland	600	200–1500
Tundra	140	10–400
Cultivated land	650	100–4000
Algai beds and reefs	2500	500–4000
Estuaries (excluding marshes)	1500	200–4000

This global relationship has many local exceptions and appears to be strongly influenced by nutrient availability and the balance between primary and secondary producers. The whole basis of agriculture is the enhancement of productivity through selection of species, nutrient enhancement and reduction of competition and secondary consumers by cultivation methods and pesticides. Such systems often have higher primary productivity than tropical rain forests, as Table 4 shows, but have much lower diversity. So have some low-diversity natural formations, even in subpolar regions. For example, Sub-Antarctic tussock grassland at South Georgia dominated by one species, *Paradiochloea flabellata*, has been recorded as having an above-ground primary production rate of around 2500 g m^{-2} yr^{-1}, and a herb-field on Macquarie Island has been recorded as producing over 1800 g m^{-2} yr^{-1}.[36] Neither formation is grazed by any vertebrate herbivore, however: the throughput of energy to higher trophic levels is very low, with much of the production going to the decomposers or accumulating in peat.

Increasing the nutrient levels in grasslands may lead directly both to enhanced production and to lower diversity. Long-term fertilization of a species-rich grassland at Rothamsted in England led, over the period 1856–1949, to a reduction from 49 to 3 plant species. Enrichment of shallow-water marine communities likewise leads to higher productivity, but lower diversity. Addition of vertebrate herbivores may reduce both biomass and diversity: if predators then in turn reduce herbivore density, plant biomass and productivity may again increase.[37,38]

It is clear that high diversity is not a necessity for high productivity. It is even lets essential if a system is geared to the yield of selected and specific products. But there is a question regarding the last of Solbrig's hypotheses—that ecosystems exhibit degrees of biodiversity several orders of magnitude higher than required for efficient trophic function. If the statement means that species-poor systems which are yet functionally complete, with an array of primary producers, herbivores, carnivores, decomposers and microbivores work as efficiently as species-rich ones, then it is likely to be true. As ecosystems develop, initially productivity, diversity and biomass will increase together. There is recent experimental support for this statement: when artifical systems of 9, 15 and 31 species were maintained in controlled-environment chambers, the more diverse systems exhibited higher productivity and respiratory activity.[39,40] However, as systems become even more diverse and complex the curves are likely then to level off, and primary production may fall in some systems as secondary consumers increase and the available nutrients are shared between a larger number of competitors and consumers. But diversity may go on increasing with time. The classic ecological explanation is that as ecosystems develop so the available energy is parcelled out between more and more species, each specialized to fill a particular ecological niche: the greater diversity of tropical systems is then explained by both their relative antiquity and by their higher levels of incident solar energy. If so, diversification is a process of elaboration within the system, not essential for its integrity as a functional unit.

First Conclusion: Diversity and Ecosystem Function

My first conclusion is that there is no close or direct relationship between the diversity of species or genes in an ecosystem and its biomass, productivity or role in the biogeochemical cycles. However, there is a general tendency for the systems with the highest biomass and productivity also to be the most diverse. This relationship need not be directly causal.

Patterns Of Diversity And Their Causes

There is a mass of ecological literature which seeks to relate species diversity in ecosystems to physical habitat diversity, competition and other interactions and environmental history. At the small scale, competition, stress and disturbance are often held to have especial importance, and in certain systems it has been argued that diversity is greatest in situations of intermediate stress.[41,42] Disturbance is often held to be important in opening up closed systems to invaders, and modifying the sequence of ecological development, so that dominant species can reproduce or other species can invade.[43] Clearly the situation is one of great complexity and the precise relationships at the level of detail are likely to vary greatly from one situation and ecosystem to another. On a broader scale, however, the pattern looks rather different.

The pattern of diversity appears to have several separate historical and biogeographical causes. The concentration of island endemics reflects the well-known processes of radiation among the small subset of species able to cross ocean barriers and the much smaller subset of these migrants able to establish themselves in new habitats.[44] The southern continents support relicts of systems once found on the cool-temperate parts of Gondwanaland including Antarctica. The centers of high diversity in areas like the Cape Peninsula are also relict. The low diversity of northern temperate regions is almost certainty a reflection of recent glaciation, and perhaps also of human impact which reduced vertebrate diversity. This latter point is significant because taking mammoths, mastodons, giant elk and North American horses and camels out of the system must have had a knock-on effect on dependent species like the sabre-toothed big cats and some plants whose seed germination was aided by passage through the guts of the big herbivores.

Within the tropical forests there is evidence that contemporary diversity has been greatly affected by events in the (geologically) recent past. At the ice maximum of northern glaciations, when sea level stood some 80 m below the present, what are now the basins of the Amazon and Zaire rivers are believed to have contained widely separated areas of forest, which served as refugia for plant and animal species.[43] In South America, there are thought to have been more, and larger, refugia, so that more species survived and the diversity in the recombinant forest after ice retreat and a return of more favorable conditions was greater in South America than in Zaire.

One consequence of the historical pattern is that several of the world's species and ecological formations are effectively functional alternatives to one another. It has been known since the early days of scientific exploration that organisms that are phylogenetically distinct may occupy similar ecological niches in different situations. The parallelism between the marsupial mammals in Australia and the placental dominants in the rest of the world is celebrated, and used to be greater before human disturbance of Australian ecosystems and introduction of eutherian competitors and predators. The same parallelism existed in Madagascar and New Zealand, where large, flightless land birds radiated to fill niches elsewhere occupied by mammals, only to fall victim to human impact and introductions.

Among plants, the arboreal habit has appeared in a very wide range of groups, and has emerged on some oceanic islands and high mountains among genera that are normally herbaceous. Ombrogenous, peat-forming mires in the Northern Hemisphere and the Atlantic sector of the Southern Hemisphere are particularly associated with bryophytes, and the genus *Sphagnum* above all. In the New Zealand—Chile—Tasmania sector of the Southern Hemisphere they are the preserve of vascular plant genera like *Astelia, Donatia* and *Oreobolus*. There are close physiographic similarities between the high tropical mountains and the Subantarctic.[46]

Sometimes the parallelism extends to whole vegetation complexes, with their attendant invertebrate faunas. For example the *Eucryphia-Laurelia-Weinmannia-Amomyrtus-Nothofagus* forests of southern Chile and their close congeners in Tasmania and New Zealand, the vascular plant carpet mires that accompany them in these regions, and the distinctive insect faunas found in association, are in many ways ecological alternatives to forest and wetland complexes that occur elsewhere in the subtropics and northern temperate zones.[47] They are distinctive, diverse, and ancient, but alternative arrays of species perform the same functions in other regions of comparable climate. The same holds for many vegetation types on oceanic islands. The success of certain invasive species moved around the world by human agency should not cause any ecological or biogeographical surprise. It is a logical consequence of the diminution of barriers to geographical dispersal, and would be expected to lead to a reduction in global biodiversity followed, in due course, by new radiation and specialization if separation between subpopulations is re-imposed.

Second Conclusion: Diversity and Biogeographical History

It is clear that much biological diversity is a reflection of history, and that today's patterns have complex roots. Many centers of high diversity are so because assemblages of species that have evolved over long periods and come to occupy a wide array of ecological niches have persisted there. Some of these diverse systems have persisted in certain parts of the world while ecosystems of comparable functional type have emerged elsewhere. Some have evolved in isolated and special conditions. Solbrig's[20] hypothesis that the species richness in an area is solely the resultant of the rate of species input (by immigration and *in situ* speciation) and species output (by emigration and *in situ* extinction) may be an oversimplification, but it provides a reasonable working model that fits with island biogeographical theory.[13] The diversification process has been guided by ecological rules, and no doubt feeds back to ecosystem function, but much observed diversity is clearly a reflection of the movement of species across a world with great variety in geography and climate, and in functional and ecological terms involves considerable duplication.

Diversity And Resilience

There is a strongly-held view that diversity in a plant or animal community confers resilience. The theory is appealing in its simplicity. If there are numerous species, many occupying closely linked ecological niches, and conditions alter to disrupt the ecosystem and eliminate key components, it is more likely that there will be replacements at hand than if the system is simple. However, the hypothesis can be read backwards: again to quote Solbrig[20] "the higher the diversity of an ecosystem, the more dependent the species are on the existence of diversity. In other words the higher the diversity, the narrower the ecological niche of each species." If this is so, diversity is a consequence of habitat complexity, biological complexity, energy and nutrient availability, and time. It does not follow that diversity is essential for resilience, and high diversity may imply high risk because of extreme specialization.

The truth is likely to lie somewhere in the middle ground. A very species-poor ecosystem may well lack resilience. For example, on the island of Tristan da Cunha there it but one native tree, *Phylica arborea* (Rhamnaceae).[48] If a specific disease, or human cutting for fuelwood, were to eliminate it, that would be the end of woody vegetation until some other species was able to cross the oceans and colonize the island. In a continental forest, on the other hand, there are tens or hundreds of tree species and reductions in the frequency of one will almost certainly promote another so that the overall forest facies, with its high biomass and productivity, remains.

This has clearly happened in history. In western Europe different forest types were prominent in past interglacial periods, and the differences probably relate to different survival patterns in different refugia. In Britain, for example, the hornbeam, *Carpinus*, was very prominent in the last interglacial but is a highly local tree today and appears only to have colonized the island in the past 2500 years.[49,50] Human impact also had a major modifying effect: in southern England, the original 'wildwood' 5000 years ago was dominated by *Tilia cordifolia*, *Corylus avellana* and *Quercus robur* (small-leaved lime, hazel and oak).[51] Human intervention eliminated lime and led to a dominance of oak, with hazel as a coppiced underscrub. There was enhanced productivity of things useful to people (timber, wattle, fuelwood, acorns and hazel nuts), and yet the retention of the forest ecosystem and of its roles in regulating water runoff. Even in apparently unaltered forest systems, such as those of Sarawak, there is evidence that forest peoples have modified the composition of the system, enhancing the proportion of the species they use.[52] It is an indication of the resilience of these diverse systems that in terms of ecological processes these changes are unlikely to have had a significant impact on overall function.

The apparent redundancy in today's biota may reflect, and be a basis for, ecological dynamics. It is now well established in ecological thinking that concepts such as the 'climax', defined as the ultimate dominant formation for a bioclimatic unit, and of 'succession' as the orderly process leading to it, need to be revised and modified. Many apparent climax systems—like oak forests—do not necessarily maintain themselves by regeneration. Many forests go through cyclical (or helical) processes in which an apparent climax is disrupted (for example by fire), followed by regeneration which may pass through several stages before something broadly like—but not necessarily the same as—the original is reconstituted.[43]

The process of dynamic change in ecosystems means that at any time patchiness is likely to be present, and that the overall character of the system is likely to alter with time. The most obvious patchiness occurs in a diverse moist forest where there may be as many as 300 canopy species, and where the fall of one individual opens the ground to regeneration, not necessarily by the same species causing a change in the small-scale vegetation mosaic and also in the associated insect and other fauna, much of which is specific to particular tree species. Patchiness is on a larger scale in formations liable to be swept by fires, and switched back to a regenerative mode over areas of hundreds of hectares. But in either case, it follows that there is opportunity for change, and that some of the apparent redundancy in the ecosystem may be expressed in the emergence of previously inconspicuous species to new prominence.

The implication is that in periods of change a different part of the 'pool' of biological diversity may well emerge and establish itself, and this is one positive advantage in the presence of apparent redundancy. It follows further that this 'redundancy' can have ecological as well as evolutionary importance in ensuring that functionally integrated ecosystems are maintained in the face of major environmental changes—including those wrought by humanity. Indeed, we have historical evidence that this is so in the evident changes in forest systems as a result of the impact of our species in the past.[51,53] It is likely that this property will have great importance in a period when climatic change induced by human action appears likely to occur at an unprecedented rate.

The projections are for an elevation in global mean temperature of maybe 1.5 to 3°C in the next 50 years, with the largest changes in polar regions. Precipitation patterns and storm intensities may also change, and sea levels will rise, probably by 15–30 cm by the end of the next century. These changes imply a displacement of the zones of tolerance of dominant trees and also of cropping zones by some 180–240 km, or 120–200 m. vertically. They may imply the virtual disappearance of the Arctic tundra (and large impacts on the migratory birds that breed there), and the northward extension of the Boreal forest to the shores of the Arctic Ocean. They may mean the dominance of Sweden by broad-leaved nemoral forests, and of southern England and northern France by species more typical of the Mediterranean regions.

Moreover, the rates of change seem certain to outstrip the maximum rates at which dominant plants can redistribute themselves by natural means. A study of events since the postglacial suggests that forest trees have spread at rates of up to 2 km a year,[49] and this implies that they are liable to lag behind habitat change if climates warm as fast as some projections imply. The result is likely to be that new ecosystems, of relatively low biological diversity, dominated by the more opportunistic and fastest dispersing species, will appear, drawing on the apparent redundancy in today's systems.

Third Conclusion: Diversity and Resilience

My third conclusion is that the principal ecological importance of biological diversity may well be its storage of a great quantity of genetic information out of which healthy ecosystems can be reconstituted under an enormous range of conditions and circumstances. As in most good information systems, there is redundancy and back-up. Not all the information will be of equal importance for the future. But in any ecosystem it is virtually certain that there will be a range of species existing in sub-optimal conditions, among but subordinate to, those best suited by current conditions. Some of these are of great potential importance as the founder stocks for new ecosystems if conditions change.

The Implications For Conservation

Conservation today has, in my view, been unduly influenced by four assumptions:

- that the concept of the climax, and of succession leading to it, is valid;
- that ecosystems (at least climax ones) can be regarded as static;
- that conserving the areas of highest biological diversity should have highest priority;
- that any loss of biological diversity is dangerous.

All four assumptions can be challenged. It now seems clear that climax systems do not persist without change, and without cyclical regenerative processes. It follows that just protecting a patch of climax forest or dry grassland may not work: the area conserved may need to be large enough to incorporate a complex of patches and stages in the cycle. Is also follows that conservation management must accept the reality of ecological dynamism. The cycle in the sense of a closed loop is less valid as a model than the helix, with time as the long axis, where there are subtle differences as cycle succeeds cycle. Invasions are natural and some species will spread, especially given human transport systems as aids in accelerated distribution.[54,55] In the past 50 years the collared dove, *Streptopelia decaocto*, has spread across Europe, for reasons that are not clear.[56]

Other birds, mammals, invertebrates and plants show constant variations in distribution. Long-range movement is not rare, and some such species do succeed as invaders. Conservation needs to maintain the potential for ecosystems to form and re-form themselves, and this will be especially important as climates change.

It seems to me to follow that conservation should be concerned with the larger landscape. Corridors and patches of diverse habitat should be encouraged within the altered and intensively managed ecosystems of agriculture, forestry and horticulture. Strict nature reserves and National Parks should be large and physically diverse. Fires and other disruptive processes may need to be tolerated rather than suppressed. Where secondary systems created by human agency, like most limestone grasslands, are deliberately maintained there may also need to be a wide range of stresses and impacts, to allow variation its opportunity. No ecosystem should be expected to remain static.

I am concerned about the notion that we should always give highest priority to the most diverse areas, even though this may be justified in some cases. It is right to seek to conserve unique centers like the Cape Peninsula in South Africa, or the richest tropical forest sites (especially those with physical diversity, like those around mountains such as Cameroon or Kinibalu or the eastern slopes of the Andes). But there is evidence that areas rich in one kind of organism are not necessarily so in others: bird diversity may not be a good indicator of insect diversity, and present correlations may be coincidental. There are certainly reasons for conserving refugia of lower diversity that may be important in a changing future—not least because we do not understand the origins of biodiversity. And we obviously need to give high priority to habitats of useful diversity like the homes of wild relatives of crop species. The methodology of conservation needs thorough evaluation, based on critical science.

The World Charter for Nature[57] and the draft International Covenant on Environment and Development[58] state that 'nature as a whole warrants respect: every life form is unique and is to be safeguarded regardless of its value to humanity.' Many conservationists argue passionately that all life forms must be conserved, and some have criticized the second World Conservation Strategy, *Caring for the Earth* [59] for tolerating some loss of biological diversity through the inevitable conversion of additional areas to support food production for increasing human populations. It has been argued that reduction of biological diversity is incompatible with sustainable development and imperils the evolutionary future. Without wanting to stir the hornet's nest too vigorously, I contend that such an extreme position is scientifically untenable.

Species Diversity, Evolution And The Human Future

I suggest that biological diversity is only partly related to ecosystem functional integrity and to the maintenance of the biogeochemical cycles. In summary:

- a high degree of biological diversity is not an essential prerequisite for ecosystem function and for the provision of the essential major services discharged by ecosystems—the cycling of carbon, nitrogen, oxygen, sulfur, phosphorus and other essential elements or the regulation of water flow and local climate;

- biological diversity is especially related to certain features of habitat structure and history, and has great biogeographical significance;
- loss of diversity in the past has not impeded the onward progress of evolution: it is in the nature of biological material that it is capable of renewed diversification to make good organic losses;
- the parallelism we see between ecosystems is an indication of some functional redundancy;
- some of this functional duplication and apparent redundancy is none the less important, and of evolutionary value.

Is there any implication for conservation priorities? It is evident that quite a number of species alive today are most unlikely to play a significant part in the evolutionary future. The coelacanth fishes may persist in the deep Indian Ocean, in the form of *Latimeria* and its relatives, and the Rhynchocephalian reptiles on three islands off New Zealand as *Sphenodon,* but the likelihood that these groups—or the Crocodilians, or the marsupial mammals, or a host of others—will be significant founder stocks in future is in my view vanishingly small. The same holds for island endemics, whose vulnerability in the face of introduced mammals, birds, invertebrates and plants from continental habitats is obvious, especially when the invasions are accompanied by habitat disturbance.[13,60] I suggest that the vast majority of the Coleoptera and arachnids that have evolved to occupy limited ranges and niches in the canopy of tropical rain forests[9,10] are also unlikely to have evolutionary significance.

That does not mean that I advocate writing such species off. Although they may not be important in future evolution, I accept that there are strong cultural and ethical reasons for seeking to conserve ancient relicts like *Sphenodon* and island endemics like the Galapagos finches. However, there are different reasons for conserving different groups and habitats, and we need to think clearly about which pertain to what. I suggest that we do need to apply evolutionary, biogeographical and ecological understanding to the analysis of which life forms are the indispensable foundations of the future. If we do, we may surprise ourselves and discover that some ecosystems of low biological diversity—like arid lands, temperate forests and grasslands, or coastal zone—are of great importance as habitats for future founder stocks. We would be wise also to recall that it is the widely dispersed and fairly abundant species that are most likely to survive stresses causing extinctions,[16] and be the founder stocks for new diversification.

We can readily conclude that a high priority must be given to conserving species of direct importance to people. The human future depends especially on the sustainability and productivity of the modified and highly productive ecosystems created by agriculture and managed forestry. The success of this process is evident in simple statistics. Wheat yields in western Europe have risen from around 0.5 tonnes ha^{-1} under the mediaeval open-field system, with poorly-yielding varieties and inefficient husbandry, to 7 to 8 tonnes ha^{-1} today.[61] But much of the improvement depends on continuing improvement by crop breeders, incorporating genes from wild relatives that are part of the general pool of biological diversity (Box 1).

The economic and social value of a wide range of wild species for food, medicines and many other products is also unquestioned. Forest people use a very high proportion of the species around them. One study of four indigenous Amazonian peoples indicated that they used between 50% and 66% of all trees either for food, construction material, raw materials for other products, and medicinal or trade goods.[22] Over 50 species of wild animal are important for food in Botswana. Traditional medicine—on which about 80% of the 3 billion people in developing countries depend—is based on wild plants. The values of many such species are only now being catalogued.[62]

The continued survival of species important, or potentially important, to people depends on the integrity of the ecosystems in which they occur. In the past this conservation used to be more or less automatic: nature maintained wild crop relatives, medicinal plants, wild foods and other products as a 'free good'. This is no longer an automatic service, because so large a part of the Earth's land surface has been occupied or modified by humanity. The maintenance of the ecosystems in which such important species occur is therefore a socially important activity and payment for habitat conservation may be a necessary feature of the implementation of the Convention on Biological Diversity.

All this will need more science. I am concerned that the contemporary debate about biological diversity lacks in-depth analysis of many crucial relationships such as:

- that between habitat diversity, habitat history, habitat size and patchiness and biological diversity;
- that between extinctions and human perturbations;
- that between ecosystem health, ecosystem integrity and biological diversity;
- that between apparent ecological redundancy and ecosystem resilience;

and as a consequence, that we lack a sound scientific basis for conservation both under the Convention on Biological Diversity and generally.

We need to study the key relationships more critically.

References and Notes

1. Reid, W. 1992. How many species will there be? In: *Tropical Deforestation and Species Extinction*. Sayer, J. A. and Whitmore, T. C. (eds.) Chapman and Hall, London.
2. Myers, N. 1979. *The Sinking Ark: A New Look at the Problem of Disappearing Species*. Pergamon Press, Oxford.
3. Lovejoy, T. E. 1980. A projection of species extinction. In *Council on Environmental Quality (CEQ): The Global 2000 Report to the President*. Council on Environmental Quality, Washington, DC.

4. Raven, P. H. 1987. The scope of the plant conservation problem worldwide. In: *Botanic gardens and the World Conservation Strategy*. Bramwell, D., Hamann, O., Heywood, V. H. and Synge, H. (eds.). Academic Press, London.
5. Raven, P. H. 1988. Biological resources and global stability. In: *Evolution and Coadaption in Biotic Communities*. Kawano, S., Connell, J. H. and Hidaka, T., (eds.). University of Tokyo Press, Tokyo.
6. Myers, N. 1988. Threatened biotas: "hot spots" in tropical forests. *Environmentalist 8*, 1–20.
7. Wilson, E. O. 1988. The current state of biological diversity. In: *Biodiversity*. Wilson, E. O. and Peter, F. M. (eds.). National Academy Press, Washington, DC.
8. Wilson, E. O. 1989. Threats to biodiversity. *Sci. Am.* September 1990, 108–116.
9. Erwin, T. L. 1982. Tropical forests: their richness in Coleoptera and other Arthropod species. *Coleoperist's Bull. 36*, 74–75.
10. Erwin, T. L. 1988. The tropical forest canopy: The heart of biotic diversity. In: *Biodiversity*. Wilson, E. O. and Peter, F. M. (eds.). National Academy Press, Washington, DC.
11. Tolba, M. K., El Kholy, O., El Hinnawi, E., Holdgate, M. W., McMichael, D. F. and Munn, R. E., (eds.). 1992. *The World Environment 1972–1992*. Chapman and Hall, London.
12. Arrhenius, O. 1921. Species and area. *J. Ecol. 9*, 95–99.
13. Williamson, M. H. (1981). *Island Populations*. University Press, Oxford.
14. Raven, P. 1990. The politics of preserving biodiversity. *BioScience 40*, 769.
15. Heywood, V. H. and Stuart, S. N. 1992. Species extinction in tropical forests. In: *Tropical Deforestation and Species Extinction*. Whitmore, T. C. and Sayer, J. H. (eds.). Chapman and Hall, London.
16. May, R. M., Lawton, J. H. and Stork, N. E. 1995. Assessing extinction rates. In: *Extinction Rates*. Lawton, J. H. and May, R. M. (eds.). University Press, Oxford.
17. United Nations. 1992. *Report of the United Nations Conference on Environment and Development, Rio de Janeiro, 3–14 June 1992*. United Nations, New York.
18. Glowka, L., Burhenne Guilmin, F. and Synge, H. 1994. *A Guide to the Convention on Biological Diversity*. IUCN, Gland, Switzerland.
19. Sittenfeld, A. 1993. INBio-Merck collaborative research agreement, Costa Rica. *Partnerships in Practice*. Department of the Environment, London.
20. Solbrig, O. 1991. *Biodiversity: Scientific Issues and Collaborative Research Proposals*. MAB Digest 9, Paris: UNESCO.
21. Grassle, J. F. 1989. Species diversity in deep-sea communities. *Trends Ecol. Evolut. 4*, 12–15.
22. United Nations. 1995. Commission on Sustainable Development. Third Session. Conservation of Biological Diversity. Report of the Secretary General, Paper E/CN, 17/1995/7. United Nations, New York.
23. Maveneke, T. 1993. CAMPFIRE, Zimbabwe. Case Study 4 in *Partnerships in Practice*. Department of the Environment, London.
24. WCMC. 1992. *Global Biodiversity: Status of the World's Living Resources*. Chapman and Hall, London.
25. Holdgate, M. W. 1979. *A Perspective of Environmental Pollution*. University Press, Cambridge.
26. Currie, D. J. and Paquin, V. 1987. Large scale biogeographic patterns of species richness of trees. *Nature 329*, 326–327.
27. Brown, J. H. 1988. Species diversity. In: *Analytical Biogeography*. Myers, A. A. and Giller, P. S. (eds.) Chapman and Hall, London.
28. Marshall, L. G. 1988. Extinction. In: *Analytical Biogeography*. Myers, A. A. and Giller, P. S. (eds.). Chapman and Hall, London.

29. Martin, P. S. 1967. Prehistoric overkill. In: Martin, P. S. and Wright, H. E. (eds.). *Pleistocene Extinctions: the Search for a Cause*. University Press, Yale.

30. Holdgate, M. W. 1964. The fauna of the Tristan da Cunha Islands. *Phil. Trans. Roy. Soc. Lond. B. 249,* 361–402.

31. Roughgarden, J. 1972. Evolution of niche width. *Amer. Nat.* 106, 683–718.

32. Roughgarden, J. 1989. The structure and assembly of communities. In: *Perspectives in Ecological Theory*. Roughgarden, J., May, R. and Levin, S. (eds.). University Press, Princeton.

33. Olson, J. S., Watts, J. A., and Allison, L. J. 1983. *Carbon in Live Vegetation of Major World Ecosystems.* Oak Ridge National Laboratory Publication 1997. Oak Ridge, Tennessee: National Laboratory.

34. Swift, M. J., Heal, O. W. and Anderson, J. M. 1979. *Decomposition in Terrestrial Systems.* Oxford: Blackwell.

35. Whittaker, R. H. and Likens, G. E. 1975. The biosphere and man. In: *Primary Productivity of the Biosphere*. Leith, H. and Whittaker, R. H. (eds.). Springer-Verlag, New York.

36. Holdgate, M. W. 1977. Terrestrial ecosystems in the Antarctic. *Phil. Trans. Roy. Soc. Lond. B 279,* 5–25.

37. Regier, H. and Cowell, E. B. 1972. Applications of ecosystem theory, succession, diversity, stability, stress and conservation. *Biological Conservation 4* , 2, 83–88.

38. Paine, R. T. 1966. Food web complexity and species diversity. *Amer. Nat. 100,* 56–75.

39. Naeem, S., Thompson, L. J., Lawler, S. P., Lawson, J. H. and Woodfin, R. M. 1994. Declining biodiversity can affect the performance of ecosystems. *Nature 368,* 734–737.

40. Kareiva, P. 1994. Diversity begets productivity. *Nature 368,* 686–687.

41. Grime, J. P. 1973. Competitive exclusion in herbaceous vegetation. *Nature 242,* 344–347.

42. Grime, J. P. 1974. Vegetation classification by reference to strategies. *Nature 250,* 25–31.

43. Connell, J. H. and Slatyer, R. O. 1977. Mechanisms of succession in natural communities and their role in community stability and organization. *Amer. Nat. 111,* 1119–1144.

44. Holdgate, M. W. 1986. Characteristics and consequences of biological invasions. *Phil. Trans. Roy. Soc. Lond. 314,* 733–742.

45. Lynch, J. D. 1988. Refugia. In: *Analytical Biogeography*. Myers, A. A. and Giller, P. S. (eds.). Chapman and Hall, London.

46. Troll, C. 1960. The relationship between the climates, ecology and plant geography of the southern cold temperate zone and the tropical high mountains. *Proc. Roy. Soc. B 152,* 529–532.

47. Godley, E. J. 1960. The botany of southern Chile in relation to New Zealand and the Subantarctic. *Proc. Roy. Soc. Lond. 152,* 457–475.

48. Wake, N. M. and Dickson, J. H. 1964. The terrestrial botany of the Tristan da Cunha Islands. *Phil. Trans. Roy. Soc. Lond. B 249,* 273–360.

49. Bennett, K. D. 1986. The rate of spread and population increase of forest trees during the postglacial. *Phil. Trans. Roy. Soc. Lond. B 314,* 523–529.

50. Godwin, H. 1975. *History of the British Flora*. Second Edition. University Press, Cambridge.

51. Rackham, O. 1986. *A History of the Countryside*. J. M. Dent. London & Melbourne.

52. Brookfield, H. 1983. Farming the forests in South East Asia. In: *Environmental Change in Rain Forests and Drylands*. UNU Global Environment Forum, II. United Nations University, Tokyo.

53. Denevan, W. 1992. The pristine myth. The landscape of the Americas in 1492. *Ann. Assoc. Amer. Geogr. 82,* 369–385.

54. Kornberg, H. and Williamson, M. H. (eds.). 1987. *Quantitative Aspects of the Ecology of Biological Invasions*. Proceedings of a Royal Society Discussion Meeting held on 26 and 27 February 1986. The Royal Society, London.

55. Williamson, M. H. and Brown, K. C. 1986. The analysis and modeling of British invasions. *Phil. Trans. Roy. Soc. Lond. B 314*, 505–521.

56. O'Connor, R. J. 1986. Biological characteristics of invaders among bird species in Britain. *Phil. Trans. Roy. Soc. Lond. B 314*, 583–598.

57. United Nations. 1982. UN General Assembly Resolution 37/7 on a World Charter for Nature.

58. IUCN/ICEL. 1995. *International Covenant on Environment and Development*. March 1995. IUCN Environment Law Centre, Bonn.

59. IUCN/UNEP/WWF. 1991. *Caring for the Earth: A Strategy for Sustainable Living*. IUCN, Gland, Switzerland.

60. Elton, C. 1958. *The Ecology of Invasions by Animals and Plants*. Methuen, London.

61. Prescott, J. 1994. *Sustaining Food Production and Our Countryside*. Wye College, Ashford: Occasional Paper 94/2.

62. McNeely, J. 1988. *Economics and Biological Diversity*. IUCN, Gland, Switzerland.

Sir Martin Holdgate
35, Wingate Way
Trumpington
Cambridge CB2 2HD
England

USING BIODIVERSITY AS A JUSTIFICATION FOR NATURE PROTECTION IN THE US

R. EDWARD GRUMBINE

Introduction

To fathom the last twenty-five years of growth in awareness of biological diversity in the US, take this simple test. Ask yourself the following questions:

1. What were two primary goals of environmental activists in 1970?
2. What were three species threatened with extinction at the time?
3. Was the term "ecosystem" in your personal lexicon on Earth Day 1970?

Your response to the first question likely includes air and water pollution, as these problems were receiving much attention at the time of the first Earth Day. Congress had already passed several laws to address such concerns and additional legislation was forthcoming. The second question was probably more difficult to answer—the Endangered Species Act as we know it today did not exist. You might have mentioned Whooping Cranes, Bald Eagles, or Bison, but most citizens were just beginning to wake up to the loss of species as a critical problem. As for the final question, you probably would not have had a solid working definition of "ecosystem" unless you had taken a college course in biology. In 1970, few activists in the nascent environmental movement had yet to embrace scientific ecology as an organizing principle.

If you were to ask yourself these same questions for 1996 your answers would be surprisingly different. Though pollution is still perceived as a threat by most Americans, many more environmentalists would now highlight the loss of biological diversity as a key problem. For question #2 you would have no trouble listing numerous species—Northern Spotted Owl, Peregrine Falcon, Grizzly Bear, Snail Darter, Mission Blue Butterfly, Kirtland's Warbler, or any of a dozen other commonly known endangered life-forms. And though your definition of ecosystem might not pass muster with a Ph.D., you would have little trouble describing it as a community in which plants and animals interact with the physical environment.

In the mid 1990s, after the 25th anniversary of Earth Day, loss of biodiversity is at center stage for many concerned citizens, activists, scientists, and managers (Grumbine 1992). This was not the case in 1970. The current emphasis on biodiversity has grown from a complex mix of cultural factors which are easy to highlight but difficult to untangle. First and foremost, there has been since the first Earth Day an unprecedented growth in scientific understanding of biological diversity, the ecological functions that diversity serves, and the biological consequences of environmental deterioration. This new knowledge has in turn been bolstered by trends in US environmentalism that reflect broad changes in American social values.

As historian Samuel Hays (1987) has observed, the first Earth Day marked the high water mark in the metamorphosis from conservation to environmentalism in the US. Americans, with greater amounts of education, disposable income, and leisure time were beginning to view *game* as *wildlife*, value *nonconsumptive* outdoor activities (e.g., photography) equally with *consumptive* pursuits (e.g., hunting), and voice stronger concern about resource *protection* as well as resource *management*. During the 1970s and 1980s, as environmental groups gained members, larger budgets, and lobbying clout, their agenda expanded from countering threats to specific parks and wildernesses to include concern for general environmental problems such as population growth, resource consumption, pollution, and energy policy. Arguments challenging human-centered values also grew stronger, to the point that a new field of a philosophical inquiry—environmental ethics—began to flourish (Nash 1989). Overall, more Americans began to actively question whether *progress*, defined simply as endless material growth, could really be sustained into the future.

Today, another phase in the evolution of American environmental values appears to be taking place, spurred on by new understanding of biodiversity. If the original Earth Day marked the beginning of a more inclusive approach to managing Nature for humans, future Earth Days may come to represent the rise in importance of biodiversity protection as the primary basis for human work with Nature.

In this paper, I trace how the concept of biodiversity has evolved toward its present position at the center of a compelling scientific framework for protecting Nature. I focus on two related trends—the development of the ecological roots of knowledge about biodiversity and the development within environmentalism of scientific justifications for protecting Nature.

Defining Biodiversity and the Biodiversity Crisis

'Biodiversity' has become a central rallying cry for a growing portion of the US environmental movement. The term and its relative, 'ecosystem management,' are referred to so often that the media portrays them as "buzzwords," empty phrases that everyone employs but few understand. Open any recent textbook, however, and biodiversity is easily defined. Noss and Cooperrider (1994, p. 5) provide a standard definition:

> *[Biodiversity is] . . . the variety of life and its processes. It includes the variety of living organisms, the genetic differences among them, the . . . ecosystems in which they occur, and the ecological and evolutionary processes that keep them functioning, yet ever changing and adapting.*

Where did the modern concept of biodiversity come from? Part of the answer is that biodiversity has appeared today because it is disappearing so rapidly. Conservation biology, the science of scarcity and diversity, would not be needed if not for significant loss of lifeforms. A science exploring extinction and habitat fragmentation, it has blossomed since the 1980s as a response to widespread destruction of species and ecosystems.

In the US, thousands of species are either listed or awaiting protection as candidates for listing under the Endangered Species Act. Estimates of species at risk over the next decade range from 2.5–15% of all lifeforms on Earth (Primack 1993). Beyond individual plants and animals, many US ecosystem types have been reduced to critical levels (Noss et al. 1994).Yet only about 6% of the US is in some kind of protected classification. Biologists are beginning to describe not only species and ecosystems at risk but also endangered biophysical processes, including large mammal and song bird migrations, river system flooding and deposition patterns, and forest nutrient cycles (Brower 1994). Some scientists warn that entire faunal groups may "all but disappear" within the next century, including primates, large carnivores, and most hoofed animals (Soulé 1986). In both direct and indirect ways, human activities are causing a biodiversity crisis—the largest mass extinction in 65 million years.

Scientific Roots of Biodiversity

The modern definition of biodiversity derives from the development of the science of ecology and its application to conservation issues. To fathom why scientists and environmentalists did not comprehend biodiversity fully in 1970 first requires a glance at how ecology has matured as a discipline. There have been at least four developmental stages in the science of ecology: formative, descriptive, quantitative, and predictive nonequilibrium. (For full treatments see Worster 1994, Golley 1993, and McIntosh 1987.)

Several people stand out as major influences on ecological thinking long before the field coalesced into a unified discipline. Charles Lyell, the father of geology, in his book *Principles of Geology* (1830) helped overturn Linnaean concepts of a static nature under strict divine rule. Lyell was among the first to understand that geologic change occurred gradually over eons, that species dispersed actively around the world, and that competition was a driving force in biotic interactions. Lyell was a major influence on Charles Darwin. In *On the Origin of Species* (1859), Darwin built upon Lyell and advanced natural selection as the primary mechanism of evolution. Contemporary with Darwin but living in the New England woods, Henry David Thoreau was one of the first naturalists to understand succession as a major pattern of change in ecosystems. Thoreau also was one of the first to glimpse the loss of species and habitat and its cultural ramifications at the onset of the Industrial Revolution. George Perkins Marsh contributed a pioneering global account of humanity's role in reducing the capacity of Earth to support life in *Man and Nature* (1864).

Ecology in the early decades of the 20th century was a descriptive, holistic science. The key themes were the balance of nature and succession toward a stable, climax state. Plant ecologist Frederick Clements dominated the field with his idea of communities as interdependent superorganisms evolving collectively.

By the time A. G. Tansley coined the term "ecosystem" in 1935, Clements's views were falling from favor. Qualitative, descriptive ecology was being superseded by a more quantitative ecology of energy and nutrient flows, food chains, and tropic levels. Natural history was out, mathematical models were in. The science of interrelationships was becoming subject to compartmentalization and reductionism. In the 1940s, Raymond Lindeman developed important theories on energy flows in ecosystems and G. E. Hutchinson refined the concept of feedback and constructed some of the first mathematical models of populations. Later ecologists built on these fundamentals with Eugene Odum (ecosystem characteristics), Frank Bormann and Gene Likens (nutrient flows), and Robert MacArthur (population models) making key contributions (see Bocking 1994).

Since the late 1970s, as knowledge of nature patterns and processes has accumulated and the biodiversity crisis has grown, a new ecological worldview has been emerging (Pickett et al. 1992, Botkin 1990). Ecology is moving away from a reductionist approach toward a more contextual, non-equilibrium perspective. Where in the past scientists (and environmentalists) characterized ecosystems as orderly and relatively balanced, current viewpoints emphasize natural systems as dynamic, changing at different space and time scales, and full of uncertainty. Nature is episodic as often as it is homeostatic. Nature is not always in "balance"; and changes are difficult, sometimes impossible, to predict.

Definitions of biodiversity reflect these latest changes. No longer is diversity just about numbers of species or types of ecosystems. The new emphasis on non-equilibrium processes (especially natural disturbances such as fires and floods) has resulted in a comprehensive definition that includes not only the diversity of life from genes to landscapes, but also the fundamental patterns and processes of Nature that weave lifeforms over time and in space.

Early Ecological Justifications For Protecting Nature

In the first decades of this century, the descriptive balance of Nature view of ecology reigned supreme for biologists and citizen alike. Two of America's greatest naturalists, Joseph Grinnell and Tracy Storer (1916, p. 377), wrote in an early *Science* article that mammals in National Parks added "the witchery of movement" to the "natural charm of the landscape." These distinguished biologists believed that the National Parks' highest purpose was to "furnish examples of the earth as it was before the advent of the white man" (p. 377).

A few years later, views were beginning to change. Beginning in the 1920s, several professional ecologists published papers calling for Nature protection for the sake of science. In 1920, Victor Shelford (1920) criticized the Park Service and the Forest Service for an unecological approach to management. Francis Sumner (1920, 1921) Called for "nature conservation" over resource management. Both Shelford and Sumner were members of the new Ecological Society of America and advocated setting aside representative examples of all US ecosystems in a comprehensive national system. Other ecologists joined them, publishing articles such as "The Preservation of Natural Areas in the National Forests" (Pearsons 1922) and "The Importance of Preserving Wilderness Conditions" (Adams 1929). In 1921 the American Academy for the Advancement of Science endorsed the Ecological Society's policies on reserves.

What sparked this outcry from a few leading scientists? Wilderness historian Craig Allin (1982) suggests that it resulted from a massive upsurge in road building on public lands by both the Forest Service and Park Service between 1916 and 1921. Also likely is that Shelford and his colleagues, on the cusp between Clementsian and quantitative ecology, recognized the need for protecting representative ecosystem types as examples of steady state conditions and as baselines for gathering new scientific data.

Shelford's efforts led to the remarkable paper "The Preservation of Natural Biotic Communities" (Shelford 1933). This visionary work (not unlike The Wildlands Project's vision today) outlined a national strategy for preserves that included protection for both species and ecosystems, expansion of park and reserve boundaries to match species habitat needs, managing for ecological "fluctuations" (i.e., natural disturbances), and a core/buffer zoning approach to planning.

Reading Shelford's paper more than 60 years later, one can only dream of what condition US public lands would be in today if policy-makers of the time had embraced Shelford's bold vision. Unfortunately, no sanctuary system was forthcoming. Shelford's work did result, though, in the beginning of the Forest Service's Research Natural Area program, where small examples of different timber types were declared off-limits to commercial logging.

While Victor Shelford developed his Nature sanctuary plan, three other biodiversity pioneers, George Wright, Ben Thompson, and Joseph Dixon, were focusing on the National Parks. As wildlife experts studying the habitat needs of park fauna, Wright and his colleagues discovered that every single park was far too small to sustain large mammal populations over time. At the conclusion of their landmark *Fauna of the National Parks of the United States* (1933, p. 37–39) they made one of the first statements suggesting biodiversity as the *raison d'etre* for parks: " . . perhaps our greatest natural heritage," rather "than just scenic features . . is nature itself, with all its complexity and its abundance of life."

Wright and his colleagues were proved correct in both their scientific and policy assessments by the debate that surrounded the creation of Everglades

National Park in 1934. Wildlands advocates count Everglades as the first park where wilderness preservation was used to justify protection. The park was established for the "preservation intact of the unique flora and fauna and the essential primitive conditions" (US Statutes at Large 1934); but this legal language obscures the true justification behind the protection of the park. The record shows that conservationists convinced Congress to accept wildlife as "scenery" since the river of grass had no magnificent mountains or gorges (Runte 1987). Though the Everglades bill does represent a statutory milestone for accepting wilderness and wildlife, grandeur, magnificence, and Romantic ideas of the balance of Nature continued to hold sway.

Biodiversity From Aldo Leopold To Earth Day

As ecology developed into a modern science in the 1930s and 1940s, there remained a need to consolidate ecological justifications for protecting Nature into a coherent whole. The person who accomplished this, as much as anyone, was Aldo Leopold. Best known for *A Sand County Almanac* (1949), Leopold wed the science of ecology with a land ethic where humans were "plain members and citizens" of Earth.

Leopold's thinking about ecology and management went through a profound transformation over several decades. In 1921, he was using recreational justifications for protecting wildlands that were radical for the time. But Leopold soon left such arguments behind. Beginning in 1933 he published a series of papers that provided the basis for much of the current definition of biodiversity as well as the ethical foundations of conservation biology. Leopold made four key contributions.

In 1939, he offered one of the first inclusive definitions of biodiversity: "... the biota as a whole is useful, and biota includes not only plants and animals, but soils and water as well" (Leopold 1939, p. 727). Leopold expanded on this in 1944 by adding the concept of health to conservation (Leopold in Flader and Callicott 1991, p. 310):

> *Conservation is a state of health in the land. The land consists of soil, water, plants, and animals, but health is more than a sufficiency of these components. It is a state of vigorous self-renewal in each of them, and in all collectively ... In this sense land is an organism and conservation deals with its functional integrity, or health.*

This commingling of biodiversity conservation and land health is the root of current attempts to define ecological health and ecological integrity.

Leopold's second contribution was to use the new ecological concepts of biotic pyramids, energy flows, and food chains to point out defects in prevailing balance-of-nature perspectives on ecosystems. He suggested that balance implies "only one point at which balance occurs, and that balance is normally static" (Leopold 1939, p. 727).

Third, Leopold used his awareness of the dynamics of Nature to provide a scientific rationale for wilderness protection. In 1941, he wrote that "all wilderness areas . . . have a large value to land-science (Leopold 1941, p. 3) and that their principal utility was as a"base-datum of normality, a picture of how healthy land maintains itself . . ."

The fourth and most important contribution of Leopold to understanding biodiversity was that he placed people squarely in Nature as"plain members and citizens of the land community" (Leopold 1949, p. 204). He had already recognized this intimate relationship as early as 1933 when he defined civilization"as a state of mutual and interdependent cooperation between human animals, other animals, plants, and soils . . ." (Leopold 1933, p. 635). With this conception Leopold became the first modern ecologist to link the health of land with the health of culture.

Like Shelford and Wright before him, Leopold had little immediate influence on policy. By the 1960s, however, as the pace of environmental deterioration quickened, other ecologists were beginning to catch up with where Leopold had been. Science came to play an increasingly key role in environmental policy debates. Rachel Carson in *Silent Spring* (1962) built her argument against pesticides by exposing their negative effects on both human health and ecosystem functioning. By 1968, the international scientific community was becoming active. The United Nations Educational, Scientific, and Cultural Organization (UNESCO) sponsored a global conference that year on the use and conservation of the biosphere. This led to the biosphere reserve model of ecosystem protection. All this work forged links between Leopold's view of science and the environmental problems of the day.

Congress and the Administration, too, were beginning to act for biodiversity. In 1966 the Endangered Species Preservation Act was passed. This prototype of more powerful laws to come protected only vertebrates and contained many other loopholes. A year before Earth Day, in 1969, as the environmental movement gathered strength, Congress extended protection to invertebrates. In 1972, President Nixon stated that "even the most recent act to protect endangered species simply does not provide the kind of management tool needed to act early enough to save a vanishing species" (Nixon 1972). Nixon signed the Marine Mammal Protection Act that year.

The groundswell of presidential and popular support for ecology and endangered species surrounding Earth Day 1970 led to Congress passing the 1973 Endangered Species Act with but 12 no-votes total in both legislative houses. The ESA, still the strongest American environmental law, validated Aldo Leopold's "ecological consciousness" toward species and ecosystems and set the stage for future policy debates.

Biodiversity and the Environmental Movement

Environmental protection did not gain a lasting place in American values merely as a result of the development of the science of ecology. Nor did biodiversity come to the fore simply because of individual biologists such as Shelford, Leopold, and Carson. It also took the concerted efforts of leaders within the environmental movement to understand the implications of ecology and render these ideas accessible to the American public.

Just as ecology has developed as a science, so have environmentalist arguments evolved for protecting Nature. When Robert Marshall, Aldo Leopold and their colleagues founded The Wilderness Society in 1935, they focused only on wilderness and roadless areas. They believed that wildlands should be protected primarily for the benefits they conferred on people. Marshall's views, in particular, were influential: wilderness offered a respite from civilization, encouraged spiritual contemplation, and offered a unique aesthetic experience (Marshall 1930). During this period Leopold was only beginning to voice his biotic view of land; and the conservation movement had, aside from Shelford's work, little scientific ecology on which to base political prescriptions.

In 1949, the year *A Sand County Almanac* appeared, the Sierra Club convened its first biennial wilderness conference. These conferences were to become the main philosophical and strategic forum for the movement. Through the first four gatherings, there was little mention of ecology as having anything to do with wilderness protection. Recreational, spiritual, and aesthetic justifications prevailed. The *Sierra Club Bulletin* from 1950–1976 had only two references to ecology, four to endangered species, and five to wildlife conservation (Sierra Club 1976).

But the power of scientific ecology in general and Leopold's ideas in particular was beginning to be felt. In 1950 journalist Bernard DeVoto, responding to the Echo Park Dam controversy in Dinosaur National Monument, proclaimed that the park deserved to be protected "as wilderness . . . for the field study of . . . the balance of Nature, the web of life, the interdependence of species . . ." (DeVoto 1950, p. 44). DeVoto's is a classic attempt to incorporate Leopoldian ecology with Romantic ideas of balance in Nature.

In 1951 and again in 1955, at the third and fifth wilderness conferences, Howard Zahniser of The Wilderness Society (TWS) unveiled a national plan for wilderness protection, based partly on Victor Shelford's original Nature sanctuary vision (Kendeigh et al. 1950–1951). Zahniser's plan became the precursor to the original Wilderness Act bill in 1956. In Congressional hearings over the new bill, Zahniser mentioned scientific baseline data arguments in favor of the legislation, but these justifications were never highlighted by conservationists during the debate.

As executive director of the Sierra Club, David Brower was as responsible as any leader for bringing science to conservation. Brower controlled the agendas of the wilderness conferences. Beginning in 1959 with Raymond Cowles, he invited several professional ecologists to address the conferences. Cowles spoke of population growth from an ecological perspective. In 1963, James Gilligan, author of the first Ph.D. dissertation on US wilderness policy, described wildlands as "essential habitat for scarce species." Slowly, biodiversity was creeping into conservation arguments.

Conservationists did not find it easy, however, to include ecology along with recreational and spiritual justifications for wilderness. Sharing the podium in

1963 with Gilligan was forest ecologist Stephen Spurr, whose view of ecology challenged Brower and the conferees. Spurr argued strongly against any wilderness preservation strategy grounded in a stable, balance of Nature view. "Stability is only relative, and only superficial," spoke the ecologist, and "natural succession will never recreate old patterns, but will constantly create new patterns" (Gilligan 1963, p. 60). Spurr used ecological theory to confront the conferees' "nostalgia" for a nature that never existed, Instead of drawing lines around roadless areas and lobbying Congress to designate new wilderness, Spurr argued for greater use of science and technology to manipulate Nature for human ends.

This conflict between ecology and preservation was manifest again in 1963 with the influential *Wildlife Management in the National Parks,* the so-called Leopold Report (Leopold et al. 1963). At the behest of Interior Secretary Steward Udall, a blue ribbon committee chaired by Aldo Leopold's son, zoologist A. Starker Leopold, was convened to review wildlife in the parks. The committee's report was both revolutionary and paradoxical. Following ecology (and the thirty year old insights of George Wright), the report concluded that "maintaining suitable habitat is the key to sustaining animal populations, and . . . protection, though it is important, is not of itself a substitute for habitat" (p. 1–2). But after verifying Spurr's assessment that ecosystems change over time, the Leopold Committee recommended that "the biotic associations within each park be maintained . . . as nearly as possible in the condition that prevailed when the area was first visited by the white man." Each park "should represent a vignette of primitive America" (p. 4). As historian Alfred Runte (1987) has noted, these scientists could not escape their cultural values. Science required them to portray Nature as dynamic, yet they advocated freezing Nature into pre-European landscapes.

The Sierra Club, Wilderness Society, and National Parks and Conservation Association were all quick to endorse the Leopold Report. These groups supported the committee's *philosophy* while avoiding the committee's *ecology.* The following year the Wilderness Act was passed by Congress. Ecological values rated all of three words in the new law.

Despite these inconsistencies, support for endangered species and broad environmental protection continued to grow. In 1968, the Sierra Club lobbied for a national ecological survey, but the bill died in Congress (McCloskey 1968). At the biennial wilderness conference in 1969, population biologist Paul Ehrlich proclaimed that human population growth and resource consumption were inextricably linked to the loss of wilderness.

Biodiversity Comes of Age: 1970–1990

During the 1970s and into the 1980s, scientific and policy conceptions of biodiversity continued to converge with environmentalist notions of ecology. Ecologists added to their knowledge of competition, diversity, stability, and community dynamics (see Cody and Diamond 1975; and on biogeography,

MacArthur and Wilson 1967). R. H. Whitaker (1972) refined and broadened the concept of diversity to include within-habitat (alpha), between-habitat (beta), and regional (gamma) diversity. The same year that Whitaker published his classic work, the United Nations Conference on the Environment was held in Stockholm. For all the impassioned debate in Sweden, though, few ecologists attended. Scientists were not ready to present their ideas in political forums.

Several national and international conferences and policy documents built upon Stockholm. In 1981 the US Council on Environmental Quality produced the *Global 2000 Report to the President,* which was the first US policy document to attempt a definition of biodiversity. The US State Department, following the Council's lead, sponsored an International Strategy Conference on Biological Diversity in 1981. A World Charter for Nature was ratified by the U.N. General Assembly in 1982. The charter included recommendations to protect parks and wildernesses, but was especially notable for its preamble which tied protecting diversity to an ethical position: "Every form of life is unique, warranting respect regardless of its worth to man, and, to accord other organisms such recognition, man must be guided by a moral code of action." The US was the only member of the General Assembly to vote against the charter.

Yet by the time UN delegates were voting on the World Charter for Nature, biodiversity protection had already been codified in US law—at least on the National Forests. The 1976 National Forest Management Act (NFMA) today remains the only US law that explicitly requires a federal agency to protect viable populations and ecosystems. As with the Everglades legislation four decades prior, however, the motives of Congress were unclear. The NFMA was a response to excessive clearcut logging on the National Forests. Forest activists and Congress were concerned about stand conversions, the forestry practice of logging native forests to replace them with preferred commercial species, creating industrial monocultures. Yet Congress did not understand biodiversity well enough to act decisively. In an extremely ambiguous section of the NFMA, legislators required the Forest Service to "provide for the diversity of plant and animal communities" (US Code 1982).

The NFMA diversity provision was clarified by a committee of scientists who wrote rules under the Code of Federal Regulations whereby the law would be implemented. These rules, completed in 1979 and revised in 1982, require the Forest Service to preserve existing variety, maintain viable populations, recognize forests as ecosystems, and base management on ecological relationships. Clear as the rules were, it would take many years and numerous appeals and lawsuits to force the agency to begin to implement them.

Along with NFMA, two additional events in the late 1970s brought ecologists and activists closer together. As the pace of development continued, concerns were raised as to how "external threats" would affect protected areas. The National Parks and Conservation Association (1979) published a national report documenting such threats. The Park Service (1980) released its own study highlighting similar problems. The following year, Congress, in response to erosion and watershed degradation on lands surrounding Redwood National

Park, amended the Park Service Organic Act to affirm park protection. While judicial interpretations have limited the effectiveness of the Redwood Park Amendments, the issue of external threats served notice that protected areas were in fact embedded in an ecological matrix that required protection as a whole.

Further illuminating biodiversity in the late 1970s were several books by prominent scientists warning of an extinction crisis. Norman Myers's *The Sinking Ark* (1979) was read widely and caused much debate. Paul and Anne Ehrlich titled a 1981 textbook *Extinction*. The Nature Conservancy, ahead of most conservation groups in understanding diversity, began to build a national database that cataloged threatened and endangered species, habitat types, and other elements of biodiversity (The Nature Conservancy 1975).

By the beginning of the 1980s a critical threshold was being reached in scientific comprehension and environmental awareness of biodiversity. The first International Conference on Conservation Biology, held at the University of California, San Diego in 1978, brought together a diverse group of geneticists, population biologists, evolutionists, and biogeographers. The conference resulted in the path-finding anthology *Conservation Biology: An Ecological and Evolutionary Approach* (Soulé and Wilcox 1980). The synthetic discipline of conservation biology was born. Soon thereafter, other books appeared linking conservation with genetics, evolution, and population biology (Frankel and Soulé 1981, Schonewald-Cox et al. 1983, Harris 1984). In 1986, a second conference of the newly formed Society for Conservation Biology was held, followed by the initial publication of a professional journal. In late 1986 in Washington, DC, the Smithsonian Institution and the National Academy of Sciences hosted the first high-profile international gathering of professionals concerned with loss of biodiversity. From this time onward there has been a great outpouring of papers and reports covering all aspects of the new field.

What was new about conservation biology? The discipline was synthetic, with island biogeography, population genetics, and habitat fragmentation studies leading the way. There was an emphasis on applying academic theory to management problems. And conservation biology was explicitly value-laden: diversity, complexity, and evolution were imbued with normative value (Soulé 1985). Many conservation biologists supported shifting the burden of proof in environmental decisions from those who wished to protect diversity to those who desired to develop Nature. While conservation biology was mission-oriented, however, the methods used to gather data were objective, peer reviewed, and as value-free as in any other scientific discipline. The field has contributed these general management goals to conservation (Meffe and Carroll 1994):

1. Critical ecological processes must be maintained.
2. Goals and objectives must come from a deep understanding of the ecological properties of the system.
3. External threats to reserves must be minimized and external benefits maximized.

4. Evolutionary processes must be conserved.
5. Management must be adaptive and minimally intensive.

The Earth First! movement was well out in front of almost all environmental groups in using conservation biology arguments as first principles in protecting Nature. Evident in the earliest volumes of the *Earth First! Journal* (1981–1982), this ecological wilderness perspective was consolidated in *The Big Outside* (Foreman and Wolke 1989): "Protecting *natural diversity*, then, must be the major goal of the wilderness movement . . . natural diversity means that all indigenous species must be free to evolve under natural conditions, in as many different natural habitats as possible" (p. 24).

Still, many environmental groups were slow to embrace the new field. It took the Northern Spotted Owl and its old-growth forest habitat to catapult biodiversity toward the forefront of environmentalism. The owl awakened activists (and managers and Congress) to critical aspects of protecting biodiversity. What began in the 1970s as an owl-only issue was transformed by 1990 into an ecosystem protection issue, in part due to EF!'s dramatization of threats to old-growth forests. Once old-growth ecosystems were adopted as a strategy focus, it became easier for activists to appreciate the role that ecosystem patterns and processes played in maintaining biodiversity. As scientific assessments on owls and old growth were joined by reports on Marbled Murrelets, salmonids, and other species, activists were pushed toward another level of sophistication, recognizing the need for regional/landscape-scale protection. In 1994, ecosystem management studies supported (in concept) by environmentalists were initiated by the federal government for the entire Columbia River Basin and the Sierra Nevada Mountains in California.

Biodiversity Beyond Earth Day 1995

Three years after Earth Day 1970, when Congress passed the Endangered Species Act, protecting diversity was perceived as "low-cost, no-lose" (Yaffee 1982, p. 57). Today the Act faces efforts to gut its most stringent provisions. Yet it is abundantly clear that the biodiversity crisis has worsened and that the law should be strengthened. The sum of our growing scientific understanding of biodiversity reveals a deep chasm between environmental policy and environmental protection (Grumbine 1994b).

The single major consequence of the revolution in awareness of biodiversity has been to deepen our appreciation of interrelationships. For ecologists, this trend has been manifest in two important ways: the evolving definition of diversity from species to the current inclusive hierarchical view, and the change from static balance to non-equilibrium theories of Nature. In the environmental movement, biodiversity has begun to nudge activists away from viewing Nature as a series of special places (parks and wilderness) embedded in developed landscapes toward the protection of regional landscapes (Noss 1983) or greater ecosystems (Grumbine 1990) where use and protection are grounded in a sense of limits on human behavior.

In American society at large the concept of protecting biodiversity continues to challenge cherished but outmoded images of people vs. Nature. While anthropocentric values and resourcism still hold sway with the majority, a growing number of citizens are asking provocative questions. Are there limits to private property rights when biodiversity is at risk? Is industrial-scale resource depletion sustainable? What should give when human activities are exposed by conservation science as endangering species, ecosystems, and landscapes? Who should decide what constitute acceptable levels of risk in losing elements of biodiversity? These questions were not part of the discussion on Earth Day 1970. We had no definition for biodiversity, no comprehensive Endangered Species Act, and property owners were unconcerned about their "rights." "Sustainability" was not part of the environmental lexicon. There were no conservation biologists to decry threats to viable populations, and "plenty" of old growth remained to be cut. Environmental protection was perceived as either/or, cut-and-dried, not replete with uncertainty and multiple levels of risk.

Conservation biologists today confront myriad complex issues. In a society that considers science to be value-neutral, how do you practice objective science and yet advocate for biodiversity (for example, see Noss 1994 and Brussard et al. 1994). And how do you create a tighter link between science and policy? In the 1930s, when Victor Shelford designed a national reserve system using science, the political response was a series of tiny Research National Areas. In the 1990s, several scientific panels recommended a moratorium on old-growth logging and the result was President Clinton's Option Nine, and Congress's "forest health" rider—the first of which failed to uphold the scientists' recommendations, the second of which runs directly counter to them.

Activists, too, are adjusting to the new world of diversity. They are becoming less hesitant to employ scientific arguments in their strategies, regardless of the perceived political costs. The history of US environmentalism reflects a tendency by activists to downplay scientific rationales in favor of ethical justifications. The normative standards of conservation biology can help to overcome this tendency.

Activists should not accept science uncritically, however. The history of the concept of biodiversity makes clear that along with evolving scientific "fact," ecological theory is also dependent on cultural context. The balance of Nature steady-state model was partly a product of Romantic values at the last turn of the century, just as chaos, uncertainty, and non-equilibrium theories are tied to current circumstances. The process of science suggests that the current definition of biodiversity provides an improved picture of how Nature works; but the views that Stephen Spurr expressed at the 1969 Sierra Club wilderness conference are alive and well today. The influential ecologist Daniel Botkin believes "We can engineer nature at nature's rate and in nature's way . . ." (Botkin 1990, p. 190). The current debate over defining ecosystem management provides another indication of American ambivalence over new concepts of biodiversity (Grumbine 1994). Is it people over Nature or people in partnership with Nature? Ecosystem management or ecosystem protection? The goal of biodiversity protection considers all human use of Nature as flowing from ecosystems *only after* basic patterns and processes are maintained and restored. If ecosystem management for native diversity is to take hold and flourish beyond 1995, the new goal of protecting biodiversity and the old standard of providing natural resources for human use must be reconciled. This is a values and political question which does not depend exclusively on science for resolution.

One hundred thirty-seven years ago, Charles Darwin ascertained that all life—including humans—is subject to the forces of evolution. Three decades before the first Earth Day, Aldo Leopold, working a cut-over sand county farm in Wisconsin, saw through the delusion that people are separated from and not responsible to Nature. In 1962, Rachel Carson published her blockbuster against pesticides. In the late 1980s, American school children became aware of the Northern Spotted Owl and its old-growth home. Yet a century and a third after *The Origin of the Species* many US citizens do not believe in evolution. Twenty-five years passed between the appearance of *A Sand County Almanac* and a law to protect species from extinction. DDT, banned in the US since 1969, is still manufactured by US companies for export. The pace of positive change is painfully slow. Sociologist Bill Devall is correct to note that though Americans have been quick to support environmental reforms, changes that require difficult behavioral and values adjustments remain incomplete (Devall 1995).

Biodiversity protection represents the core idea that may bring Americans to support protecting all species, human and nonhuman together. Noting the tension between scientific and environmentalist views of ecosystems, ecologist Frank Golley (1993, p. 205) remarked, "It is not clear to me where ecology ends and the study of the ethics of nature begins, nor is it clear to me where biological ecology ends and human ecology begins. These divisions become less and less useful." Moving from a 19th century model of preserving Nature toward a 21st century image of protecting biodiversity will help break down further the delusion that people and Nature can be separate. The hope is that adjusting management goals to reduce extinction and habitat destruction will not only end the present biodiversity crisis but also provide the opportunity for people to forge a new relationship with Nature. Hope and time are intertwined—moist biologists do not believe that we have the luxury of an additional 25 years to wait for biodiversity to become accepted by society as ecology was from 1970 to 1995. Long before Earth Day's 50th anniversary, Americans must learn that there can be no alternative to protecting the native sources of life—healthy, functioning wild ecosystems.

Acknowledgments

Thanks go to Curt Meine, Reed Noss, and John Davis for their helpful comments on earlier drafts of this paper. Readers should be aware that this paper represents preliminary work in a neglected area that could use further critical scholarship. An earlier version of this paper appeared in the 1995 Humboldt Journal of Social Relations 21(1): 35–59.

R. Edward Grumbine is director of the Sierra Institute (University of California Extension, Santa Cruz, 740 Front St. Suite 155, Santa Cruz, CA 95060), author of Ghost Bears *(Island Press, 1992), and editor of* Environmental Policy and Biodiversity *(Island Press 1994).*

Literature Cited

Adams, Charles, 1929, "The Importance of Preserving Wilderness Conditions," *New York State Museum Bulletin* 279: 37–34.

Allin, Craig, 1982. *The Politics of Wilderness Preservation.* Westport, CT; Greenwood Press.

Botkin, Daniel, 1990. *Discordant Harmonies.* New York: Oxford University Press.

Brower, Lincoln P. 1994. A new paradigm in conservation biology. Endangered biological phenomena: Pages 104–106 *in* Gary Meffe, and C. R. Carroll, eds. *Principles of Conservation Biology.* Sunderland, MA: Sinauer Associates.

Brussard, Peter et al. 1994. "Cattle and Conservation Biology—Another View," *Conservation Biology* 8(4): 919–921.

Carson, Rachel, 1962. *Silent Spring.* Boston: Houghton Mifflin.

Cody, Martin L. and J. Diamond, 1975. *Ecology and Evolution of Communities.* Cambridge, MA: Belknap Press of Harvard University Press.

Darling, F. F. and N. Eichhorn. 1969. *Man and Nature in the National Parks: Reflections on Policy.* Washington, DC: The Conservation Foundation.

Darwin, Charles (1859). *On the Origin of Species.* New York: Modern Library, n.d.

Devall, Bill. 1995. "Twenty Five Years Since Earth Day: Reflections of a Sometime Social Activist," *Humboldt Journal of Social Relations.*

DeVoto, Bernard. 1950. "Shall We Let Them Ruin Our National Parks?," *Saturday Evening Post* 223: 44.

Ehrenfeld, David. 1970. *Biological Conservation.* New York: Holt, Rinehart, and Winston.

Ehrlich, Paul, and Anne Ehrlich. 1981. *Extinction: The Causes and Consequences of the Disappearance of Species.* New York: Random House.

Foreman, Dave and Howie Wolke. 1989. *The Big Outside.* Tucson: Ned Ludd Books.

Frankel, O. and M. Soulé. 1981. *Conservation and Evolution.* Cambridge: Cambridge University Press.

Giligan, James. 1963. "Summary of the Outdoor Recreation Resource Review Committee," Pages 51, 52 *in* Francois Leydet, ed. *Tomorrow's Wilderness.* San Francisco: Sierra Club.

Golley, Frank. 1993. *A History of the Ecosystem Concept in Ecology.* New Haven, CT: Yale University Press.

Grumbine, R. Edward. 1994a. "What is Ecosystem Management?," *Conservation Biology* 8(1): 27–38.

Grumbine, R. Edward. 1994b. *Environmental Policy and Biodiversity.* Washington, DC: Island Press.

Grumbine, R. Edward. 1992. *Ghost Bears: Exploring the Biodiversity Crisis.* Washington, DC: Island Press.

Grumbine, E. 1990. "Protecting Biological Diversity through the Greater Ecosystem Concept," *Natural Areas Journal* 19(3): 114–120.

Grinnell, Joseph and Tracy Storer. 1916. "Animal Life as An Asset of National Parks," *Science* 44 (September 15, 1916).

Harris, Larry. 1984. *The Fragmented Forest.* Chicago: University of Chicago Press.

Hays, Samuel. 1987. *Beauty, Health, and Permanence.* Cambridge: Cambridge University Press.

Kendeigh, SC et al. 1950–1951. "Nature Sanctuaries in the United States and Canada: A Preliminary Inventory," *The Living Wilderness* 15(35): 1–45.

Leopold, Aldo. 1949. *A Sand County Almanac*. New York: Oxford University Press.

Leopold, Aldo. 1944. "Conservation: In Whole or In Part?" Pages 310–319 *in* Susan Flader, and JB Callicott, eds. 1991. *The River of the Mother of God and Other Essays* by Aldo Leopold. Madison, WI: University of Wisconsin Press.

Leopold, Aldo. 1941. "Wilderness As a Land Laboratory," *The Living Wilderness* 6:3.

Leopold, Aldo. 1939. "A Biotic View of Land," *Journal of Forestry* 37(9): 727–730.

Leopold, Aldo. 1933. "The Conservation Ethic," *Journal of Forestry* 31(6): 634–643.

Leopold, A. Starker et al. 1963. *Wildlife Management in the National Parks*. Washington DC: US Department of the Interior.

Lyell, Charles. (1830). *Principles of Geology*. Chicago: University of Chicago Press, 1990–1991.

Marsh, George Perkins (1864). *Man and Nature: Or, Physical Geography As Modified by Human Action*. Cambridge, MA: Cambridge University Press, 1965.

Marshall, Robert. 1930. "The Problem of the Wilderness," *Scientific Monthly* 30, 142–143.

MacArthur, Robert and E. O. Wilson, 1967. *The Theory of Island Biogeography*. Princeton: Princeton University Press.

McCloskey, Michael. 1968. "A Conservation Agenda for 1969," *Sierra Club Bulletin* 53(12): 5.

McIntosh, R. P. 1985. *The Background of Ecology: Concept and Theory*. Cambridge MA: Cambridge University Press.

Meffe, Gary and C. R. Carroll, eds. 1994. *Principles of Conservation Biology*. Sunderland, MA: Sinauer Associates.

Myers, Norman. 1979. *The Sinking Ark*. Oxford: Pergamon Press.

Nash, Roderick. 1989. *The Rights of Nature*. Madison, WI: University of Wisconsin Press.

NPCA, 1979. "Adjacent Lands Survey: No Park Is an Island," *National Parks and Conservation Association Magazine* 53(3): 4–9.

Nixon, Richard. 1972. Weekly Compilation of Presidential Documents 8 (February 8, 1972), 218–224.

Noss, Reed et al. 1994. *Endangered Ecosystems of the United States: A Preliminary Assessment of Loss and Degradation*. Washington, DC: US Fish and Wildlife Service.

Noss, Reed. 1994. "Cows and Conservation Biology," *Conservation Biology* 8(3): 613–616.

Noss, Reed. 1983. A Regional Landscape Approach to Maintain Diversity, *BioScience* 33: 700–766.

Noss, Reed and Alan Cooperrider. 1994. *Saving Nature's Legacy*. Washington DC: Island Press.

Pearsons, G. A. 1922. "The Preservation of Natural Areas in the National Forests," *Ecology* 3: 284–287.

Pickett, S. T. A. et al. 1992. The new paradigm in ecology: Implications for conservation biology above the species level. Pages 65–88 *in* Peggy Fiedler and S. Jains, eds. *Conservation Biology*. New York: Chapman and Hall.

Primack, Richard. 1993. *Essentials of Conservation Biology*. Sunderland, MA: Sinauer Associates.

Runte, Alfred. 1987. *National Parks: The American Experience*. Lincoln, NB: University of Nebraska Press.

Schonewald-Cox, Christine et al. eds. 1983. *Genetics and Conservation: A Reference for Managing Wild Animal and Plant Populations*. Menlo Park, CA: Benjamin/Cummings.

Shelford, Victor, 1920. "Preserves of Natural Conditions," *Transactions of the Illinois State Academy of Science* 13: 37–58.

Shelford, Victor. 1933. "The Preservation of Natural Biotic Communities," *Ecology* 14(2): 240–245.

Sierra Club Bulletin. 1976. *Sierra Club Bulletin Index 1950–1976*. San Francisco: Sierra Club.

Soulé Michael et al. 1986."The Millennium Ark: How Long a Voyage, How Many Staterooms, How Many Passengers?" *Zoo Biology* 5: 101–113.

Soulé Michael. 1985."What is Conservation Biology," *BioScience* 35: 727–734.

Soulé Michael and Bruce Wilcox, eds. 1980. *Conservation Biology: An Ecological and Evolutionary Approach*. Sunderland, MA: Sinauer Associates.

Spurr, Stephen. 1963. "The Value of Wilderness to Science," Pages 57–64 *in* Francois Leydet, ed. *Tomorrow's Wilderness*. San Francisco: Sierra Club.

Sumner, Francis. 1921. "The Responsibility of the Biologist in the Matter of Preserving Natural Conditions," *Science* 54: 39–43.

Sumner, Francis. 1920. "The Need for a More Serious Effort to Rescue a Few Fragments of Vanishing Nature," *Scientific Monthly* 10: 236–248.

The Nature Conservancy. 1975. *The Preservation of Natural Diversity: A Survey and Recommendations*. Washington DC: The Nature Conservancy.

U.S. Code. 1982. 16, Sec. 1604 (g) (3) (B).

U.S. Statutes at Large. 1934. 48: 817.

U.S. Council on Environmental Quality. 1981. *The Global 2000 Report to the President*. 3 vols. Washington DC: U.S. Council on Environmental Quality.

U.S. National Park Service. 1980. *State of the Parks—1980*. Washington DC: U. S. Department of Interior.

Whitaker, R. H. 1972. *Communities and Ecosystems*. New York: MacMillan.

Worster, Donald. 1994. *Nature's Economy*. New York: Cambridge University Press.

Wright, George, Thompson Ben, and Joseph Dixon. 1933. *Fauna of the National Parks of the United States: A Preliminary Survey of Faunal Relations in National Parks*. Washington DC: Government Printing Office.

Yaffee, Stephen. 1982. *Prohibitive Policy: Implementing the Federal Endangered Species Act*. Cambridge, MA: MIT Press.

BURNING THE LIBRARY OF AMAZONIA

RICHARD EVANS SCHULTES

The encyclopedic botanical knowledge of the Amazon Indians is in danger of being lost. A student of that knowledge for forty-seven years argues that it must be preserved.

As every food shopper knows, vegetal chaos has come to reign in the grocery stores of the United States. Clementines are rung up as tangerines, plantains as bananas, collard greens as mustard greens and butternut squash as acorn squash. Most of us urban Americans live lives of extreme floristic deprivation. We encounter few plants, recognize fewer and make use of fewer still. So sparse is our botanical knowledge that only with great difficulty can we grasp the deep botanical awareness of peoples who live in intimate association with the flora that surrounds them and who are utterly dependent on it for the necessities and ameliorants of life.

Consider the Amazonian Indians. They have made a poisonous tuber—the root of the yucca, or tapioca plant—a staple of their diet by devising ways of preparing it that remove the cyanic material it contains. They make an alcoholic drink of the pomegranate-like fruit of the moriche palm, a valuable source of vitamin B^1, which is otherwise in short supply in their diets. They cleanse their bodies and wash their clothing with plants containing saponines, compounds that foam in water. They harvest the tough, alga-like leaves of an abundant river weed and burn them to make a kind of table salt. They make a stimulating morning drink from the caffeine-rich bark of a forest liana, or woody vine. They apply drops of a tea made from the leaves of a vine to cure eye infections. They employ a curious parasitic plant to stanch the blood from a wound.

The forest peoples' acquaintance with plants is subtle as well as extensive. The Indians often distinguish "kinds" of a plant that appear indistinguishable, even to the experienced taxonomic botanist. Moreover, they frequently identify such kinds from a considerable distance by sight alone. None of the several "explanations" offered for that taxonomic acuteness satisfactorily explains the consistency of the Indians' identifications. One would suppose that for plants used as foods, medicines or hallucinogens, the Indians are probably distinguishing between chemovars, or chemical varieties of a species. But the subtle signs of these kinds seem to have escaped the notice of Western botanists.

It is well known that in many regions of the Amazon Basin, native plant and animal species are being destroyed faster than they can be recorded, classified and studied. So dire has the plight of the forests become that Conservation International, a nonprofit organization based in Washington, D.C., has organized rapid-assessment teams to make quick surveys of the large organisms in unexplored areas. The hope is that the surveys will give a rough index of the diversity of each area and that particularly rich zones will be set aside as preserves. The plant and animal species they harbor could then be available for study at the slower pace required by scientific and technological investigation.

But the earth is losing not only the biodiversity of the forest; it is also losing what I call its crypto-diversity, the hidden chemical wealth of the plants. A few Western botanists have a "taxonomic eye," the gift of recognizing significant morphological variation in a plant. But only the Indians are well acquainted with the various properties of the plants as well as with their external appearance. Why not enlist that vast, intimate knowledge? Why not regard the Indians in the Amazon Basin as a kind of phytochemical rapid-assessment team already on the ground, which could help locate the most promising plants for chemical and pharmacological evaluation?

The task is enormous. No one knows for certain how many species of plants the great Amazon Valley harbors. Richard Spruce, the nineteenth-century English explorer of the Amazon, estimated that there might be as many as 80,000 species of higher plants, and many botanists who have worked in the Amazon since Spruce's time are inclined to agree. Otto Gottlieb, a noted Brazilian chemist specializing in tropical plants, has stated that of the species inhabiting the Amazon forest 90 percent have not been subjected to even superficial chemical analysis.

Time is running out. Yet here is the rub. The Indians' botanical knowledge is disappearing even faster than the plants themselves. What we in the developed world call civilization is rapidly encroaching on indigenous communities, just as it is encroaching on the plants, and native botanical lore is usually an early casualty. In only one generation, acculturation can lead to the disappearance of botanical knowledge acquired over millennia. Ethnobotanical conservation—the preservation of the indigenous knowledge of plants and their properties and uses—is therefore as urgent as environmental conservation.

My own acquaintance with the promise of ethnobotanical conservation began in 1941, when I first went to the Amazon Basin. I had just earned my Ph.D. at Harvard, and I had been offered two jobs. One was as biology master at a private school in New England; the other was a ten-month grant from the National Research Council to go to the Amazon region to identify the plants employed in the many kinds of curare the Indians use for hunting. I decided on the Amazon—which is fortunate, because otherwise, I would probably still be a biology master. From my forty-seven years of ethnopharmacological research in the northwestern Amazon, primarily the Colombian region, I have field notes on more than 1,300 species valued by the Indians for their biological activity as medicines, narcotics or poisons. Those include species and even genera and entire families of plants that have never been even superficially examined by phytochemists.

The northwestern Amazon is the least studied but also the most interesting part of the basin. The richness of the flora there is a consequence of the area's complex geological structure. The region extends from the eastern slopes of the Andes, at an altitude of 12,000 feet, to forested plains as low as 300 feet. And the plains are not an unbroken carpet of green; they are spotted with isolated outcrops of quartzitic and granitic mountains.

The quartzitic sandstone mountains scattered throughout the western part of the Colombian Amazon are low, flat and often grotesquely eroded. They are the remnants of a once continuous mountain range that stretched from the Guianas and southern Venezuela westward. The ancient mountains are permeated with large tunnels and caves, and the walls of some of them are covered with ancient engraved or painted images of animals.

On the flat-topped heights of the sandstone mountains the botanist is in seventh heaven; the flora is strange, endemic and still largely unstudied. The Cerro de La Pedrera, a mountain a few miles above the mouth of the Río Apaporis, was first "collected" by the German botanist Karl Friedrich Philipp von Martius in 1820; and the work of several more recent plant explorers notwithstanding, some of the species von Martius discovered and described have not been re-collected in the past 174 years.

To the east, near the frontier with Brazil, the mountains change from metamorphosed sandstone to granite. They are mostly rounded and less subject to complex erosional patterns than are the sandstone mountains, and they have steep sides streaked with colors. The flora of the granitic mountains, in contrast to that of the sandstone mountains, is limited and not so rich in endemic plants. In between the scattered mountains stretches the great flat *planada*, a plain created when alluvial silt was deposited from both the quartzitic and granitic formations during the Cretaceous and Tertiary periods, when what is now Amazonia was a great bay. Most of the planada is clothed with dense forest, but in many places there are extensive white-sand patches, the remnants of eroded quartzitic and granitic mountains. The white-sand patches also support a highly endemic vegetation.

The Colombian Amazon is home to a large and highly diverse Indian population. Seventy thousand Indians belonging to fifty ethnic groups speak a mosaic of languages or dialects that have been classified, often with uncertainty, into fourteen or more linguistic groups. Acculturation has fortunately been less rapid in the Colombian Amazon than it has in most other parts of the basin. The region is protected from encroachment from the east by its rivers, all but one of them (the Río Putumayo) unnavigable. And the people living in the malaria-free and fertile Andes have not been anxious to invade the Colombian lowlands, where malaria flourishes. According to a recent census, the average population density in the Colombian Amazon is thirty-three inhabitants for every hundred square miles.

Although the area is still settled mostly by indigenous peoples, many of the tribes have been exposed to outsiders seeking to exploit one or another product of the forests. The histories of two tribes with which I worked closely, the Kofán and the Witoto, suggest how devastating contact with other cultures can be for such peoples. The Kofán, who live in Ecuador and adjacent parts of Colombia, were first contacted in 1536 by Catholic missionaries. Later they were enslaved by the Spanish, and their numbers greatly decreased. In the 1940s the number of Kofán in Colombia was estimated to be fewer than 300. Their language, which I never did learn to understand, has not yet been assigned to any large linguistic group and may be the sole survivor of a distinct linguistic group.

The discovery of oil in the regions inhabited by the Kofán has led to severe disruptions in their traditional way of life. But, at least until recently, the Kofán excelled in the preparation of many kinds of curare and made arrow poisons from a large number of plants not previously known as sources of poisons. They also possess perhaps the most extensive plant pharmacopoeias of all the tribes in the Colombian Amazon.

The Witoto, who live near the border between Colombia and Peru, were almost unknown until the rubber boom at the end of the nineteenth century and the beginning of the twentieth. Thereafter they were mercilessly exploited by Casa Arana, a Peruvian rubber company then in possession of the Colombian state of Amazonas, one of six administrative units into which the Colombian Amazon is divided. The brutal treatment accorded the Witoto and related tribes is reported to have reduced their numbers from some 50,000 in 1910 to only 7,000 a decade later. Although the Witoto are now highly acculturated, many old customs persist and their acquaintance with medicinal plants is still quite extensive.

The Indians of the Colombian Amazon distinguish two general classes of bioactive plants. The first are regarded as sacred, and they are used exclusively by or under the strict supervision of the *payé*, or medicine man. Most of the sacred plants are hallucinogens. The plants in the second class are known and used by the general population, and they are often, but not always, known and collected by the women.

The Indians are extremely friendly and cooperative, at least until they acquire the sometimes unsavory veneer of Western culture. The naturalist interested in plants and animals, both of which are close to the Indian's heart, is usually immediately accepted and may even meet with excessive collaborative attention. The Indian leaders are gentlemen, and all that is required to bring out their gentlemanliness is reciprocal gentlemanliness.

The story of arrow poison, which was my own initial study, provides a convincing argument for the urgency of ethnobotanical conservation. The poison

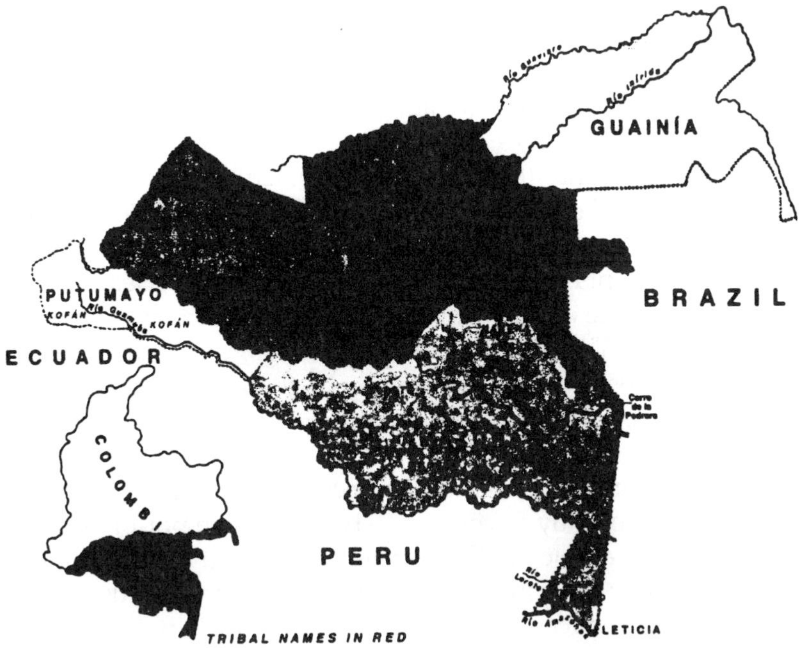

The Colombian Region of Amazonia

was first known to Europeans more than three centuries ago as a preparation for hunting in tropical South America, and it was reported, frequently in greatly exaggerated terms, by the early chroniclers. But it was not until the 1930s that curare attracted the serious attention of modern medicine. It soon became evident that tubocurarine, an alkaloid from one of the many plants used in arrow poisons, acts as a muscle relaxant. Tubocurarine comes from *Chondrodendron tomentosum,* a forest liana of the moonseed family that serves as an ingredient of many kinds of curare.

Today tubocurarine is widely used before deep surgery. By relaxing skeletal muscle, tubocurarine makes it easier for the surgeon to gain access to internal organs. Attempts to synthesize tubocurarine itself have not been successful: to produce only the active isomer of the molecule is difficult. For that reason pharmaceutical companies still purchase curare syrup from the Indians of Colombia, Ecuador and Peru. The liana is extremely slow-growing, and the Indians must fell it to harvest its bark. It is becoming rarer, and each year the Indians must go farther afield to meet the pharmaceutical demand.

My field research showed that *C. tomentosum* is only one of many ingredients used in Indian curares. Whereas some kinds of curare are made with only one plant species, other curares may call for as many as fifteen ingredients. Even today perhaps only a fraction of what the Amazonian Indians know about arrow-poison plants is known to modern science. Each tribe and even each payé may have a different recipe for curare.

What Western science does not know about curare preparation indicates how much there is still to learn from the so-called primitive peoples of Amazonia. An excellent example is the Western lack of understanding of the possible synergistic effects of many vegetal ingredients in the curares. Which of the additives increase the toxicity of the main ingredient? Which enable the poison to adhere better to the darts? Which help the bioactive ingredients enter the bloodstream more easily? Which are added for magic or superstitious reasons? We simply do not know.

Most recent research has focused on widely used arrow poisons, the majority of which are extracted from species of the genera *Chondrodendron* and *Strychnos*. But the Kofán, the master poison makers of the Colombian Amazon, employ many other plants in addition to the widely used species, either alone or in complex recipes. Some of the plants are not even known as sources of poison by most other tribes. The Kofán, like other tribes, make arrow poisons for specific hunting purposes, depending mainly on the kind of animal the individual wants to shoot. Few of the minor arrow poisons have ever been chemically studied.

One intriguing example is *Schoenobiblus peruvianus*, a member of the daphne family. The plant was recently discovered to be both a fish poison and the source of a curare. The Kofán employ both fruits and roots of the plant, almost always without admixture, in preparing a kind of curare for killing birds. Another tribe dries the leaves and reduces them to a powder, which is then applied as a poultice to persistent infected cuts. The chemical makeup of the species has not been analyzed, but it is known that the family is rich in coumarin derivatives, which act as anticoagulants.

The amazing phytochemical skill of the Indians is hardly limited to applications in hunting and medicine. Vegetal intoxicants and hallucinogens are also abundant in the forest, and their use is highly sophisticated. The Indians prepare a hallucinogenic drink, variously known as *ayahuasca, caapi, yajé, pindé* or *natém,* from the bark of *Banisteriopsis caapi,* "the vine of the soul," a forest liana that trends to grow in charming double helixes. The plant contains a group of alkaloids collectively known as beta-carbolines, of which the alkaloid harmine is the major component. The beta-carbolines belong to a class of compounds known as monoamine oxidase inhibitors, which prevent the normal metabolism of certain neurotransmitters. Thus the neurotransmitters tend to accumulate in the central nervous system. One effect of ayahuasca is to overstimulate brain functions that depend on the neurotransmitter serotonin, a biochemical process that can become manifest as a hallucination. I have taken ayahuasca with the Indians during ceremonies and have experienced its extraordinary effects.

A historical accident demonstrated that the alkaloids in *B. caapi* are extraordinarily stable. The plant was identified and named by Richard Spruce in 1853. He collected a bundle of *caapi* stems and shipped them to the Royal Botanic Garden in Kew, England, for chemical analysis. In transit from the jungle the specimens were left in a damp abandoned Indian home for several years. They were finally located at Kew in 1969 and analyzed. One hundred-sixteen years after collection, the specimens yielded 0.4 percent by weight of beta-carboline alkaloids, which compares favorably with the 0.5 percent yield from a recently collected specimen.

To increase the intensity and duration of intoxication with ayahuasca, the Indians in the Colombian and Ecuadorian parts of the Amazon Basin sometimes add the leaves of another liana, of the Malpighiaceae family, *Diplopterys cabrerana,* or the leaves of a bush belonging to the madder, or coffee, family, *Psychotria viridis.* Both plants contain another type of psychoactive alkaloid: tryptamines. Tryptamines also prolong the effects of serotonin in the body, generating hallucinations at sufficiently high dosages.

But the tryptamines are inactive when they are taken orally, unless monoamine oxidase inhibitors are present. The harmine and its derivatives in *B. caapi* are inhibitors of that kind and thus enable the tryptamines to act. The net hallucinogenic effect is that *B. caapi* by itself usually engenders visions in dull colors, but those induced when tryptamines are added to *B. caapi* can be brightly colored, in reds and yellows. How did the Indians find two plants that would act synergistically with *B. caapi* among the 80,000 species in the forest? Trial and error hardly seems an adequate explanation.

The same chemistry figures in the story of the hallucinogenic snuff prepared from four or five species of *Virola*. More than thirty species of *Virola,* a genus of the nutmeg family, occur in the forests of northern South America and adjacent Central America. The Indians strip bark from the lower part of the tree trunk before the heat of the sun penetrates the forest. A colorless resin-like exudate appears on the inner surface of the bark. The liquid rapidly turns reddish brown, darkens and dries to a hard, gummy mass, which is scraped off. The scrapings are kneaded in water and then reduced by boiling, to a thick syrup. The syrup is dried, pulverized and mixed with the ashes of the bark of a wild cacao species. The product of all this labor is then snorted through snuffing tubes made from hollow bones of birds, either U-shaped for self-administration or straight for blowing the powder into the nostrils of a friend.

The psychoactive ingredients in *Virola* are tryptamine alkaloids, present in the exudate of the bark in concentrations as high as 11 percent by weight. The alkaloids are bioactive when they enter the bloodstream directly through the nasal membranes.

I discovered, however, that the Bora and the Witoto of Colombia and Peru use *Virola* as an oral hallucinogen. They boil slivers of the bark for at least an hour, until a thick paste forms. The paste is then rolled into pellets, and the pellets are coated with an alkaline ash prepared from one of several plant species and set out to dry. Why should the tryptamines, taken orally, be active at all? Further chemical examination disclosed that the exudate contains trace amounts of beta-carboline alkaloids, which serve as built-in monoamine oxidase inhibitors.

The genus *Virola* is of interest for another reason. Many tribes in the Colombian Amazon use the exudate to treat fungal skin infections, such as ringworm. Every day for ten or fifteen days, the sufferer goes into the forest in the early morning, strips off a piece of the bark and paints the affected area with the resin. In my experience, the treatment appears to cure or at least suppress the fungal infections. Twice I dried bark specimens and sent them to laboratories to be checked for fungicidal properties; none were found. When fresh material was recently analyzed in Brazil, however, it turned out that some secondary plant compounds in *Virola* are undoubtedly fungicidal.

But what explains the earlier failure to detect a fungicidal activity? When I dried the bark under the tropical sun, I may have unwittingly allowed its chemical composition to be altered, perhaps through the activity of some enzyme. If the fungicidal properties of *Virola* are borne out by further research, it will be one more indication that Western science ought never treat the Indians' knowledge lightly or ignore their familiarity with their plants.

One disarming example of the Indians' uncanny ability to distinguish subtle variations between plants comes from the yoco liana, *Paullinia yoco,* of the soapberry family. The bark of the vine is nearly 3 percent caffeine, and when its scrapings are kneaded in cold water, the result is a stimulating drink. The Indians normally take no food in the early morning. Instead, they drink a gourdful of yoco when they get up and another gourdful perhaps a half-hour later. Pieces of the trunk of the liana are kept in the malocas, or communal houses, ready for use. When men embark on a canoe trip or on a lengthy trip through the forest, they carry pieces of the trunk instead of food. In effect, the liana is a kind of portable coffee stick.

The Indians distinguish some fourteen varieties of yoco. Among them are yoco blanco (white yoco), tigre yoco (yoco of the jaguar) and yoco de brufo (sorcerer's yoco). Some are said to be stronger than others, some to be best suited for particular purposes. I have worked on yoco for many years, and other botanists also have collected specimens of different varieties, but none of us has been able to find any morphological differences between them. They may be forms that vary by age or by ecosystem, or they may be chemovars. But note that Western coffee drinkers, too, make distinctions as fine or finer between the varieties of the bean to which they are accustomed.

The numerous varieties of yoco have not yet been analyzed for chemical differences. But even if chemical differences are discovered, an intriguing question will remain. The Indians often identify the lianas on the basis of its trunk alone, because the liana's leaves are usually a hundred feet high or higher in the forest canopy. Furthermore, the identification is often made many yards from the trunk. What could enable such quick-sighted identification? Only the natives' long familiarity with their flora.

The narcotic liana *B. caapi* provides an even more astonishing example of the Indians' botanical perspicuity. The Indians of the Colombian state of Vaupés recognize many "kinds" of *B. caapi*. Again, botanists can find no morphological differences between those kinds, although the Indians can identify them by sight. The Indians insist they make an identification by taking into consideration many factors, such as the kind of soil and the age and size of the vine. They differentiate the kinds, however, primarily according to their biodynamic effects: the predominant color of the hallucinations, they maintain, is one important consideration. Moreover, some kinds are held to be best for communication with the spirit world, whereas others are held to be best for diagnosing illness or for prophecy.

Of course, not all this Indian lore is necessarily translatable into the vocabulary of Western science. Still there is enough consistency in the classification and use of the vines and enough agreement among tribes to suggest that investigation of the kinds of *B. caapi* might be well worth the effort to try to understand it.

Most urban Americans know of the Indian curares, if only through detective novels, in which they sometimes figure as poisons exotic enough to elude the local coroner. Less known are the barbascos, or fish poisons.

The Indians of the northwestern Amazon have discovered many plant species that can be used to stupefy fish. Most commonly employed is a forest liana known as timbo, a species of the genus *Lonchocarpus,* which belongs to the bean family. Its active ingredient is rotenone, which is easily biodegradable and is now used in the U.S. as an insecticide. The bark of the liana is scraped off and macerated, then thrown into still water or placed in a porus bag or basket and dragged through the water. Most fish poisons effect the function of the gills. The fish come to the surface stunned and seeking oxygen, but their flesh is fully edible, since none of the toxic ingredients have been absorbed.

Timbo grows wild in the forest, but many of the plants used as fish poisons are cultivated. Two of the most common cultivated plants are a member of the daisy family and a bushy species of the spurge family. Those plants are employed when only a few fish are needed. But when a tribal festival is planned and there will be many mouths to feed, there is no substitute for timbo.

In addition to those few fish poisons, there are many others, which are employed in remote areas or in times of emergency, when the major kinds are not readily available. Indeed, it seems every botanical expedition to the northwestern Amazon Basin discovers a new source of fish poison. Recently a fish-poisoning plant was discovered along the Río Loretoyacu in the Colombian state of Amazonas. It is a species new to science, *Patinoa ichthyotoxica,* a member of the Bombacaceae family. The Indians of the area gather the pulp from the large fruits and dry it for use on short expeditions. The seeds, roasted, are edible. The pulp, dried and powdered, is carried in pouches. It is cast into stagnant pools and within twenty minutes fish appear at the surface, where they are easily caught.

It is a common misconception that the indigenous peoples of the Amazon use relatively few plant species as medicines. Some tribes, it is true, use only a few plants as curatives. But I have found that the tribes have pharmacopoeias that differ greatly in size, and some, including the Kofán, the Witoto, the Makuna and the Yukuna, have extensive pharmacopeias. Throughout the region most bioactive plants are employed as simples; mixtures and complex prescriptions are rare. By far the commonest method of preparing plant medicines is to make a tea.

One disease prevalent in the Amazon is conjunctivitis—an inflammation of the tissues lining the eye. It is highly contagious and frequently affects all the people living in a maloca. Of the several plants used against the condition, the one said by the Indians of Vaupés to he most effective is a vine of the gourd family, *Cayaponia ophthalmica,* which is cultivated by the medicine men for this purpose. The leaves are gently boiled, and the tea is dropped into the eye over a period of three or four days.

Not surprisingly, given the prevalence of fevers in the tropics, the largest category of medicinal plants is the febrifuges, agents that reduce fever. One interesting febrifuge is the woody shrub *Brunfelsia chiricaspi*, a member of the nightshade family. The plant's specific name comes from its vernacular name in the Colombian state of Putumayo, *chiricaspi*, which means cold tree or tree of chills. A small amount, prepared as a tea, gives the sensation of chills. Whether it actually lowers body temperature is not known.

In large doses, the plant acts as a hallucinogen rather than as a febrifuge. I have tried several of the Indian hallucinogens, in part because the Indians consider them sacred plants, and it would have been an unpardonable rudeness to refuse them when the Indians were kind enough to offer them to me during a ceremony. I never ventured to try *B. chiricaspi*, however. Its effects are much more unpleasant than those of *B. caapi* and include—in addition to the sensation of chills—vertigo, nausea and a loss of muscle coordination. The chemistry of the species, which is used as a slow-acting curare as well as a febrifuge and a hallucinogen, is extraordinarily complex and still far from being understood.

Ethnobotany is sometimes considered by people who have no contact with Indians as a rather sentimental discipline, a form of scientifically sanctioned nostalgia for simpler ways of life. I would argue, however, that it is a thoroughly practical discipline. The human species depends on plant species for its own welfare and ultimately for its survival. Each living species is the repository of organic molecules that are the products of the plant's irreproducible evolution. Science should intensify its study of these chemicals, because there is little time left to learn; when a plant becomes extinct, the opportunity to learn is lost forever. It is only common sense that we who can apply technical analyses to problems should learn from peoples who are intimately familiar with their floral environment and its useful properties for the benefit of all mankind.

Richard Evans Schultes *is Jeffrey professor of biology and director of the Harvard Botanical Museum (emeritus).*

BIODIVERSITY IN SOUTHERN AFRICA:

An Overview

JOSEPH P. DUDLEY

Introduction

The "southern Africa region" is defined herein as the continental mainland extending southward from 10° south latitude, approximately coinciding with the northern limits of the Zambezi River watershed. Southern Africa as defined includes the political domains of Mozambique, Malawi, southern Tanzania, central Angola, southern Zaire, Zambia, Zimbabwe, Botswana, Namibia, South Africa, Swaziland, and Lesotho. The offshore island of Madagascar, with its extraordinarily high levels of endemism in plants and vertebrates, is not included. (Madagascar has been isolated from the African continent so long that it ranks as an evolutionarily independent continental biota.)

The continental mainland of Southern Africa is a region of extraordinarily high biological diversity. Biodiversity in Southern Africa is remarkable not only for its species diversity (alpha/beta diversity), but as much or more so for its high degree of evolutionary (phyletic) and landscape (gamma) diversity. Few if any equivalent continental regions of the Earth possess comparable degrees of biological diversity at macroevolutionary and landscape scales of analysis.

Regional Landscapes

On an east to west transect, biotopes within Southern Africa range from deepwater tropical marine, subtropical coral reef and coastal mangrove forest to

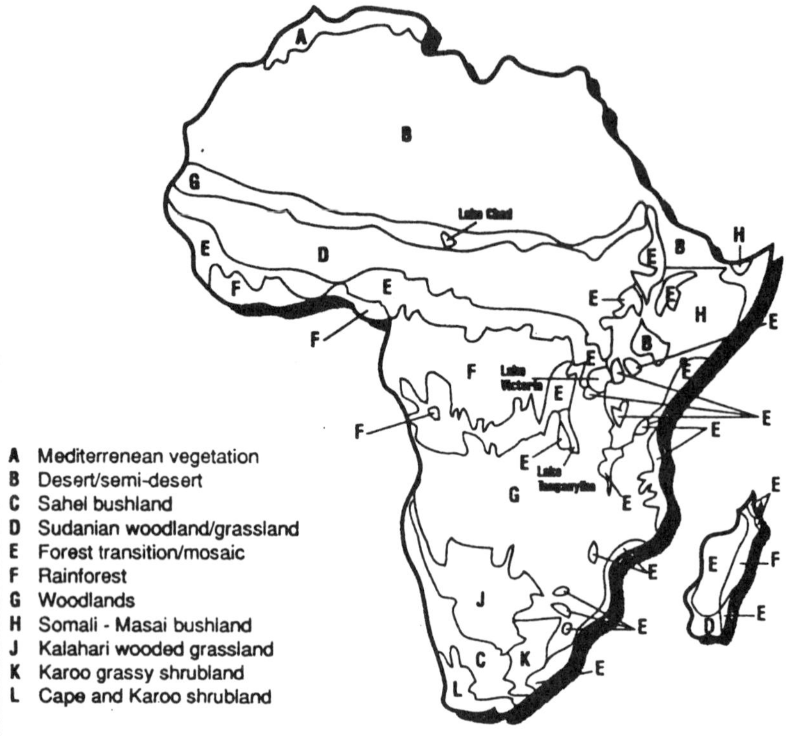

*vegetation zones of Africa**

montane rainforest, semi-arid woodland, arid grassland and scrublands, and near-absolute coastal desert. The nutrient-rich antarctic Benguela current flowing northward along the western (Namib) coast historically supported a rich coastal and coldwater deep-sea marine fauna of fishes, fur-seals, whales, and seabirds (including an endemic penguin). Beaches and tropical waters of the Mozambique current along the east coast provide nesting and foraging habitat for Indian Ocean sea turtle and pelagic fish populations. Most inland areas are subject to periodic severe droughts, which have been a major evolutionary force in the region since time immemorial. Many of the region's endemic plants and animals are inhabitants of arid desert or semi-desert landscapes (Karoo, Kalahari, Namib).

The Zambezi River drains the region's largest watershed, an area extending nearly coast to coast between 10°S and 2°S latitudes. The Zambezi system, which includes Lake Malawi (southern-most of the Rift Valley Great Lakes), has a highly diverse endemic fish fauna. The headwaters of the Zambezi's western tributaries rise from central Angola, draining southward and eastward into the Mozambique Channel and the Indian Ocean. The Zambezi's Victoria Falls, one of the world's greatest natural wonders, are located at almost the geographical center of region: 18°S by 26°E. Average flow volume is 550,000 m^3/minute but the river reaches 5,000,000 m^3/minute in flood stage. The Zambezi gorge constitutes the southern extremity of the great African Rift Valley.

Adjacent to the Zambezi's westernmost tributaries is the Okavango watershed, which flows southeastward into a vast seasonally-inundated inland marsh on the northern fringes of the Kalahari desert. Southeast of the Okavango Delta, and into which it drains during exceptionally high flow periods, lie the great Makgadikgadi salt flats, a major flamingo breeding area. The Kalahari is a vast arid wooded grassland covering much of the south-central part of the region. Mountain ranges in the southern and eastern Cape, Angolan/Namib coastal region, and Mozambique/Zimbabwe border region support archipelagos of moist temperate montane and subtropical cloud and rainforest habitats exhibiting high levels of endemicity. The Cape flora of the southern tip of the African continent is so unique that it ranks as one of the planet's six major floristic provinces (Cape, Paleotropical, Neotropical, Holarctic, Australian, Antarctic).

Flora

Southern Africa is particularly noted for high levels of plant diversity and endemicity, particularly among proteas, cycads, aloes, and leguminous trees (Coates-Palgrave 1991). The southern Cape *fynbos* flora alone supports an estimated 8500 indigenous vascular plant species—70% endemics—within an area of only 71,000 km^2 (Stuart et al. 1990). The nearby Drakensberg Mountains of the eastern Cape region are also a very important area for plants, with some 300 known endemic species. The country of South Africa contains probably the highest concentration of endemic flowering plant species of any region on this planet (Pomeroy 1993).

The Namib-Karoo flora contains some 3500 species, over half of which are endemic forms (Stuart et al. 1990). The Namib Desert of the southwestern coast supports some particularly distinctive taxa. The Skeleton Coast region (so-called because of the beached whale carcasses once common there) is famous for its cartwheeling dune spiders. Perhaps the Namib Desert's most enigmatic organism is the endemic Wonderplant *Welwitschia mirabilis*. *Welwitschia* produces only two true leaves during its lifetime, which reach up to 3 meters in length and become shredded into leathery, thong-like strands by sun and wind. *Welwitschia* is actually a massive but almost entirely underground "tree," individuals of which (like Baobabs) are known to be among the oldest living organisms on earth (2000+ years old: Coates-Palgrave 1991).

The Zambezian flora which dominates much of southern central Africa is thought to be the continent's richest flora with more than 8500 species, about 54% endemic. The western Angola highlands and the outlying mountains ("inselbergs") of the Chimanimani Range in western Mozambique are suspected to be reservoirs for numerous still unidentified endemic plant species. Boabab trees (*Adansonia digitata*), a distinctive inhabitant of arid savanna landscapes within the region, have been carbon dated to ages of at least 4000 years (possibly just two generations post-Pleistocene time for some individuals!).

Macrofauna

Birds

More than 920 bird species occur in the region, of which some 140 species may be regional endemics (Sinclair 1987). Some species found locally (Arctic Skua, *Stercorarius parasiticus*, European Stork, *Ciconia ciconia)* are Palearctic seasonal migrants, others (e.g., Rockhopper Penguin, *Eudyptes chrysocome;* South Polar Skua, *Catharacta maccormicki*) occur as Antarctic vagrants. Raptor diversity within the region is truly bewildering—Hwange National Park (12,953 km^2) hosts fifteen species of "eagles" alone: *Aquila* (5 spp.), *Hieraaetus* (3 spp.), *Circaetus* (3 spp.), and one species from each of four other genera. There is also astounding regional diversity among falcons (16 spp.), vultures (8 spp.), hawks (23 spp.), and owls (12 spp). Other notably diverse groups include warblers (73 spp.), thrushes/robins (43 spp.), weavers (34 spp.), shrikes (26 spp.), swallows (22 spp.), larks (26 spp.), quails/francolin (19 spp.), doves (14 spp.), herons/storks (28 spp,), rails/coots (21 spp.), and teeter-assed** shorebirds (60 spp.).

Mammals

The Southern Africa region supports substantial populations of many of the continent's most characteristic and charismatic large mammal species (African Elephant, rhinos, Cheetah, Wild Dog, Lion, Gemsbok, Giraffe, zebras, hippos, wildebeests . . .). More than half of Africa's surviving Savanna Elephant (*Loxodonta africana africana*) population inhabit Zambesian dry subtropical woodland and scrub landscapes (Cumming et al. 1990). Endemism among large mammal species is limited, although distinctive regional types or subspecies of various cosmopolitan species are recognized—some of which are now threatened or extinct. The extinct Quagga (*Equus quagga*) was an endemic zebroid horse once common and widely distributed within what is now the country of South Africa; the last specimen died in captivity in 1883 (Haltenorth & Diller 1994).

Large mammals brought back from the verge of extinction by conservation efforts include White Rhinoceros (*Ceratotherium simum*), Black Wildebeest (*Connochaetes gnou*), and Bontebok (*Damaliscus dorcas dorcas*) (Smithers 1986). Black and White Rhinos (*Diceros bicornis, Ceratotherium simum*) had been extirpated within much of their former range by the mid-20th century, when conservation efforts began to make headway in protecting and restoring relictual populations. Translocations were used to re-establish rhinos in a number of areas from which they had been extirpated before the turn of the century (e.g., Hwange National Park, Zimbabwe).

Human Impacts

Southern Africa supports a large, culturally diverse, and very rapidly growing (2–3% per annum) human population. Human population densities within parts of the region are estimated to have increased ten-fold or more during the past century. Botswana, Malawi, Zimbabwe, and Zambia currently exhibit human population growth rates that are among the highest on the planet, with projected doubling times of about twenty years. Human impacts within the region are consequently considerable and increasing rapidly. Deforestation of dry woodlands for conversion to croplands is proceeding at a rate of more than 700,000 ha/year. Pumping of underground water for agricultural, industrial, and domestic uses may be depleting underground aquifers. Gold, diamond, coal, copper, and heavy-metal mining are diminishing surface water quality and availability within watersheds in many parts of the region. Human impacts are a serious threat to most ecosystems and biotas within Southern Africa (BSP 1993).

Illegal killing of wild animals for subsistence or commercial purposes is rife in many areas. With some notable exceptions (Mozambique, Angola), elephant poaching levels in Southern Africa have generally remained low relative to those in East Africa (Kenya, Tanzania, Sudan, Somalia, Ethiopia). Rhino populations, however, have been decimated in recent years by poaching and virtually extirpated from many areas. The reintroduced White Rhinoceros population of Hwange National Park was nearly wiped out between December 1992 and January 1994: 80 of 82 known individuals (many dehorned) were lost to poachers during this brief period. Although rhino poaching now appears to be under better control within Zimbabwe, Botswana, Namibia, and South Africa (at least partly because the more vulnerable populations have been wiped out) it remains to be seen whether White and Black Rhinoceros populations in Southern Africa can be sustained indefinitely against a poaching tide which has in the past twenty years extirpated these species throughout most of their restored range (TRAFFIC 1993).

The southern Cape of Africa has become one of the planet's premier agricultural regions, as its Mediterranean climate and good soils proved ideal for conversion to orchard crops and viticulture. Little remains of the original Cape fynbos habitats, and only a tiny fraction is protected within reserves (0.4% of its original 31,800 ha habitat area). The species-rich temperate montane moist forests, which once blanketed the slopes of the southern Cape's coastal mountain ranges, have now been largely replaced by farms and exotic pine (*Pinus*) and eucalyptus plantations. Formerly cultivated weedy exotic tree and plant species such as Blackwood (*Acacia melanoylon*) constitute a serious threat to the integrity of surviving areas of indigenous forest habitat (Vermeulen 1993). Industrial/mechanized agriculture in much of the inland region is largely irrigation-dependent due to limited and erratic rainfall. Fertile soils of the temperate/subtropical highland plateau regions of Zimbabwe and South Africa support extensive areas of commercial agriculture. Drier regions and areas of infertile soils are dominated by commercial livestock operations and subsistence communal farming and pastoralism.

Industrial trawling, sealing, whaling, and guano harvesting have severely damaged Namib marine ecosystems. Major hydroelectric/irrigation dams in the

Zambezi River system have radically altered floodplain, riverine, and estuarine ecosystems along the Mozambique coast. Dynamite fishing is devastating coral reef ecosystems of the Mozambique coast. In the Zambezi Delta, destruction of estuarine mangrove forests and lowered nutrient inputs due to upstream dams are threatening the continued viability of the Mozambique prawn fishery (the country's most important export industry).

Cattle, goats and sheep are present almost everywhere outside of strict wildlife reserves across most of Southern Africa—even hyperarid desert areas, where pumping of underground water allows livestock production. Heavily-subsidized livestock disease and parasite control programs facilitate expansion of livestock populations to levels beyond long-term carrying capacities. European-based international development agencies directly and indirectly subsidize livestock production programs within the region. Cattle, sheep and goats severely overgraze rangelands in drought years before destocking is effected through drought mortality or salvage marketing.

Cultural norms in Southern Africa complicate livestock problems. Traditional cultures esteem *and communal land-occupation systems enforce* the role of livestock as the principal source of real property assets. One artifact of the"livestock as real property" syndrome is that 10 skinny cows contain greater value than 8 sleek cows; communal pastoralists are loath to sell off potential breeding stock except under extreme duress. Hence, overstocking becomes chronic within communal areas whenever conditions permit, and overgrazing/overbrowsing by cattle, sheep, and goats is often the norm even within years of adequate rainfall. As is the case with the degradation of arid rangelands in western North America due to overgrazing, livestock carrying-capacity issues are not resolvable in terms of either ecological or economic cost-benefit approaches.

Wildlife reserves within Southern Africa are for the most part restricted to areas where tsetse-fly infestation and/or lack of perennial surface water sources historically precluded large-scale human settlement or agriculture. Fuelwood cutting and conversion to agriculture is rapidly decimating woodland landscapes throughout much of the region. Vast areas of dry woodlands newly-cleared of tsetse fly are being targeted for urban population resettlement programs and accommodation of rapidly expanding rural populations.

Animals within many of the region's largest and most important wildlife reserves (e.g., Kruger National Park, South Africa; Hwange National Park, Zimbabwe) depend on man-made water supplies during the peak dry season months. As is the case with domestic livestock, water provisioning allows water-dependent ungulates (e.g., elephant, buffalo, hippo, wildebeest) to permanently occupy vast areas of former seasonal habitat and encourages the build-up of artificially high densities of these species.

Temperature and rainfall data from Hwange National Park in western Zimbabwe show trends consistent with the predictions of global warming scenarios. Interior regions of Southern Africa may be threatened by wide-spread desertification due to the combined effects of global climate change and deforestation.

Geopolitical Context

The southern Cape of Africa was first occupied by Europeans some three centuries ago. European settlements spread westward along the coast and northeastward into the interior during the eighteenth and nineteenth centuries. Nonetheless, European domination of the region was not consolidated until the first decades of the twentieth century. Most available agriculturally-productive lands within the region were expropriated by European settlers, and indigenous black Africans often relegated to semi-serfdom (or outright slavery prior to the late nineteenth century) on the farms of European colonialists and/or subsistence farming in tribal or communal lands ("Tribal Trust Lands," "Tribal Homelands," following "Indian Reservation" paradigms used in the United States of America).

With few exceptions, ethnic African populations in colonial South Africa/Rhodesia/Namibia retained or were retroactively granted land tenure of sorts only in areas with limited agricultural potential and no mineral or timber resources of known or anticipated value. The greatest single difference between the European colonial occupation of Southern Africa and of what became the United States was that deliberate systematic extirpation of indigenous populations was rather more the exception than the rule in Africa, partly because the historically agro-pastoralist indigenous Bantu and "Hottentot" peoples provided a valuable source of labor for the farming (and later mining) enterprises of ethnic European colonials. European movement northeastward from the Cape into the interior of the present-day republics of South Africa and Zimbabwe (formerly Southern Rhodesia) was first delayed, but ultimately probably facilitated, by the decimation and displacements of indigenous peoples resulting from internecine warfare, raiding forays, and migrations of Bantu "Nguni" tribes (e.g., Zulu, Matabele) between 1810 and 1860. (Tribal warfare and antipathies similarly facilitated European territorial conquests in several regions of North America.)

Political boundaries drawn by European colonial powers have left many African nations with a complex legacy of unresolved ethnic and ideological conflicts. Some stresses may be seen as exotic in origin (e.g., political polarization of Marxist/Maoist Socialist versus "other" paradigms), others are of endemic and precolonial ethno-historical origins (e.g., Zulu/Xhosa of South Africa; Hutu/Tutsi of Rwanda/Burundi), and some are synergisms of both spheres of influence (military versus civilian executive control in national governments). Such existing or potential conflicts were often exacerbated or abetted by the political intervention of outside powers during the post-colonial Cold War era. Political instability, leading to displacements of regional populations and their concentration in refugee camps, is rapidly becoming one of the most significant environmental threats to African landscapes (cf Mozambique, Liberia/Ivory Coast, Rwanda/Zaire/Tanzania/Burundi, Somalia/Ethiopia/Eritrea, Uganda/Sudan . . .).

Conclusion

There appears to be a growing consensus that in order for the large mammals of Southern African landscapes to escape the fate of their North American and European equivalents, rural communities will have to derive direct economic benefits from the management of local wildlife populations. The social and economic costs of free-ranging African wildlife species (especially elephants) outside reserves are difficult for Americans and Europeans to comprehend. Elephants outside reserves often raid croplands and will even enter settlements and destroy village granaries. Human deaths and injuries result from confrontations or even chance encounters between farmers and crop-raiding elephants.

The "CAMPFIRE" and "ADMADE" programs of Zimbabwe and Zambia were developed to give local communities vested economic interests in managing wildlife on communal lands, and have drawn considerable financial support and backing from both international conservation NGOs (e.g., IUCN, WWF, WCI) and international development agencies. The CAMPFIRE program, which derives 90% of its returns from the trophy hunting of elephants, has been remarkably successful in promoting the protection of wildlife on communal lands in southern and western Zimbabwe.

Wildlife conservation in Southern Africa will not succeed if promoted along the strict preservationist paradigms typically regarded by rural Africans as unwarranted and undesirable artifacts of colonial rule. Tourism provides an economic justification for wildlife conservation only within the bounds of national parks, and even so becomes viable only within those parks amenable to large-scale tourism. The importance of habitats outside reserve areas to regional biological diversity requires the development of alternative programs for enlisting support and cooperation from local human communities in protection of local habitats and wildlife. Resolving this dilemma will require tremendous inputs of energy and effort at all levels (local, national, and international), but the potential returns should more than justify the investment. The biological diversity of Southern Africa is a heritage of outstanding global significance, and needs much wider recognition as an international conservation priority.

References

Biodiversity Support Program. 1993. African Biodiversity: Foundation for the Future. WWF/TNC/WRI/USAID, Beltsville, Maryland.

Coates-Palgrave, K. 1988. Trees of Southern Africa. Struik, Cape Town.

Cumming, D., R. F. Du Toit, and S. N. Stuart. 1990. African Elephants and Rhinos: status survey and Conservation Action Plan. IUCN/SSC African Elephant and Rhino Specialist Group, Gland, Switzerland.

Haltenorth, T. and H. Diller, 1994. Mammals of Africa including Madagascar. Harper Collins, Hong Kong.

Pomeroy, D, 1993. Centers of high biodiversity in Africa. *Conservation Biology* 7(4): 901 -0 907.

Sinclair, I. 1987. Field guide to the birds of southern Africa. Struik, Cape Town.

Smithers, R. H. N. 1986. Land mammals of Southern Africa: a field guide. Macmillan, Johannesburg.

Stuart, S. N., R. J. Adams, and M. D. Jenkins. 1990. Biodiversity in Sub-Saharan Africa and its Islands: Conservation, Management, and Sustainable Use. IUCN-Species Survival Commission No. 6. Gland, Switzerland.

Vermuelen, W. J. 1993. Multiple-use management of the Southern Cape Forests. *In* The Ecology and Management of Indigenous Forests in Southern Africa (D. Pierce & D. J. Gumbo, eds.). Zimbabwe Forestry Commission, Harare.

Joseph Dudley is a conservation biologist in the Department of Biology and Wildlife at the University of Alaska-Fairbanks. He is currently working in Zimbabwe's Hwange National Park (Main Camp, Private Bag DT 5776, Dete, Zimbabwe, Africa).

CULTURE AND BIODIVERSITY IN THE ANDES

GRIMALDO RENGIFO VÁSQUEZ

Introduction

According to scientific definitions biodiversity or biological diversity is the variety of life in all forms, levels and combinations: includes biodiversity of ecosystems, of species and genetic biodiversity (IUCN, PNUMA, WWF, 1991) and refers to all existing species of plants, animals and microorganisms interacting within a given ecosystem (Altieri, 1995) i.e., all living beings.

According to Maturana and Varela (1995, p. 73) living beings are self-sustainable and capable of reproduction:

> 'if they are machines, living systems are self-sustainable machines in that they are able to transform themselves in order that their own product corresponds to their own organization. If we consider the opposite statement: a system which is self-sustainable is therefore a living system. In other words, self-sustainability characterizes even the "organization" of living beings'.

Such a definition, excludes stones and clouds. But now let us turn to an Andean vision of the world to see how this world view explains 'biodiversity'.

In the Andean culture, diversity is holistic. Basically, it includes all living beings. *Runakuna en quechua* (human beings) perceive *alpacas* or sweet corn as living beings. However, they also consider rivers, stones, stars, wind and all other biological elements incapable of reproduction, as forms of life. And in this, the notion of self-sustainability is totally meaningless.

The activities which characterize any form of life[1] are not seen as the expression of individual acts. Rather, they are the sum of reciprocal and equitable participation of all reproductive activities. In Andean culture, there is no difference between what is 'living' and what is 'lifeless'. For example, Andean people do not consider a new potato crop to be the result of hybridization of different living organisms in a given environment. Instead, they perceive it as a symbiotic process among all other existing things. In other words, the life of *papa* seeds does not only derive from the seeds themselves. It also involves, equally importantly, the participation of earth, water, stars, *Apus* and *runas* (people). In such a vision of life, everything is linked with everything else and they are all together, engaged in the 'reproduction' of the world.

Nowadays, this culture of biodiversity is threatened. Reports abound of the extinction of species, varieties and cultures due to industrial pollution, unrestricted development, trade, and modern agriculture. Many programmes on biodiversity have been launched in response to this problem. Yet, despite these efforts still little is known about methods of breeding and training of different 'life forms' within local communities. Even less understood is how these local communities regard the loss of seeds. Little is known about how seeds are grown by peasants and even less information is available about the so-called 'seed culture'.

Breeding and growing of diversity

What is really puzzling in a peasant's crop is not only the diversity of species, but the great variety within the species themselves, be it *papa* (sweet potato), *frijol* (bean) or *quínua* (a tree with edible seeds and leaves). The same may be said of the breeding of animals such as *alpacas, llamas, ocas* (geese) and *cuyes* (sort of guinea pigs). Andean agriculture and livestock are extremely varied. The species diversity in the *charcas* (small farms) is accounted for by the interaction between ecological phenomena (such as climate variations, the tropical location of these areas, the historical geological changes and plentiful sun) and human interaction (horticulture, and animal husbandry). As it was pointed out by François Greslou (1989) only few studies acknowledge the Andean vision of the world as a driving force in the orientation of a 'breeding culture' in these tropical areas.

Following western explanations, interaction between humanity and nature is perceived as a conflictual relationship in that human survival is guaranteed only if humans are capable of imposing themselves on nature through technology intervention. A cultivated plant is deprived of its natural attributes so that its reproduction is only guaranteed through the process of horticulture. A cultivated plant is a subdued 'form of life', whose once complex relationship with the natural environment (transformed now to a 'cultural landscape') is simplified and even eliminated.

With such a process, the cultivated plant no longer belongs to the realm of nature (vegetable kingdom), but to the 'realm of culture.' In this framework, culture and its related techniques are seen exclusively as a human attribute. Agriculture is, therefore, perceived as humanity's triumph over nature in order for people to be free from nature's whims.[2]

In the Andes, their cosmic vision leads to a completely different approach. From an Andean viewpoint, the breeding of so many forms of life is considered to be a dialogue, even a loving conversation. It is not a question of conflictual relationships between humans and nature striving for dominance. Instead they operate the notion of a 'living world', mutual equitable relationships between humans and nature, breeding and cultivation, agricultural centrality and life in an *ayllu* (community).

A living world to be respected

In Andean belief *runas* (humans) understand that all things which exist are 'forms of life'. Earth is not a lifeless receptacle or a mere base for growing plants. Earth is a living being, it is our *Pachamama* (Mother Earth), mother of all, including *runas*. Andean people have conversations with her. The same can be said for water, seeds, stones, hailstones and springs. In this sense, biodiversity of life forms is not just a privilege of'living organisms'. As don Calixto Huacca from the Community of Maka Qollo, Puno would put it:

> 'Papa seed is like a living being, sometimes it can even get tired after a certain time. For example, I grow *luk'i* that last for ten years, then disappear, this too depends on the degree of love one is able to give them. The more love you give, the more the *papa* produces. We say'conmigo se congenia y cuando no siempre se congenia, no congenia conmigo' (the *papa* is as vital for me as I am for him). That's how we say it. . . .' (Chambi, 1996)

I would like to point out that when Andean people refer to things as living things, it does not mean that they perceive a life-giving and trascendental spirit in stones, *papas* and rivers; a spirit which makes itself visible in a rite, as if by magic, changes its lifeless nature into a living substance.

If we assume that there exist living beings which are invisible to *runas'* eyes but become 'visible' in a rite (the way Christ is transformed into the host), it means that we are applying the Christian notion of ideal and matter, spirit and body, transcendence and immanence to the Andean world. Nor is this just a metaphorical way to refer to things. When a peasant says that *Pachamama* is her or his mother she or he is not making an analogy between the earth and her or his biologiocal mother because for her or him the Earth has the attributes of her or his mother. In practice, he or she just means that *Pachamama* is also his or her mother. The concept of identity does not make much sense to a *runa,* since it does not reflect an Andean point of view on how to understand forms of life.

To Andean people, the world is like a person characterized by interrelated forms of life which create a living texture whose 'physiology' depends directly from the conversation held between its 'forms of life'. Each part, like organs in one body, in fact, mutually interact, have conversations, thus participating in the activities that contribute to the reproduction of the world. Everything which occurs inside this living animal, or *Pacha,* is immanent. That is, everything springs directly from its inner workings.

Mutual equitable relationship between human beings and nature

When *runas* say '*Pachamama* is my mother' they mean that they put the earth and humans on the same level. In the same way, when they say 'I breed *alpacas* just as they bring me up' they show that among all, an activity is not the exclusive privilege of one form of life but the result of relationships among all of them. This is a concept which is totally foreign to a western concept of hierarchy. [. .]

Hierarchy is based on the idea that all species evolve and develop from less complex into more complex forms, the former being inferior to the latter. Hierarchy also implies a specific unit of measure which is used to classify and organize.

This is the researcher's assumption, for example, when he or she comes across a *papa* whose features are, to a certain extent, considered better and superior (just because this *papa* is frost-resistant) to others. The new unit of measure based on its frost resistance will be used to classify *papas* closer to the frost-resistant prototype identified as the most promising, whereas the others are dropped from consideration. In the attempt to find the best variety, the scientist makes use of pre-existing genetic material (either cultivated or not) and from this, a new variety is created. In this process we fail to learn from the peasant's knowledge which has contributed to the creation of the genetic material. Rather, *papas* which do not possess these required attributes are redundant.

In this manner, scientists transform natural conditions with little consideration for nature's own rhythms and conditions. Their only interest is to classify the *papas*. Once the new and stronger *papa* has been created, its task will be to substitute the other traditional and now obsolete varieties of *papas*. In many Andean areas, several *papa* varieties grown by *runas* have been labeled as obsolete. They only survive because the Andean peasants continue to grow them with love and care. We can only blame the scientists for their contradictory behaviour: they pretend to be puzzled by the decrease and extinction of numerous species and varieties but it is the scientific approach which contributed to species extinction. It is also why the new ethic of biodiversity is not convincing.

The Andean way of life is in harmony with rhythms and cycles of nature and the activities carried out in the *chacra* (small farm) by the *chacareros* (farmers). It does not violate nature. Reproduction of species and varieties is not seen as the search of miracle varieties. Reproduction, breeding and cultivation follow the rhythm of the *chacra* and its inner system of life. Nor is it an 'isolated activity' geared towards scientific discovery.

Peasant agriculture tends to reproduce certain species, creating at the same time crops with specific forms, with the potential to develop other species under certain circumstances. In the Andes, the creation of new crops and varieties, in contrast to modern science does not involve the redundancy of the varieties already existing. They, in fact, contribute to the creation of new ones. Therefore, the new varieties are not seen as better or worse than the previous ones. Peasants take care of them as if they were new members of the family whose life comes from *Apus* or *Pachamama*.

The new crops are grouped together in families. Thus crops are made up of variegated groups so that the new species can easily harmonize with all the other members of the family. The new varieties coexist and do not substitute the others.

Breeding and cultivation

In the Andes it is common to hear *alpaca* breeders say, as stated above:

> 'Just in the same way we breed alpacas, they bring us up.' In the ceremonies for celebrating the first produce which are displayed in the so called *Fiesta de la Mamacha Candelaria, aymaras* from Conima, Puno, the wives of the local authorities receive the very first *papa* tubers out of the furrows and put them together with the *papas* of previous rites. The Andean 'priest', playing the role of the *papa* itself, then addresses the new *papas* with the following words: "In the same way we have brought up these people, now it is your turn to bring them up." (Chambi y Chambi, 1995. p. 60)

In the report by Mauricio Alvarez from Paucartambo, Cusco, as it was told to Luis Perez Baca from Centro de Servicios Agropecuarios (CESA):

> 'Turnip is a mother which brings us up during times of food shortage when harvests rot in the fields because of the *jatun poccy* (rainy season). Together with the *arac papa* turnips are the only edible goods which human beings could feed themselves on. During the *Fiesta de la Virgen Purificación* the so called *Mamacha Nabo,* a maiden with a yellow dress, or *yuyu,* dances and sings a sad song before leaving the community. In this song she tells the crops that their mission is accomplished, since they saved people from starvation. And now it is their turn to take care of the *Kcara huicsas* (turnips). That is why from this day onward, turnips lose their flavour, because there is other produce humans will use to feed themselves. With the new crop, we will have a new type of food. When we are nearing carnaval everybody gathers new produce *(Chaccos)* of the *chacra*. We celebrate this by kissing these new fruits and by kindly thanking *Pachamama* for the *miccuna* (food) she supplies us with.' (Perez Baca, 1994)

These are but few examples which help further explain that in the Andes breeding and cultivation are not just privileged activity of people but of every 'form of life'. To further clarify what we mean when we refer to breeding we will tackle the example of a 'conversation held among equal beings'. From a western point of view, we see language as peculiar to human beings and to some species of animals—yet, stones and stars do not have speech ability. If someone were to say 'the river is talking to me' or 'the river is creating a poem', most probably we would think that this person is speaking metaphorically. But, in the Andes this is not so.

From an Andean point of view, each form of life is provided with a voice of its own, and expresses itself through *señas* or *lomasas* (signs).

Given the diversity of human communities, of *chacras* and *microclimas,* no conversations are alike. Everything differs from everything else. The variety of plants grown by peasants is accounted for by their long and loving conversations with the plants.

To be able to understand the 'sign' of a plant means being able to have a conversation with it, to let oneself 'be brought up' by the plant. In turn, being brought up is being capable of 'feeling' the moment when any form of life is talking to you, namely letting the forms of life bring you up. Those who let themselves be brought up are those who know that they are incomplete. If not so, humanity would impose themselves on nature. This is what occurs with scientific cultivation, horticulture and animal husbandry. People impose themselves on plants or animals to such an extent that animals and plants lose all their characteristics. In the West humans have lost the ability of letting nature bring them up. They just impose on nature without ever engaging her in conversation.

In the Andean conception of the world the notion of 'ancestry' for plants and animals does not exist in the way agronomists understand it. Indian corn and *papas* have other *mamas, mamatas* or *maman* (mothers) which are other sorts of corn with specific characteristics. For Andean people, the 'mother' of corn participates in corn reproduction, in the sense that she lets this happen. To them, corn does not derive from any previous biological 'species', but from the corn *maman*. As long as *maman* exists there will always be corn. Therefore Andean people are not only growing corn, but 'corn families' and their *mamas*.

Aymaras breeders say: 'The day in which there will no longer be *alpacas* the world will disappear.' *Alpacas* were given by *Apus —cerros tutelares* (guardian hills)—to *ayllus* (Indian communities) so that they may breed them. If *aymaras* do not breed them, *alpacas* go back to *puqyos* (the place where they came from), and the end of the world has come. Breeding is at the core of biodiversity in *Pacha*.

Centrality of nature

Farmers refer to *chacras* as the setting in which every form of life is bred and grown. *Sallqa* (nature) which comprises meadows, mountains, birds, deer and stag, *vicunas,* wild *papas,* is not perceived by Andean people as something which is wild and untamed, but rather as a *chacra* bred by *Apus.* The word *Sallqua* in Peruvian may be translated into 'wildlife' in English, the contrary of what is tamed. Such a dichotomy, however, does not make sense when it is applied to the Andean world.

The human community helps *Apus* in breeding and cultivating 'nature' through the activities carried out in a *chacco.* The *chacco* is a sort of wild livestock roundup, including animals such as *vicunas,* partridges and foxes, which are rounded together in order for people to collect and hunt all the 'produce' nature provides. But for Andean people it represents also another form of ritual breeding through which *sallqa,* after a formal request to the *Apus,* is 'pruned' in order to encourage each form of life to bloom again.

By convention, this type of activity is considered to precede any form of agriculture. But in the Andes cultivation goes hand in hand with this hunting and gathering activity. *Chacra* and *chacco* are two complementary ways of breeding different forms of life. In this regard, *chacra* is but another way of raising the varied forms of life in nature. It is not a human invention, but goes with all that was and is done by the *Apas;* the breeding of *sallqua* increases in certain areas and it involves all those elements which 'demand' a different treatment. The *chacra* can be either inside or outside *sallqua* and in specific areas, since not all plants and areas need a *chacra. Chacra* has its own area. Whereas in the *chacra* humans take care of breeding, in *sallqua* this is a privilege of the *Apus.*

Humans are brought up both in the *chacro* and *chaco.* As biodiversity applies to all forms of life, people are also different from each other. Each ethnic group distinguishes itself from the others through their customs. Those from Quispillacta, Ayacucho, say that their customs 'come from nature'. Humans are connected to their place of origin, which accounts for the differences between people.

When humans create a *chacra* they do the same as the *Apus.* Agriculture is one of the forms of life. Foxes, which belong to nature, have their own *chacras* of geese and *papas,* which are called *atoq oca* and *atoq papa,* which science considers as wild forms of *papas* and geese. *Vicunas* like foxes are raised in the *Apus' chacra.* In the humans' chacra *runa* do not limit themselves to grow plants. Peasants use to say: 'the llama is my *chacra'.* They talk about *chacra de oro, chacra de sal, chacra de sachas* and so forth. Therefore, *chacra* is not only the place in which agriculture takes place, but it is any place of breeding, since different forms of life talk to one other in order to contribute to the *Pacha* reproduction—as the *chacra* as such presupposes *runas'* engagement, it belongs to everything and is for everything.

The *ayllu* (Indian community)

As the peasants would say, not all *papas* can take root in all ground, nor can every plant mix with all plants, nor a *runa* become accustomed to all places. The *chacra* is the expression of empathies, of harmony among different forms of life which respect one another and give one another shelter. The community in which all forms of life live together in harmony with the sole aim to reproduce life is called *ayllu.*

The *ayllu* is not the community of blood-related human beings. Andean people refer to Apus as their grand-parents, to *Pachamama* as mother. Corn species are considered as their offspring, the water of certain springs in some rites is addressed as a son-in-law. In addition, any corn species has its own mother, the so called *Saramama, papas* have their *Papamama,* rivers their own *Yacumama* and their *pacha.* Peasants consider corn and beans as if they were brothers since they grow without 'sticking together'. And in other cases, they say 'La papa y quínea son familias porque no les afecta las plagas' (*Papas* and *quíneas* are 'families' because they are not infested by blights).

Each form of life is limitless in an *ayllu* (be it a *runa, waca* or *sallqa*). *Runas* contain the *waca* and the *sallqua* forms of life within themselves, as well as in each *Apus* dwell the forms of *sallqa* and *runa.* This explains why in certain rites *runas* dress up as corn or condors. To them, in that moment, those disguised represent the corn itself or the condor itself. Performance differs from the West where nature is an observed object.

Runas are performed as if they were 'animals themselves'. (Rengifo et al., 1993, p. 37) When the ceremony is over, the so called form of a *runa* prevails again. The same happens when the *alpacas* mate. It does not mean that male and female *alpacas* are performing the mate following *runa's* customs, nor does it means that the *runas* are playing a part, it is a human performance, it does not mean they are humanizing or embodying the *alpacas*. It just means that the *alpacas,* as any other form of life, mate when it is deemed to be the right time in a livestock cycle. In the *ayllu* everybody has the same attributes: humans are not the measure of all things.

This explains how natural it is to a runa to talk to a plant or to *alpacas*. Within the *runas* themselves there are other forms of life. That is why in the Andes the conception of *runa* cannot be absolutely compared to the same conception of human being for western societies. The pre-condition on which a western society is grounded was that any sense of nature is expelled from the concept of a human being. The Aristotelian concept of 'man' as rational animal' represents a way to distinguish and, to a certain extent, elevate humanity relationship to nature. Modern society was built on such a conception, humanity being created in order to dominate nature.

In the case of *runas,* the viewpoint is different. Mutual breeding with plants, animals, and *wacas* is one of the most natural things. *Runas'* behaviour and conduct are not guided by the utilitarian conception of what is useful to the *chacra*. Andean reciprocity is the 'pleasure of giving and breeding with love'. It is not an obligation within a given framework. Breeding is not an action based an domination, give and take. In the *ayllu* everybody breeds and at the same time is brought up.

Translation by Elisabetta Vaccaro.

Grimaldo Rengifo Vásquez works at PRATEC, Lima, Peru. This free translation is based on a longer article written in Spanish.

Notes

* This English version of the article is a greatly reduced text of the original Spanish. It is also what one must call a 'free translation'. It was difficult to capture the exact meaning of Peruvian and local Andean dialect within the rigidity of the English language particularly when discussing such holistic concepts presented by the author. Spanish readers are encouraged to contact the author in order to obtain a copy of the original version.

1. With 'forms of life' we refer to the single expression which characterizes a living being and differentiates but does not oppose it to any other form of life, so that one can easily tell the difference between, say a *llama* and an *alpaca,* but this difference completes them. We are not referring here to external as opposing an internal figure. The expression 'form of life' implies/includes internal and external elements.

2. Robin Pellew director of WWF-RV states that 'A global assessment of diversity, coordinated by PNUMA and carried out by 1,500 scientists, found that biological elements of the Earth are in serious danger. The evolution of biodiversity—the multitude of genes, species and ecosystems that collectively form what we call nature—despite its 4,000 years of existence, is doomed to be destroyed, in the most part, within four generations . . .' (Robin, 1996, p. 36).

References

Altieri, M., 'La biodiversidad en Agroecosistemas', *II Curso de Educación a distancia: Agroecologia y desarrollo rural II,* Cied, Lima, Clades, (Agosto, 1995).

Chambi, Pacoricona, N. y Chambi, Pacoricona W., *Ayllu y papas. Cosmóvision, religiosidad y agricultura en Conima,* Puno, Lima, Chuyma Aru, (diciembre, 1995).

Chambi Pacoricona, N. et al., *La crianza de las yoqchas en los distritos de Conima, Tilali e llave,* Puno, Asociación Chuyma se Apoyo Rural, (1996).

Greslou, F., 'Manejos Campesinos de los Recursos Fitogenéticos Andinos', *Manejo Campesino de Semillas en los Andes,* Lima, PPEA, PRATEC, (1989).

IUCN, PNUMA, WWF, 'Cuidar la tierra: estrategia para el futuro de la vida', Suiza, Gland, (1991)

Maturana, H. and Varela, F., *De máquinas y seres vivos,* Segunda Edición, Santiago de Chile, Edit. Universitaria, (1995).

Pellew, R., 'La evaluación mundial de la diversidad biológica', *Nuestro Planeta,* Tomo 7, no. 5, (1996).

Pérez Baca, L., *Crianza de la papa en la localidad de Paucartambo,* Qosqo, CESA, (1996).

Rengifo et. al., *Chacras y chacareros. Ecología, demografía y sistemas de cultivo,* Tarapoto, Cedisa, Fondo de Contravalor Perú-Canadá, (1993).

PLANT MANAGEMENT AMONG THE NAHUA AND THE MIXTEC IN THE BALSAS RIVER BASIN, MEXICO:

An Ethnobotanical Approach to the Study of Plant Domestication

ALEJANDRO CASAS,[1] MARÍA DEL CARMEN VÁZQUEZ,[1] JUAN LUIS VIVEROS,[1] AND JAVIER CABALLERO[1]

Different forms of management of wild, weedy, and domesticated plants carried out by the Nahua and the Mixtec in the Balsas River Basin, Mexico, are described. Along with cultivation of domesticated plants, these forms of plant management include gathering from wild populations; in situ *tolerance of plant individuals during clearings of natural vegetation;* in situ *enhancement and protection of particular plants among populations of some species; as well as sowing or planting of propagules and transplantation of complete individuals of weedy and wild plants in controlled* ex situ *environments. Processes of artificial selection and possible routes of domestication occurring in these forms of plant management are discussed.*

Key Words: ethnobotany; plant management; domestication; Nahua; Mixtec.

Introduction

The cultural area known as Mesoamerica, in southern Mexico and Central America, has been considered one of the most important centers of plant domestication in the world (Harlan, 1975; Hawkes, 1983; Vavilov, 1926). This can be explained by the co-existence of two factors: a great natural biodiversity and a long social and cultural history. Toledo (1988) estimates that approximately 30,000 vascular plant species exist in Mexico. This country is therefore considered one of the richest floristic regions of the world (Rzedowski, 1978). In addition, there are 56 indigenous ethnic groups living today in the Mexican territory (Toledo, 1988) whose subsistence is based on the use and management of no fewer than 7000 plant and animal useful species (Cases *et al.*, 1994).

Ethnobotanical studies such as those of Caballero and Mapes (1985), Casas *et al.* (1994), Vázquez (1986), and Zizumbo and Colunga (1982), among others, have shown that at present a large segment of the indigenous Mexican population subsists by cultivating corn, beans, chili, squash, and other crops, and through gathering a wide number of plant resources. According to Caballero (1984), indigenous people in Mexico use and manage more than 5000 species of wild, weedy, and cultivated plants. Indigenous people maintain different forms of interaction with all these plants. Such interactions are a valuable source of information for the study of the evolution of plants under domestication.

Domestication is an evolutionary process, in which human intervention through artificial selection operates as the principal evolutionary force. This process may cause changes in morphology, physiology, phenology, life cycles, or reproductive systems of plants. All these changes are designed and regulated according to the different requirements of human cultures.

In order to understand the process of domestication of plants, it is necessary to analyze both biological variation in plant populations as well as social and cultural aspects involved in the use and management of this variation. Disciplines such as population genetics, ecology, and systematics may generate important information for the analysis of genotypic and phenotypic variation caused by domestication in plant populations, while disciplines such as ethnobotany and human ecology, which analyze the interactions and interrelationships between humans and plants, make possible the study of the different forms of manipulation of plants. Thus, ethnobotany and human ecology may make significant contributions to the understanding of how artificial selection is occurring, and its consequences for the evolution of plants. In addition, information generated in these areas may be helpful in explaining how domestication occurred in the past.

When the process of plant domestication is analyzed, a problem that commonly arises is how to determine if the plant populations under study are wild or domesticated. Also, when a domesticated plant is studied it is difficult to determine the degree of domestication (incipient or advanced) the plant has. This issue becomes more problematic when populations of a given species are distributed in a continuum from natural wild sites to man-made habitats. In other cases, relatives of cultivated plants may be either wild plants, weeds, or hybrids between the crop and its sympatric wild or weedy relatives. The resolution of these problems may involve complex systematic and ecological studies. However, in the search for answers it may be helpful to consider the general characterization of wild, weedy, and domesticated plants proposed by De Wet and Harlan (1975). These authors classified plants according to their ecological responses to different environments. Plants were classified as:

1. Wild plants, as plants growing naturally outside habitats disturbed by humans, and which cannot successfully invade habitats permanently disturbed by human beings (De Wet and Harlan, 1975).
2. Weeds, as plants growing in permanently human-disturbed habitats but which do not depend on human beings for the reproduction and survival (De Wet and Harlan, 1975). In this paper, we include under this category both true weeds, plants growing in agricultural fields and gardens, and ruderal plants, those growing along roadsides.
3. Domesticated, as plants growing in human-made habitats, and having a strong dependence on human beings to reproduce and survive. As a consequence of artificial selection, these plants usually present gigantism in the parts used by humans, a wide range of morphological variation, phenological changes, suppression of natural mechanisms for dispersion and protection, reduced competitivity, and other features of domestication (see De Wet and Harlan, 1975; Harlan, 1975; Hawkes, 1983; Schwanitz, 1967).

Evolutionary changes can be recognized as changes in phenotypic and genotypic frequencies in populations. Human interactions with plants may influence not only populations but also communities. In all these cases, the rates of evolutionary change in plants depend upon the nature of the relationships between them and humans. These forms of interaction determine the level of intensity at which allelic frequencies are manipulated and modified.

Based on studies of plant manipulation among Mesoamerican peoples such as those of Alcorn (1981), Bye (1985, 1993), Caballero (1994), Caballero and Mapes (1985), Casas (1992), Casas *et al.*, (1987, 1994), Colunga *et al.*, (1986), Davis and Bye (1982), Vázquez (1986, 1991), Williams (1985), and Zizumbo and Colunga (1982), it is possible to distinguish two main forms of interaction between human beings and plants: *in situ* and *ex situ* management.

In situ management involves interactions that take place in the spaces occupied by populations of weeds or wild plants. At this level, humans may take products from nature without significant perturbations, as in some forms of gathering. But also they may alter, consciously or unconsciously, the phenotypic or genotypic structure of plant populations in order to increase numbers of particular desirable plant phenotypes. The main *in situ* management strategies are:

1. Gathering, which is the taking of useful plants or plant parts directly from natural wild plant populations or weedy plant populations. Harvest of plant products from domesticated cultivated plants is not considered here

as gathering. Gathering is the most elementary form of interaction between humans and plants in order to obtain useful products. Under this interaction, manipulation of plant populations or communities is not always obvious. However, gatherers sometimes may carry out gathering strategies such as rotation of gathering areas in order to prevent the decrease or loss of some resources. Sometimes, gatherers also practice activities such as digging, clearing of vegetation around desirable plants, and cutting of branches or other parts of the plants whose products are gathered that may cause alterations in plant population structure (see Bye, 1985; Casas and Caballero, in press; Hallam, 1989). This could be also the case when there is massive collection of particular plants or plant parts or the selective collection of products of particular phenotypes, as practiced in some forms of gathering observed in this study. For these reasons, gathering may be considered a form of plant management.

2. Tolerance or sparing, including practices directed to spare, within human-made environments, useful plants that existed before the environments were transformed. Plant communities managed in this way may change strongly their original composition and structure. In some areas, this practice may favor an increase of number of individuals of a particular plant species or particular phenotypes. Examples of tolerance among edible weeds or "quelites" has been reported by several ethnobotanical studies in Mexico (Caballero and Mapes, 1985; Casas et al., 1987; Davis and Bye, 1982; Vázquez, 1986, 1991; Williams, 1985). Among studies of perennial weeds and wild plants those with *Opuntia* spp. conducted by Colunga *et al.* (1986), and with *Leucaena esculenta* (Moc. et Sessé ex A.DC.) Benth. by Casas (1992) are notable.
3. Enhancement consists of different strategies directed to increase the density of population of useful plants. This type of management includes the sowing of seeds or the intentional propagation of vegetative structures in the same places occupied by populations of wild or weedy plants. Steward (1938) recorded a good example of enhancement of useful plants *in situ* among the Paiute in California. These people used channels to inundate meadows aimed at increasing productivity of a wild cereal. Similar examples can be found in Schneider (1972), Bye (1985), Groube (1989), Shipek (1989), Hallam (1989), Hillman (1989), and Pearsall (1989).
4. Protection includes conscious care activities such as the elimination of competitors and predators, fertilization, pruning, protection against frosts, etc., to safeguard critical wild or weedy plants. Bye (1985) describes an example of this interaction in the management of wild onions practiced by the Tarahumara in the north of Mexico.

Ex situ management, on the other hand, includes interactions taking place outside spaces occupied by natural wild populations or weedy plant populations, in habitats created and controlled by man. It usually occurs with domesticated plants, but it is possible to find it also with wild, weedy, and ruderal plants. There are two main forms of *ex situ* management of weeds and wild plants:

1. Sowing or planting includes artificial propagation of sexual or vegetative reproductive structures taken from wild or weedy plant populations.
2. Transplantation of complete individuals taken from wild or weedy populations.

Practices such as enhancement and protection of plants as described above are included in these two forms of *ex situ* management.

This article describes different forms of management of plants by the Nahua and the Mixtec in the basin of the Balsas river in Mexico. These forms of plant management are discussed as examples of current routes of plant domestication.

Study Area

Ethnobotanical studies were carried out among the Nahua in two villages, Xochitepec and Mitepec of the Jolalpan municipality, in the southwest of the state of Puebla; and among the Mixtec in four villages, Alcozauca, Amapilca, Ixcuinatoyac, and San José Laguna of the Alcozauca municipality, in the northeastern region of the state of Guerrero (Fig. 1). The two municipalities are

Figure 1. Study area: Alcozauca, Guerrero and Jolalpan, Puebla in the Balsas river basin, Mexico.

Table 1. General Environmental Information of Alcozauca, Guerrero and Jolalpan, Puebla, Mexico

Site	Altitudinal range (m)	Annual mean temperature	Annual mean precipitation	Dry season	Rainy season	Vegetation
Alcozauca	1200–2900	20.4°C	846mm	Nov.–Apr.	May–Oct.	(1) Riparian vegetation (2) Tropical deciduous forest (3) Oak forest (4) Oak-pine forest
Jolalpan	800–1900	25.6°C	840mm	Oct.–Apr.	May–Oct.	(1) Riparian vegetation (2) Tropical deciduous forest

characterized by mountainous landscape, with deep slopes, numerous small intermontane valleys, and narrow fluvial plains. Such irregular topography is due to the location of these sites between two important geographical systems: the Balsas river depression in the north, and the Sierra Madre del Sur in the south. Table 1 summarizes environmental information about Alcozauca and Jolalpan.

Approximately 70% of the human population in Jolalpan are Nahua people, while in Alcozauca 90% of the people are Mixtec. In the two municipalities, indigenous subsistence depends basically on seasonal cultivation of corn, beans, and squashes. This is complemented by the raising of goats, cattle, pigs, and barnyard fowls in home gardens. Other important subsistence practices are gathering, hunting, fishing, and handicraft activities based on forest exploitation and annual migration to cities and agroindustrial centers in Mexico and the United States for seasonal employment.

Methods

The study was carried out in three main phases. First, inventories of plant resources were made in the two sites during 1 year of fieldwork. Specimens of useful plants were collected in natural vegetation, markets, crop fields, and home gardens. During the collecting, 40 Nahua and Mixtec *campesinos* provided ethnobotanical information about plant uses, forms of management, ethnoclassification, social and cultural role of plants, and some ecological data such as distribution, phenology, competitors, predators, etc. All this information was complemented with interviews and direct observations. Plant specimens were deposited in the National Herbarium of Mexico (MEXU).

The second phase comprised more specific interviews to obtain detailed information about the forms of management of the most important edible plants, and some evaluations of their role in campesino subsistence. This information was obtained during a second year of fieldwork. Details of these studies can be found in Casas *et al.* (1994) for the Mixtec region and Vázquez (1986) for the Nahua region.

Finally, in the third phase, particular studies were carried out with *Porophyllum ruderale* subsp. *macrocephalum* (A.DC.) R.R. Johnson, and *Leucaena esculenta* (Moc. et Sessé ex A.DC.) Benth., to analyze examples of phenotypic variation in wild, weedy, and cultivated plant populations managed in different ways. Studies of this phase were carried out within a period of 2 years of field and experimental work. Some results of these studies are summarized in different parts of this paper but details on methods and results can be found in Casas (1992), Casas and Caballero (in press) for the case of *L. esculenta*, and Vázquez (1991) for the case of *P. ruderale*.

Results

In the two sites studied, a total of 180 species of edible plants was registered. Sometimes the same species was observed under different ecological and cultural conditions. Table 2 shows the total number of the edible plant species observed in each condition. Wild, weedy, and domesticated plants are managed in different ways, as it is described below.

Ways of Management of Wild Plants

Gathering. Wild plants are mostly gathered in natural vegetation areas. Many of them, mainly annual plants, are gathered without special preferences among individuals. However, with some others, especially perennial plants, indigenous gatherers usually distinguish individuals that are different in features related to qualities of food, such as flavor, texture, size, color, presence, or absence of toxic substances, etc. Gatherers choose the best plants to use. For example, within populations of the "guamúchil" (*Pithecotlobium dulce* (Roxb.) Benth.), people differentiate between trees with sweet and bitter fruits and they collect the sweet ones.

A similar situation was found in the case of the "guaje" (*Leucaena esculenta*). The Mixtec distinguish, through trial and error and through the association of morphological characters, three types of trees: (1) the "guaje de vasca" ("vomitory guaje"), a toxic variety; (2) the "guaje amargo" ("bitter guaje"), with bitter seeds that are edible only after being roasted; and (3) the "guaje dulce" ("sweet guaje"), whose seeds are edible raw. People selectively gather the "sweet guajes," taking the "bitter guajes" only when the sweet are scarce (Casas, 1992; Casas and Caballero, in press).

The Nahua and the Mixtec also selectively gather fruit from trees such as *Psidium* spp., *Spondias mombin L.*, *Byrsonima crassifolia* (L) H.B.K., and other species mentioned in Table 3.

Toleranced or Sparing. Edible wild plant species and especially the preferred varieties, are tolerated when they are found during the opening of forest areas to cultivate corn. Those individuals whose edible parts are not preferred by people are eliminated. Table 3 indicates species of wild plants tolerated selectively by people in this manner. This form of management may have consequences on the phenotypic structure of populations, because some phenotypes are increased in frequency and others are eliminated. For example, in the case of *Leucaena esculenta,* Casas and Caballero (in press) found that this process has produced significant morphological differences between wild and tolerated populations. Thus, in tolerated populations, trees producing larger seeds and pods were observed to be more frequent than in wild populations.

Enhancement. The management of the palm *Brahea dulcis* (H.B.K.) Mart. by the Mixtec is an example of enhancement of wild plants. In this case, fruits (called "capulines") and "palmetto" are edible, although the main use of the plant is the manufacturing of handicrafts such as hats and traditional mats called "petates." This palm has a vegetative reproductive system and its young ramets are resis-

Table 2. Number of Edible Plant Species Within Different Ecological and Cultural Status

Ecological status	Cultural status					
	Gathered	Tolerated	Enhanced	Protected	Sowed	Transplanted
Wild	67	37	2	10	16	5
Weed	48	41	7	15	14	3
Domesticated	—	—	—	—	83	—

tant to fire. People use these characteristics to increase the numbers of this palm. They fell trees and shrubs and burn the area in order to eliminate competitors. A similar form of management is practiced to create artificial grasslands in order to increase the availability of some species used as forage for animals.

Protection. Campesinos occasionally prune and take actions against pests of the "sweet" phenotypes of *Pithecollobium dulce* (Roxb.) Benth. *Psidium* spp., *Leucaena esculenta,* and other species in wild or tolerated populations. These protected wild species are also shown in Table 3.

Sowing or Planting. Some wild plants are also cultivated. Their seeds are sowed mainly in home gardens. Most of them are woody plants such as *Leucaena esculenta,* wild avocado (*Persea americana* Mill.), and several fruit trees. Nevertheless, *Euphorbia graminea* Jacq. and *Allium glandulosum* Link. and Otto can also be mentioned as two relevant examples of cultivated herbaceous wild plants. The wild plants cultivated by sowing or planting observed in this research are listed in Table 3.

Transplantation. Examples of transplantation of wild plants were found in the agave "maguey mezcalero" (*Agave cupreata* Trel. et Berger) and in the guava "guayaba tlahuanca" (*Psidium guajava*). In the first case, seedlings of agave are transplanted by people from natural populations to corn fields, where they grow as living fences or as soil barriers in traditional terraces. In the latter case, young ramets of guavas are usually transplanted from wild individuals to home gardens. Peasants occasionally transplant seedlings of pine, oak, and *Bursera* species to the fences of corn fields.

Management Of Weeds

Forty-eight species of edible weeds were found growing within and around the crop fields (Table 4). Among them, there are some plants traditionally used as edible greens. These plants are the "quelites" that have been characterized by Bye (1981). The "quelites" are called *quilitl* among the Nahua, and *yiwa* among the Mixtec. Table 4 indicates the 24 "quelite" weedy species registered in this study. All of them are important for the diet of indigenous people. Thus, Viveros *et al.* (1993) report that the "quelites" constitute around 10% of the total annual food consumption in the municipality of Alcozauca.

Gathering and Tolerance or Sparing. All edible weeds reported in this study are gathered in agricultural fields and disturbed areas. However, 27 of them were

Table 3. Wild Edible Plants

Family	Species	Site[a]	Gathered	Tolerated	Enhanced	Protected	Sowed	Transplanted
Agavaceae	*Agave* sp.	A	X					
	Agave cupreata Trel. et Berger	A	X	X	X	X		X
Anacardiaceae	*Comocladia mollisima* H.B.K.	A	X					
	Cyrtocarpa procera H.B.K.	A,J	X	X				
	Spondias mombin L.	A,J	X	X			X	
Annonaceae	*Annona cherimola* Mill.	A	X	X			X	
	Annona purpurea Moc. et Sessé	J	X	X			X	
	Annona reticulata L.	J	X	X			X	
Asclepiadaceae	*Cynanchum jaliscanum*	A	X					
	Matelea crenata (Vail) Woodson	A	X					
Bignoniaceae	*Parementiera edulis* DC.	A	X	X			X	
Bombacaceae	*Ceiba parvifolia* Rose	A	X					
	Ceiba aesculifolia (Kunth) Britton and Baker	J	X					
Boraginaceae	*Cordia cylindrostachya* (Ruiz and Pav.) Roem and Shult	A	X					
	Cordia dentata Poiret							
Cactacea	*Opuntia* sp.	A	X					
	Stenocerus stellatus (Pfeiffer) Riccob.	A	X	X				X
Compositae	*Porophyllum ruderal* (Jacq) Cass.	A,J	X	X				
	Porophyllum nutans Rob. and Greenm.	A,J	X	X				
	Tagetes filifolia Lag	A	X					
	Tridax mexicana A. Powell	A	X	X				
Convolvulaceae	*Operculina* sp.	A	X					
	Exogonium bracteatum (Cav.) Choisy	J	X					
Euphorbiaceae	*Euphorbia graminea* Jacq.	A,J	X	X			X	
	Euphorbia macropus (Klotz and Gaerke) Boiss.	A	X					
Fagaceae	*Quercus glaucoides* Mart and Gal.	A	X	X				
	Quercus obtusata Humb. et Bonpl.	A	X	X				
	Quercus urbanii Trel.	A	X	X				
Gramineae	*Zea mays* L. subsp. mexicana (Schrad.) Iltis	A	X					
Juglandaceae	*Juglans mollis* Engelm.	A	X	X				
Lauraceae	*Litsez glaucescens* H. B. K.	A	X					
	Persea americana Mill.	A	X	X			X	
Leguminosae	*Acacia angustissisma* (Mill) Kuntze	A	X					

	Canavalia villosa Benth.	A	X					
	Desmodium sericophyllum Schldl.	A	X	X				
	Desmanthus virgatus (L.) Willd.	A	X	X				
	Eysenhardtia sp.	A	X	X			X	
	Gliricidia saepium (Jack.) Steud.	A	X	X				X
	Leucaena esculenta (Moc. et Sessé ex A.DC.) Benth. subsp. *esculenta*	A,J	X	X				X
	Leucaena esculenta (Moc. et Sessé ex A.DC.) Benth. subsp. *paniculata*	A,J	X	X			X	X
	Leucaena macrophylla Benth.	A,J	X	X				
	Pachyrrhyzus erosus (L.) Urban.	A	X	X				
	Phaseolus coccineus L. subsp. *formosus* (Kunth) Mare., Masch. and Stain.	A	X	X				
	Phaseolus vulgaris L. var. *mexicanus* A. Delgado	A	X	X				
	Nissolia sp.	J	X					
	Pithecollobium dulce (Roxb.) Benth.	A,J	X	X		X	X	
	Prosopis laevigata (Humb. and Bompl. ex Eilld.) M. D. Johnston	A	X	X		X		
	Sena occidentalis (L.) Link	A	X					
Liliaceae	*Allium glandulosum* Link. and Otto.	A	X	X				X
Malpighiaceae	*Bunchosia palmerii* Wats.	A,J	X					
	Byrsonima crassifolia (L.) H.B.K.	A,J	X	X		X	X	
	Malpighia glabra L.	J	X					
Malvaceae	*Malva parviflora* L.	A	X					
Moraceae	*Ficus cotinifolia* H.B.K.	A,J	X					
Myrtaceae	*Psidium guajava* L.	A,J	X	X		X	X	X
	Psidium sp.	A	X	X		X	X	X
	Psidium sartorianum (Berg.) Nied.	A	X	X		X	X	X
Palmae	*Brahea nitida* André	A	X					
	Brahea dulcis (H.B.K.) Mart.	A,J	X	X	X	X		
	Pseufophoenix sargentii Wend.	J	X	X		X		
Passifloriaceae	*Passiflora* sp.	A	X					
Phytolaccaceae	*Phytolacca icosandra* L.	A	X	X				
	Phytolacca octandra L.	A	X	X				
Rutaceae	*Casimiroa edulis* Llave and Lex.	A	X	X				
Sapotaceae	*Sideroxylon capiri* (DC.) Pittier	J	X					
Sterculiaceae	*Guazuma ulmifolia* Lam.	A	X					
Verbenaceae	*Vitex mollis* H.B.K.	A	X					

[a] = Alcozauca, Guerrero; J = Jolalpan, Puebla.

Table 4. Weeds and Ruderal Edible Plants

Family	Species	Site[a]	Gathered	Tolerated	Enhanced	Protected	Sowed	Transplanted
Amaranthaceae	*Amaranthus hybridus* L. (*)	A,J	X	X	X			
Asclepiadaceae	*Asclepias* sp. (*)	A	X	X				
Bignoniaceae	Parmentiera edulis DC.	A	X	X			X	
Boraginaceae	*Cordia cylindrostachya* (Ruiz and Pav.) Roem and Shult	A	X					
Chenopodiaceae	*Chenopodium ambrosioides* L. (*)	A	X	X	X	X		
	Chenopodium berlandieri Moq. (*)	A	X	X	X	X		
Compositae	*Galinsoga paviflora* Cav. (*)	A	X	X				
	Porophyllum ruderale (Jacq) Cass. (*)	A,J	X	X	X	X		
	Porophyllum nutans Rob. and Greenm. (*)	A,J	X	X				
	Porophyllum tagetoides D.C. (*)	A,J	X	X				
	Tagetes filifolia Lag.	A	X					
	Tridax mexicana (*)	A	X	X				
Cruciferae	*Brassica campestris* L. (*)	A	X	X				
Cucubitaceae	*Melothria pendula* L.	A	X	X				
	Sicius sp. (*)	A	X					
Labiatae	*Hyptis suaveolens* (L.) Poir	A	X	X				
	Lippia palmeri S. Watson	J	X					
Leguminosae	*Crotalaria pumila* Ortega (*)	A	X	X	X		X	
	Crotalaria longirostrata Hook. and Arn. (*)	J	X	X			X	
	Desmodium sericophyllum Schldl.	A	X	X				
	Desmanthus virgatus (L.) Wild.	A	X	X				
	Eysenhardtia sp. (*)	A	X	X				
	Gliricidia saepium (Jack.) Steud.	A	X	X				
	Leucaena esculenta (Moc. et Sessé ex A.DC.) Benth. subsp. *esculenta* (*)	A,J	X	X	X	X	X	X

	Phaseolus coccineus L. subsp. *formosus* (Kunth) Mare., Masch. and Stain.	A	X	X				
	Phaseolus vulgaris L. var. *mexicanus* A. Delgado	A	X	X				
	Pithecollobium dulce (Roxb.) Benth.	A,J	X	X		X	X	
	Prosopis laevigata (Humb. and Bonpl. ex Wild.) M. C. Johnston	A	X	X		X		
	Sena occidentalis (L.) Link	A	X					
Malpighiaceae	*Byrsonima crassifolia* (L.) H.B.K.	A,J	X	X		X	X	
Malvaceae	*Anoda cristata* (L.) Schl. (*)	A,J	X	X			X	
	Malva parviflora L. (*)	A	X					
Martyniaceae	*Martynia annua* L.	A	X					
Myrtaceae	*Psidium* guajave L.	A,J	X			X	X	X
	Psidium sp.	A	X	X		X	X	X
	Psidium sartorianum (Berg.) Nied.	A	X	X		X	X	
Nyctaginaceae	*Salpiathus purpurascens* (Cav. ex Lag) Hook & Arn. (*)	A	X	X				
Oxalidaceae	*Oxalis* sp. (*)	A	X	X				
Phytolaccaceae	*Phytolacca icosandra* L. (*)	A	X	X				
	Phytolacca octandra L. (*)	A	X	X				
Polygonaceae	*Rumex* sp. (*)	A	X	X				
Portulacaceae	*Portulaca oleracea* L. (*)	A,J	X	X	X			
Solanaceae	*Capsicum frutescens* *L. var. baccatum* Irish.	A	X	X	X	X	X	
	Lycopersicon esculentum Mill.	A	X	X		X		
	Physalis philadelphica Lam.	A,J	X	X	X	X	X	
	Solanum mitlense Dunal (*)	J	X	X		X		
	Solanum nigrum L. (*)	A,J	X	X		X		
Verbenaceae	*Vitex mollis* H.B.K.	A	X	X				

[a] = Alcozauca, Guerrero; J = Jolalpan, Puebla; (*) Quelites.

Table 5. Relative Abundance of Edible Weeds in Agricultural Systems of Alcozauca, Gro., Mexico (I = individuals; B = Biomass)

	Agricultural systems							
	Sh		S		I		Hg	
Species	I/m2	B/ha	I/m2	B/ha	I/m2	B/ha	I/m2	B/ha
Amaranthus hybridus	0.42	39.48	0.03	2.82	0.17	15.98	3.01	282.94
Anoda cristata	0.02	2.38	0.07	8.33	0.8	95.2	0.07	8.33
Crotalaria pumila	–	–	0.91	74.22	0.01	0.816	0.02	1.14
Oxalis sp.	0.42	0.42	0.93	0.93	0.06	0.06	–	–
Physalis philadelphica	0.015	13.095	0.04	34.92	0.12	104.76	0.57	497.61
Phytolaca icosandra	0.03	9.15	0.11	33.55	–	–	–	–
Porophyllum nutans	–	–	0.01	0.157	–	–	–	–
Porophyllum ruderale	–	–	0.01	0.163	–	–	–	–
Porophyllum tagetoides	–	–	0.03	0.426	–	–	0.01	0.142
Portulaca oleracea	–	–	0.02	0.46	0.09	2.07	0.2	4.6
Solanum nigrum	0.11	9.57	0.03	2.61	0.07	6.09	–	–

[a]Sh = shifting agriculture sowing with indigenous traditional stick; S = seasonal agriculture sowing with plow; I = irrigated agriculture sowing with plow; Hg = home gardens

plants that peasants tolerated during the weeding of corn fields. Table 5 shows the relative abundance of "quelite" species registered during a sampling of cornfields in Alcozauca after the weeding.

A certain selection is carried out on intraspecific forms during the gathering and weeding. For example, with *Anoda cristata* (L.) Schl. (Malvaceae), a herb called "alache," people distinguish between two main variants: (1) the "alache macho" ("male alache"), which have slender and pubescent leaves, with a high fiber content, and which is not palatable; and (2) the "alache hembra" ("female alache") which has broader and nonpubescent leaves, a lower fiber content and a good flavor. People gather young individuals or stems and leaves of the "female alache." When individuals are removed, the populations are directly thinned, and with such an action it could be expected that populations of this variety might be easily displaced by those of the "male alache" and eventually disappear. However, this does not occur because not all individuals of the "female alache" populations are removed and, in addition, many individuals of the "male alache" populations are eliminated during weedings.

A similar situation occurs with *Crotalaria pumila* Ortega (Leguminosae), which is called "chipile," in which people distinguish the "chipile macho" ("male chipile") which have small leaves and a bad taste, and the "chipile hembra" ("female chipile") with bigger leaves and good taste. Also in the case of *Porophyllum ruderale* subsp. *macrocephalum* (Asteracea), the "papaloquelite" or "quelite de mariposa" ("butterfly quelite"), two main variants are distinguished: (1) the "papaloquelite hediondo" ("stinking papaloquelite"), which have a stronger flavor and bad odor, and (2) the common "papaloquelite," which have good flavor and odor. In these cases, the desirable variants are gathered and the nondesirable variants are eliminated during weedings (Vázquez 1991).

Enhancement. Peasants disperse seeds of the desirable weeds in order to increase their population density. These practices are ways of enhancement of weedy plants. Examples of these interactions were observed in seven species in Alcozauca (Table 4), being especially important in *Porophyllum ruderale* subsp. *macrocephallum*, *Amaranthus hybridus* L., *Anoda cristata*, *Crotalaria pumila*, and *Physalis phildelphica* Lam. Also, in Table 5 it can be observed that *Amaranthus hybridus* shows a particularly high relative abundance in home gardens because people intentionally disperse seeds of this plant in the field.

Protection. Physalis philadelphica and the weedy form of *Lycopersicon esculentum* Mill. are not only tolerated and enhanced; they are also fertilized, and protected against frost and pests. It is important to note that these practices usually occur on cultivated plants and only rarely on weeds and wild plants.

Sowing or Planting. Seeds of some weeds are collected to cultivate them in the crop fields. This occurs, for instance, with the species *Anoda cristata; Crotalaria pumila, Brassica campestris* L., *Amaranthus hybridus*, and *Porophyllum ruderale* subsp. *macrocephallum*. All these species are sold commercially in the traditional markets, mainly during the dry season.

Ways of Management of Domesticated Plants

The Mixtec and the Nahua know, use, and manage many native or introduced plants in advanced degree of domestication (Table 6). These species are cultivated and receive attention from sowing to harvest. Based on variation generated by crossing between domesticated and wild, weeds and other domesticated plants, people carry out consecutive selection, generation after generation. This selection directs the enhancement of phenotypic features that are favorable to the changing cultural and environmental conditions. Such processes exemplify domestication as a continuous process occurring even on already domesticated plants.

For example, in the municipality of Alcozauca there are 12 local variants of corn, each with some specific characteristics selected according to different utilitarian purposes (Casas et al., 1994). Variants with the sweetest flavor are consumed as "elotes" (green corn). Red colored corns are used to make "tortillas," "atole" (corn beverage), and "pinole" (roasted corn flour with sugar and cinnamon) for ceremonial practices, while black and white varieties are valued for the appearance given to the "tortillas." Seed size is selected also as a characteristic associated with higher yields. In a similar way, cob size and number of seed rows are selected to improve productivity. Furthermore, seed size is associated with specific uses; for example the biggest seeds are destined to prepare "pozole" (corn grains and meat soup). Seed hardness is selected to improve resistance to storage losses. However, there are also some soft variants selected for their faster cooking and easiness to mill, that are consumed soon after harvesting. Easiness to shell is also selected. Length of life cycle, phenology, drought resistance, and pest and disease resistance are selected according to different environmental conditions.

Year after year, successive generations of corn crops are submitted to artificial selection pressures. Such a process is modulated by social and cultural changes, for example, changes in customs, in patterns of diet, etc. Social and cultural changes are determined strongly by seasonal migration of people to other parts of Mexico and the U.S. for employment. Also, the increasing of marketing and the introduction of mass media of communication such as T.V. and radio are important (Viveros et al., 1993). In addition, artificial selection pressures are also modulated by technological and environmental changes such as changes in agricultural systems and the impoverishing of soils. A good example of this interaction can be found in the changes brought about with the use of chemical fertilizers to produce corn in Alcozauca. In this municipality, the campesinos began to use chemical fertilizers in 1974. Since then, they have selected seeds of corn variants that grow better and produce more under such technological innovation. As a consequence, many peasants today agree that corn production is impossible without chemical fertilizers.

Beans also present different routes of selection by people. For example, in Alcozauca there are 14 variants of *Phaseolus vulgaris* L., and four of *P. coccineus* L. (Casas et al., 1994). Peasants select on length of life cycle to ensure availability of immature and mature beans at different times. Vine growing beans (the "frijoles enredadores") are selected to be cultivated together with corn in the traditional corn fields called "milpas," while the bushy beans (the "frijoles mateados") are selected to be cultivated in monospecific fields. These two characteristics, as well as drought, pest, and disease resistance, are selected to exploit different environmental conditions. A reduced content of fiber in pods is selected in those variants that are eaten as green beans (the "ejoteros" beans). As with corn, peasants select beans with hard seeds to resist storage pests, but they also select beans with soft seeds that are easily cooked.

It is important to mention that in the two study areas, there are wild and weedy plant populations of *Phaseolus coccineus* subsp. *formosus* that are usually close to agricultural fields located between 1700 and 2500 meters above sea level. In theory, this situation makes possible a continuous exchange of genes between wild, weed, and cultivated varieties, which are outbreeders. This is an extraordinary source of variation which has not yet been evaluated. However, according to information from peasants, these crosses may occur frequently, because it is common to find "frijoles quemados" ("burned beans"), which are possibly hybrids between cultivated and wild beans.

A similar situation may be occurring between wild and cultivated trees such as *Psidium guajava*, *Byrosonima crassifolia* (L.) H.B.K., *Annona cherimola* Mill., *Spondias mombin* L., *Persea americana*, and *Leucaena esculenta* and between other weedy or wild plants and cultivated plants such as *Lycopersicon esculentum*, *Pachyrrhyzus erosus*, and *Physalis philadelphica*. In all these cases, the exchange of alleles might make possible, at least hypothetically, the existence of a wide morphologica and genetic variation providing a rich raw material for the operation of artificial selection.

Table 6. Domesticated Edible Plants

Family	Species	Site[a]	Native species	Exotic species
Amaranthaceae	*Amaranthus hypocondriacus* L.	A		X
Anacardiaceae	*Magnifera indica* L.	A,J		X
	Spondias mombin L.	A,J	X	
	Spondias purpurea L.	A		X
	Spondias sp.	A		X
Ammonaceae	*Annona squamosa* L.	A		X
	Annona cherimola Mill.	A	X	
	Annona diversifolia Saff.	A		X
	Annona muricata L.	A		X
	Annona purpurea Moc. et Sessé	J		X
	Annona reticulata L.	J		X
Cactaceae	*Opuntia ficus-indica* (L.) Mill	A		X
	Optuntia sp.	A		X
	Stenocereus stellatus (Pfeiffer) Riccobono	A	X	
Caricaceae	*Carica papaya*	A		X
Chenopodiaceae	*Chenopodium berlandieri* Moq.	A		X
Combretaceae	*Terminalia catappa* L.	A		X
Compositae	*Porophyllum ruderale* (Jacq) Cass.	A,J	X	
Convolvulceae	*Ipomoea batatas* Lam.	A		X
Cruciferae	*Brassica oleracea* L.	A		X
	Raphanus sativus L.	A		X
Cucurbitaceae	*Cucumis melo* L.	A		X
	Cucumis sativus L.	A		X
	Cucurbita ficifolia Bouche	A		X
	Cucurbita mixta Pang.	A		X
	Cucurbita moschata Duch.	A		X
	Cucurbita pepo L.	A		X
	Citrullus vulgaris Sch.	A		X
	Sechium edule Sw.	A		X
Ebenaceae	*Dyospiros digyna* Jacq.	A		X
Gramineae	*Saccarum officinarum* L.	A		X
	Sorghum vulgare Pers.	A		X
	Zea mays L.	A,J	X	X
Juglandaceae	*Juglans regia* L.	A		X
Labitae	*Hyptis suaveolens* (L.) Poir	A	X	X
Lauraceae	*Persea americana* Mill.	A	X	X
Leguminosae	*Arachis hypogaea* L.	A		X
	Erythrina americana Mill.	A		X
	Leucaena esculenta (Moc. et Sessé ex A.DC.) Benth. subsp. *esculenta*	A,J	X	X
	Leucaena leucocephala (Lam.) de Wit. subsp. *leucocephala*	A,J		X
	Leucaena leucocephala (Lam.) de Wit. subsp. *glabrata* Zárate	A	X	
	Pachyrrhyzus erosus (L.) Urban.	A	X	X
	Phaseolus coccineus L.	A	X	X

Table 6. Domesticated Edible Plants

Family	Species	Site[a]	Native species	Exotic species
	Phaseolus vulgaris L.	A,J	X	X
	Pithecollubium dulce (Roxb.) Benth.	A,J	X	
	Tamarindus indica L.	A,J		X
Liliaceae	*Allium cepa* L.	A		X
	Allium sativum	L.	A	X
Malpighiaceae	*Byrsonima crassifolia* (L.) H.B.K.	A,J	X	
Moraceae	*Ficus carica* L.	A		X
	Morus sp.	A		X
Musaceae	*Musa paradisiaca* L.	A		X
Myrtaceae	*Psidium guajava* L.	A,J	X	X
	Psidium sp.	A	X	
	Psidium sartorianum (Berg.) Nied.	A	X	X
Oleaceae	*Olea europaea* L.	A		X
Palmae	*Phoenix datylifera* L.	A		X
Passifloraceae	*Passiflora choconiana*	A,J		X
Piperaceae	*Piper auritum* H.B.K.	A,J		X
Punicaceae	*Punica granatum* L.	A		X
Rosaceae	*Crategus pubescens* (H.B.K.) Steud.	A		X
	Cydonia oblonga Mill.	A		X
	Eriobotrya japonica Lind.	A		X
	Malus pumila Mill.	A		X
	Prenus persica L.	A		X
Rubiaceae	*Coffea arabica* L.	A		X
Rutaceae	*Casimiroa edulis* Llave and Lex.	A	X	
	Citrus aurantifolia (Chirst) Swingle	A		X
	Citrus limetta Risso	A		X
	Citrus maxima (Brum) Merr.	A		X
	Citrus novilis Lour.	A		X
	Citrus auriantum L.	A,J		X
	Citrus sinesis Osbeck.	A		X
Sapotaceae	*Pouteria sapota* (*Jacq.*) H. B. Moore and Stearn	A,J		X
Solanaceae	*Capsicum annum* L. var. *acuminatum* Fring.	A	X	
	Capsicum annum L. var. *grossum* Sendt.	A		X
	Capsicum annum L. var. *longum* Sendt.	A		X
	Capsicum frutescens L. var. *baccatum* Irish.	A	X	X
	Lycopersicon esculentum Mill.	A		X
	Lycopersicon esculentum Mill var. *piriforme* Hort.	A		X
	Physalis philadelphica Lam.	A,J	X	X
Umbelliferae	*Apium leptophyllum* (Pers.) F. Muell.	A		X
Vitaceae	*Vitis vinifera* L.	A		X

[a] = Alcozauca, Guerrero; J = Jolalpan, Puebla.

Discussion

Cases of plant management analyzed in the present study reveal the existence of different forms and intensities in which the Mixtec and Nahua campesinos manage and select their plant resources. Gathering and other ways of management of weeds and wild plants are practices that have developed along with agriculture, and that at present are integrated in subsistence patterns based on strategies of diversified ways of use and management of natural resources. These subsistence patterns are common among Mexican indigenous peoples (Viveros et al., 1993).

Changes determined by artificial selection on already domesticated plants are examples of domestication as a continuous process. This process occurs not only on plants such as corn and beans, which traditionally have been the food products for the indigenous subsistence, but also on plants introduced from the Old World after the conquest.

It has been mentioned that artificial selection is the crucial evolutionary force in domestication. This kind of selection basically implies that human beings enhance desirable phenotypes and eliminate the undesirable ones in a plant population. Artificial selection has usually been associated with cultivation. Under cultivation, manipulation of successive generations of plants makes evident both artificial selection and its results. Equally, plant domestication has been considered a process intimately associated with sowing or planting. For example, De Wet (1992) and Harlan (1992), from studies on evolution of cereals, state that domestication is a natural evolutionary process resulting from selection pressures associated with sowing and harvesting occurring in successive generations. However, in cases presented here, mechanisms of artificial selection seem to be present in wild and weedy plant populations under the described *in situ*

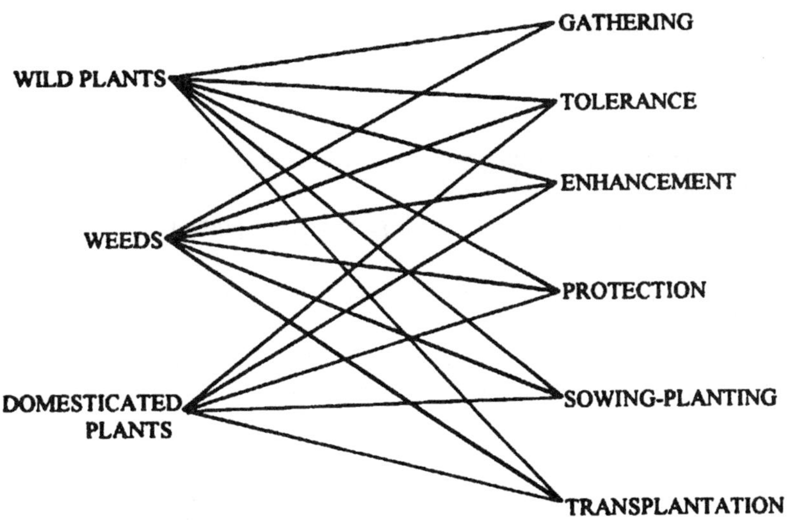

Figure 2. Forms of management of edible wild, weedy, and domesticated plants among the Nahua and the Mixtec. Possible routes of plant domestication.

and *ex situ* forms of plant management, although results are not so evident as in true cultivation. This observation suggests that evolution of plants, as directed by human being, may be a consequence of not only sequences of planting and selection, but also of a wide spectrum of human-plant interactions.

In Tables 3, 4 and 6, it is possible to appreciate that in some species different forms of management are acting at the same time. This situation suggests that domestication is not a linear sequence of forms of management but may take different routes.

Characteristics of interactions between humans and plants are determined by biological features of plants and their ecological conditions (wild, weedy, or domesticated plant). However, the cultural context also plays a determinant role. According to Hernández-X & Ramos (1977), "Ethnic groups have maintained much plant species in different domestication levels according to the uses and achievements of the desired features. . . ."

The types of plants managed and the ease of management, the requirements satisfied by those plants, the morphologic features of plants used by humans and the ease of artificial selection, as well the preferences for particular colors, flavors, odors, shapes, or textures, are all aspects modulated by culture and therefore influence the degree of intensity of the human-plant interaction. Changes affecting human culture may also influence domestication trends through time.

Acknowledgments

This research was supported by the Programa de Aprovechamiento Integral de los Recursos Naturales en la Montaña de Guerrero (P.A.I.R.-Facultad de Ciencias, U.N.A.M.) and the Jardín Botánico, Instituto de Biología, U.N.A.M. We also thank Dr. Robert Bye, Dr. Efraim Hernández-X., Dr. Barbara Pickersgill, MSc. Miguel Angel Martínez-Alfaro, and MSc. Cristina Mapes for their critical comments to the manuscript.

References

Alcorn, J. B. (1981). Huastec noncrop resource management: Implications for prehistoric rain forest management. *Human Ecology* 9(4): 395–417.

Bye, R. A., Jr. (1981). Quelites—ethnoecology of edible greens—past, present and future. *Journal of Ethnobiology* 1:109–123.

Bye. R. A.. Jr. (1985). Botanical perspectives of Ethnobotany of the greater southwest. *Economic Botany* 39(4):375–386.

Bye, R. A.. Jr. (1993). The role of humans in the diversification of plants in Mexico. In Rammamoorty, T. P., Bye, R., Lot, A., and Fa, J (eds.), *Biological Diversity of Mexico.* Oxford University Press, New York, Oxford, pp. 707–731.

Caballero, J. (1984). Recursos comestibles potenciales. In Reyna, T. T. (ed.), *Seminario Sobre la Alimentación en México,* Instituto de Geografia, Universidad Nacional Autónoma de México, México, pp. 114–125.

Caballero, J. (1994). La dimension culturelle de la diversité végétale au Mexique. *Journal d'Agriculture Traditionelle et de Botanique Appliqué, Novelle Série* 36(2): 145–158.

Caballero, J., and Mapes, C. (1985). Gathering and subsistence patterns among the P'urhepecha Indians of Mexico. *Journal of Ethnobiology* 5(1): 31–47.

Casas, A. (1992). Etnobotánica y Procesos de Domesticación en *Leucaena Esculenta* (Moc. et Sessé ex A. DC.) Benth. MSc. dissertation, Facultad de Ciencias, Universidad Nacional Autónoma de México, México, 233 p.

Casas, A., and Caballero, J. (in press). Traditional management and morphological variation in Leucaena esculenta (Moc. et Sessé ex A. DC.) Benth. (Leguminosae: Mimosoideae) in the Mixtec region of Guerrero, Mexico. *Economic Botany.*

Casas, A., Viveros, J. L., Katz, E., and Caballero, Y. J. (1987). Las plantas en la alimentación mixteca: Una aproximación etnobotánica. *América Indigena* 47(2): 317–343.

Casas, A., Viveros, J. L., Katz, E., and Caballero, Y. J. (1994). *Etnobotanica Mixteca: Sociedad, Cultura y Recursos Naturales en la Montana de Guerrero.* Instituto Nacional Indigenista-Consejo Nacional Para la Cultura y las Artes, Mexico, 366 p.

Colunga, P., Hernández, X., and Castillo, A. (1986) Variación morfológica, manejo agrícola y grado de domesticación de *Opuntia* spp. en el Bajio Guanajuatense. *Agrociencia* 65: 7–49.

Davis, T., and Bye, R. A. (1982). Ethnobotany and progressive domestication of *Jaltomata* (Solanaceae) in Mexico and Central America. *Economic Botany* 36(2): 225–241.

De Wet, J. M. J. (1992). The three phases of cereal domestication. In Chapman, G. P. (ed.), *Grass Evolution and Domestication.* Cambridge University Press, Cambridge. pp. 176–198.

De Wet, J. M. J., and Harlan, J. R. (1975). Weeds and domesticates: Evolution in the man-made habitat. *Economic Botany* 29(2): 99–107.

Groube, L. (1989). The taming of the rain forest: A model for late Pleistocene forest exploitation in New Guinea. In Harris, D. R., and Hillman, G. C. (eds.), *Foraging and Farming. The Evolution of Plant Exploitation.* Unwin Hyman, London. pp. 292–304.

Hallam, S. J. (1989). Plant usage and management in southwest Australian aboriginal societies. In Harris, D. R., and Hillman, G. C. (eds.), *Foraging and Farming. The Evolution of Plant Exploitation.* Unwin Hyman, London. pp. 136–151.

Harlan, J. R. (1975). *Crops and Man.* American Society of Agronomy, Madison, Wisconsin. 295 p.

Harlan, J. R. (1992). Origins and processes of domestication. In Chapman, G. P. (ed.), *Grass Evolution and Domestication.* Cambridge University Press, Cambridge. pp. 159–175.

Hawkes, J. G., (1983). *The Diversity of Crop Plants.* Harvard University Press, London. 184 p.

Hernández-X., E., and Ramos, A. (1977). Metodología para el estudio de agroecosistemas con persistencia de tecnología agrícola tradicional. In Hernández-X., E. (ed.), *Agroecosistemas de México.* Colegio de Postgraduados, Chapingo, México. pp. 321–333.

Hillman, G. C. (1989). Late palaeolithic plant food from Wadi Kubbaniya in upper Egypt: Dietary diversity, infant weaning, and seasonality in a riverine environment. In Harris, D. R., and Hillman, G. C. (eds.), *Foraging and Farming. The Evolution of Plant Exploitation.* Unwin Hyman, London. pp. 207–239.

Pearsall, D. M. (1989). Adaptation of prehistoric hunter gatherers to the high Andes: The changing role of plant resources. In Harris, D. R., and Hillman, G. C. (eds.), *Foraging and Farming. The Evolution of Plant Exploitation.* Unwin Hyman, London. pp. 318–334.

Rzedowski, J. (1978). *Vegetación de México.* Limusa, México. 432 p.

Schneider, T. C. (1972). A Land Management System Used by the Mesolithic Communities of Southern England. Masters thesis, Anthropology Department, Pennsylvania State University, Pennsylvania.

Schwanitz, F. (1967). *The Origin of Cultivated Plants.* Harvard University Press, Cambridge, Massachusetts. 175 p.

Shipek, F. C. (1989). An example of intensive plant husbandry: The Kumeyaay of southern California. In Harris, D. R., and Hillman, G. C. (eds.), *Foraging and Farming. The Evolution of Plant Exploitation*. Unwin Hyman, London, pp. 159–170.

Steward, J. (1938). Basin Plateau Aboriginal Socio-political Groups. Bureau of American Ethnology, Bulletin 12.

Toledo, V. M. (1988). La diversidad biológica de México. *Ciencia y Desarrolol* 8(14): 17–30.

Vavilov, N. I. (1926). Studies on the origin of cultivated plants. *Bulletin of Applied Botany, Genetics and Plant Breeding* 16: 139–248.

Vázquez, R. M. C. (1986). El Uso de Plantas Silvestres y Semicultivadas en la Alimentación Tradicional en Dos Comunidades Campesinas del Sur de Puebla. BSc. dissertation, Facultad de Ciencias, Universidad Nacional Autónoma de México, México. 104 p.

Vázquez, R. M. C. (1991). Tendencias en el proceso de domesticación del papaloquelite (*Porophyllum ruderale* (Jacq.) Cass. subsp. *macrocephalum* (DC.) R. R. Johnson ASTERACEAE). MSc dissertation, Facultad de Ciencias, Universidad Nacional Autónoma de México, México. 153 p.

Viveros, J. L., Casas, A., and Caballero, J. (1993). Las plantas comestibles y la alimentación entre los mixtecos guerrerenses. In Leff, E., and Carabias, J. (eds.), *La Dimensión Cultural del Desarrollo Sustenable*. Centro de Investigaciones Interdisciplinarias en Humanidades, Universidad Nacional Autónoma de México-Editorial Porrúa, México. pp. 625–670.

Williams, D. E. (1985). Tres Arvenses Solanáceas Comestibles y Su Proceso de Domesticación en el Estado de Tlaxcala, México. MSc. dissertation, Colegio de Postgraduados, Chapingo, México. 173 p.

Zizumbo, V. D., and Colungo, P. (1982). *Los Huaves: La Apropiación de los Recursos Naturales*. Departamento de Sociología Rural, Universidad Autónoma de Chapingo, Chapingo, México. 277 p.

Footnote

1. Jardín Botánico, Instituto de Biología, U.N.A.M., Apartado Postal 70–614, México, D.F. 04510, México.

THE SEEDS OF OUR FUTURE

VANDANA SHIVA

The Cultivation of Diversity

For ten thousand years, agriculture has been based on the strategy of conserving and enhancing genetic diversity.

According to former FAO genetic resources expert Erna Bennett,

> 'The patchwork of cultivation sown by man unleashed an explosion of literally inestimable numbers of new races of cultivated plants and their relatives. The inhabited earth was the stage for 10,000 years, for an unrepeatable plant breeding experiment of enormous dimensions. (Bennett, 1978, p. 1 14)'

In this experiment, millions of peasants and farmers participated over thousands of years in the development and maintenance of genetic diversity. The experiment was concentrated in the so-called developing world where the greatest concentrations of genetic diversity are found, and where humans have cultivated crops the longest. The traditional breeders, the Third World peasants, as custodians of the planet's genetic wealth, treated seeds as sacred, as the critical element in the great chain of being. Seed was not bought and sold, it was exchanged as a free gift of nature.

Throughout India, even in years of scarcity, seed was conserved in every household, so that the cycle of food production was not interrupted by loss of seeds.

Replacement from Diversity to Uniformity

With the Green Revolution, peasants were no longer to be custodians of the common genetic heritage through the breeding, improvement and conservation of seeds. The 'miracle seeds' of the Green Revolution transformed this common genetic and intellectual heritage into private property. Peasants as plant breeding specialists gave way to scientists of multinational seed companies and international research institutions like CIMMYT and IRRI. Plant breeding strategies of maintaining and enriching genetic diversity and self-renewability of crops were substituted by new breeding strategies of uniformity and non-renewability, aimed primarily at increasing transnational profits and First World control over the genetic resources of the Third World. The Green Revolution changed the ten thousand year evolutionary history of crops by changing the fundamental nature and meaning of 'seeds'.

Farmers' seeds were now declared 'primitive cultivars' and 'land races', suggesting no intellectual work had gone into their evolutions (see figure 1). The Green Revolution varieties were called 'elite' and 'modern' varieties. They were even called 'miracle' varieties. These varieties, themselves evolved from a narrow genetic base, were spread throughout the world, thus displacing the millions of crop varieties which farmers had evolved over millenia.

The Report on the State of World's Plant Genetic Resources has identified replacement of local varieties as the most frequent cause for genetic erosion (see Figure 1). The Green Revolution has been recognized as the most significant reason for the displacement of agricultural biodiversity by the IV Technical Conference on Plant Genetic Resources in Leipzig in June 1996. Green Revolution monocultures are a form of cultural invasion which displace local cultural and biological diversity. (Vandana Shiva, 1991) Food cultures based on diverse cereals, millets, oil seeds and pulses were displaced by rice and wheat varieties released from international agricultural centres. Thousands of local rice varieties were displaced by a few IRRI varieties. A culture of growing crops to feed the earth, animals and humans was replaced by the culture to ignore the needs of nature and other species, and to meet human needs of nutrition only through long distance market chains.

The crop and varietal diversity of indigenous agriculture was replaced by a narrow genetic base and monocultures. The focus was on internationally traded grains, and a strategy of eliminating mixed and rotational cropping, and diverse varieties by varietal simplicity. While the new varieties reduced diversity, they increased resource use of water, and of chemical inputs such as pesticides and fertilizers.

The monocultures of the Green Revolution were a form of cultural invasion which resulted in the displacement of biological diversity. The new crop varieties were themselves not biologically invasive. Crops bred through conventional means have lost their competitive capacity and their natural ability to spread. They depend on humans to be sown and cultivated.

However, while the plants introduced as Green Revolution monocultures are not biologically invasive in themselves, they do set off a second order species invasions in the form of pests and diseases. The genetic uniformity of monocultures creates the ecological conditions for transforming harmless pests and diseases into devastating epidemics as has become evident with the creation and spread of new pests and diseases for wheat and rice.

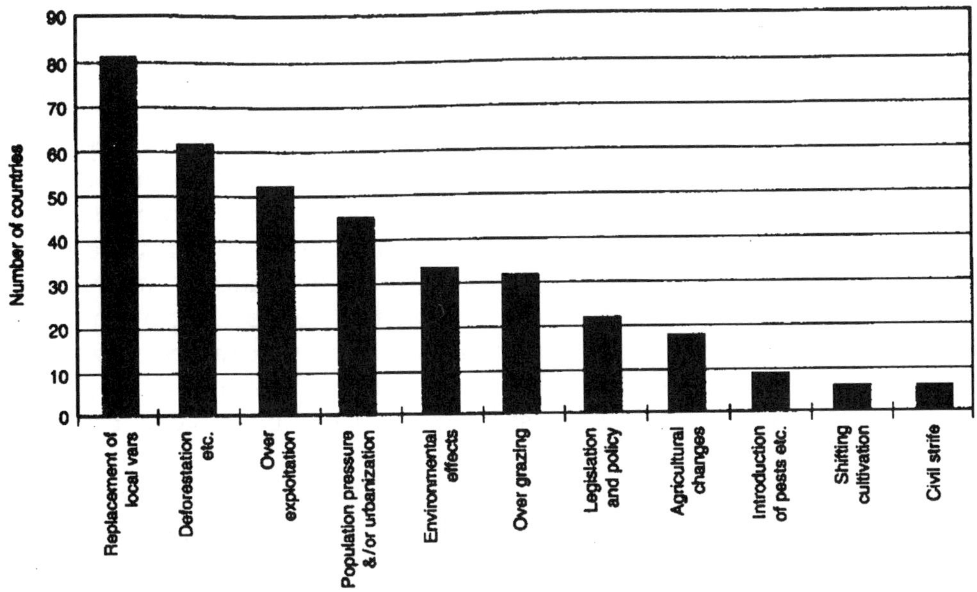

Figure 1. Main causes of genetic erosion as cited in country reports.
Source: Report on the State of the World's Plant Genetic Resources.

In 1966, the International Rice Research Institute released a 'miracle' rice variety—IR-8 was particularly susceptible to a wide range of disease and pests: in 1968 and 1969 it was hit hard by bacterial blight and in 1970 and 1971 it was ravaged by another tropical disease called *tungro.* In 1975, Indonesian farmers lost half a million acres of Green Revolution rice varieties to leaf hoppers. In 1977, IR-36 was developed to be resistant to eight major diseases and pests including bacterial blight and *tungro.* However this was attacked by two new viruses called 'ragged stunt' and 'wilted stunt'.

The vulnerability of rice to new pests and disease due to monocropping and a narrow genetic base is very high. IR-8 is an advanced rice variety that came from a cross between an Indonesian variety called 'Pea' and another from Taiwan called 'Dee-Geo-Woo-Gen'. IR-8, Taichung Native (TN1) and other varieties were brought to India and became the basis of the All India Coordinated Rice Improvement Project to evolve dwarf, photoinsensitive, short duration, high yielding varieties of rice suited to high fertility conditions. The large scale spread of exotic strains of rice with a narrow genetic base was known to carry the risk of the large-scale spread of disease and pests. As summarized in a publication titled 'Rice Research in India—An Overview' from CRRI,

> 'The introduction of high yielding varieties has brought about a marked change in the status of insect pests like gall midge, brown planthopper, leaf folder, whore maggot etc. Most of the high yielding varieties released so far are susceptible to major pests with a crop loss of 30 to 100% . . . Most of the HYVs are the derivatives of TN1 or IR-8 therefore, have the dwarfing gene of Dee-Geo-Woo-Gen. The narrow genetic base has created alarming unifor-

> mity, causing vulnerability to disease and pests. Most of the released varieties are not suitable for typical uplands and lowlands which together constitute about 75% of the total rice area of the country. (Moore Lappé and Collins, 1982, p. 114)'

The 'miracle' varieties displaced the diversity of traditionally grown crops, and through the erosion of diversity, the new seeds became a mechanism for introducing and fostering pests. Indigenous varieties or land races are resistant to locally occurring pests and diseases. Even if certain diseases occur, some of the strains may be susceptible, while others will have the resistance to survive. Crop rotations also help in pest control. Since many pests are specific to particular plants, planting crops in different seasons and different years causes large reductions in pest populations. On the other hand, planting the same crop over large areas year after year encourages pest build ups. Cropping systems based on diversity thus have built in protection.

The wheat seeds that spread worldwide from CIMMYT through Borlaug and his 'wheat apostles' were the result of nine years of experimenting with Japanese 'Norin' wheat. 'Norin' released in Japan in 1935 was a cross between Japanese dwarf wheat called 'Daruma' and American wheat called 'Fultz' which the Japanese government had imported from the US in 1887. The Norin wheat was brought to US in 1946 by Dr. D. C. Salmon, an agriculturist acting as a US military adviser in Japan, and further crossed with American seeds of the variety called 'Bevor' by USDA scientist Dr. Orville Vogel. Vogel in turn sent it to Mexico in the 1950s where it was used by Borlaug, who was on the Rockefeller Foundation staff, to develop his well-known Mexican varieties. Of the thousands of dwarf seeds created by Borlaug, only three went to create the 'Green Revolution' wheat plants which were spread worldwide. On this narrow and alien genetic base, are the food supplies of millions precariously perched.

Large scale monocultures of exotic varieties have turned minor diseases such as Karnal Bunt into epidemic proportions. Green Revolution wheat varieties came to India as a project of cultural invasion. As if in an ecological backlash, the Karnal Bunt has invaded the wheat crop of the US. In combination with drought, the fungal disease has wiped out more the 50 percent of the wheat crop in the bread basket of the world.

Uniformity and the Ownership of Seed

Diversity and sharing of seed have always gone hand in hand. Uniformity and exclusive rights have similarly been associated. In fact, Plant Breeders' Rights are rights based on the production of uniformity.

The existing international agreement that covers Plant Breeders' rights is the International Convention for the Protection of New Varieties of Plants—UPOV Convention. The UPOV Convention was adopted initially by five European countries and membership was restricted to European countries till 1968. At that time the Convention was revised and membership opened to all countries.

The UPOV Convention is rigid, requiring that members adopt its standards and scope of protection as national law. The UPOV Convention, has resulted in a high degree of standardization, goes against the reality of biological diversity and the socioeconomic diversity of different countries. It is therefore inappropriate as a *sui generis* system evolved to protect plants, people and creativity in diverse realities.

The standardization is built into the way plant varieties are defined. To be eligible for protection, a variety must be:

- new—the variety must not have been exploited commercially;
- distinct—it must be clearly distinguishable from all other varieties known at the date of application for protection;
- uniform—all plants of the variety must be sufficiently uniform to allow it to be distinguished from other varieties taking into account the method of reproduction of the species;
- stable—it must be possible for the variety to be reproduced unchanged.

This definition by its very nature rules out farmers' varieties and destroys biodiversity, and produces uniformity as necessity. The reward under such Plant Breeders' Rights (PBR) systems does not go for breeding to maintain and enhance diversity and sustainability, but to the destruction of biodiversity and creating uniform and hence ecologically vulnerable agricultural systems.

While UPOV fails to recognize and therefore protect farmers' rights as positive rights, UPOV 1978 does have a farmers' exemption which gives the farmer the right to save seed of protected varieties. Similarly, the breeders' exemption allows researchers and breeders free access to a protected variety to use for breeding other varieties.

However, UPOV 1991 has removed these exemptions. Breeders and researchers will have to pay royalty to the PBR holder to use the protected variety for breeding other varieties. The farmers' exemption has been made optional. Art 15 of UPOV 1991 states:

> 'Each contracting party may within reasonable limits and subject to the safeguarding of the legitimate interests of the breeder restrict the breeder's right in relation to any variety in order to permit farmers to use for propagating purposes on their own holdings the product of the harvest which they have obtained by planting, on their holdings, the protected variety.'

It is the breeders who will decide their legitimate interests and enforce this on the state. Since the breeders are multinational seed companies in this case, more powerful than most Third World governments, 'reasonable limits' will be set by these corporations and not by individual governments. Breeders' authorization will therefore be the final determinant in respect to sale and marketing of harvested material. UPOV 1991 is therefore as monopolistic as patent regimes.

The Convention on Biological Diversity and the conservation of agricultural biodiversity

The ecological vulnerability of agricultural monocultures has made the conservation of agricultural biodiversity an environmental imperative. The Convention on Biodiversity Conservation (CBD) has been one of the responses of the world community to conserve the ecological basis of biological production through biodiversity conservation.

There are two new political conditions to which CBD has given rise. First it has recognized the national sovereign right of countries to their biological wealth. Secondly it has recognized the contribution of indigenous communities to knowledge about the utilization of biodiversity. The recognition of sovereignty and indigenous knowledge create a major shift in the political context of the ownership, use and control of genetic resources, especially in the area of agricultural biodiversity, including seeds and plant genetic resources. The seeds of the Third World can no longer be treated as the 'common heritage of mankind', freely accessible to all, including northern seed corporations. The seeds sold by transnational seed corporations can no longer be regarded as the only ones embodying an intellectual contribution. Farmers' seeds also embody significant intellectual contribution by Third World farmers. A third major shift in recent years has been the recognition that 'improvement' of seed is not absolute and context independent. Improvement is a contextual category. 'Improvement' of crops from the viewpoint of agribusiness is often a regression and loss from the point of view of farmers. Thus agribusiness 'improves' crops for industrial processing, or for increased use of chemical inputs, whereas farmers need crops easy to process at home and crops that decrease the dependence on external inputs.

Conventional breeding only looks at yield potential and yield potential is defined as 'the yield of a crop when growth is not limited by water or nutrients, pests, diseases or weeds'. In the realities of farmers' fields these are precisely the limitations that farmers face, and their breeding is responsive to the environmental stress and the ecological diversity within which they must practice the agriculture. (Trygve Berg, 1995) Corporate breeding strategies therefore cannot be treated as the only direction of evolution in breeding and research, and 'improvement' from the perspective of transnational corporations' (TNCS) interests cannot be translated into an overall societal benefit. TNCs' monopoly in seed through IPRs is therefore neither desirable nor necessary from the public interest perspective.

That is why we need to question the assumptions of the US Secretary of State who says that the Government of India should produce legislation to protect US seed corporations so that they can provide the 'best seeds for Indian farmers'. What is good for Pepsi, Cargill, Monsanto and Pioneer is not necessarily good for Indian farmers and Indian consumers. Pepsi needs tomatoes and potatoes for processing into tomato ketchup and potato chips for its fast food chains like KFC and Pizza Hut. Cargill needs sorghum and maize for its feed industry. Monsanto is interested in selling more of its herbicide Round Up through herbicide resistant rape-seed. These are not the characteristics of seed that Indian farmers would choose in their breeding strategies or characteristics of food that Indian consumers would choose for cooking qualities in their diverse food systems.

Farmers' rights reflect the recognition of sovereignty in ownership and creativity in traditional breeding by farmers as well as alternative breeding strategies for protection of the biodiversity base of agriculture. Without farmers' rights the biodiversity rich Third World countries cannot assert their sovereign rights to their agricultural biodiversity or in their agricultural policy. Further, without the ownership rights of farming communities, biodiversity cannot be conserved.

The Biodiversity Convention provides an opportunity to governments to protect agricultural biodiversity, farmers' livelihoods, and sovereign rights to biodiversity.

In the preamble the Convention states that contracting parties recognize 'the close and traditional dependence of many indigenous and local communities embodying traditional lifestyles on biological resources, and the desirability of sharing equitable benefits arising from the use of traditional knowledge, innovations and practices, relevant to the conservation of biological diversity and sustainable use of its components'.

However, no article in the Convention directly addresses farmers' rights or mechanisms for the compensation of indigenous knowledge. Articles 10(c) and 18.4 refer to indigenous practices, but not to rights of farmers or local communities. Article 10(c) states:

> 'Each contracting party shall as far as possible and as appropriate protect and encourage customary use of biological resources in accordance with traditional cultural practices that are compatible with conservation or sustainable use requirements.'

Article 18.4 states:

> 'The contracting parties shall in accordance with national legislation and policies, encourage and develop methods of co-operation for the development and use of technologies, including indigenous and traditional technologies in pursuance of the objective of this Convention.'

The Biodiversity Convention thus offers avenues for the protection of farmers' rights and national rights to biodiversity. An important issue for government agencies is to explore ways in which the Biodiversity Convention can be used in the TRIPs negotiations. Government and non-government bodies also need to explore mechanisms for the recognition and implementation of farmers' rights vs IPRs.

In situ conservation of cultivated biodiversity makes it imperative that the concept of farmers' rights be evolved and implemented to enable farm communities to effectively conserve local biodiversity. However, IPRs as being defined in other negotiations such as the GATT are based on a denial of farmers' rights and the intellectual and material contribution of Third World farmers to the global production and knowledge systems.

The challenge for operationalizing farmers' rights is therefore to give them a due place in the implementation of IPR legislation.

IPRs and seeds: protection of transnationals vs protection of farmers

New Intellectual Property Rights (IPRs) legislation is being introduced in the area of plant genetic resources (PGRs) under pressure of the US government as well as the requirements of the TRIPs agreement of the World Trade Organization (WTO). While WTO gives a five year transition period to introduce PGRs legislation the US pressure is to introduce such legislation immediately. Further, the US is demanding monopoly protection for TNCs which control the seed industry. On the other hand people's organizations are fighting to protect farmers' rights to their biodiversity and their right to survival as well as the freedom of scientists to work for the removal of hunger rather than corporate profits. Farmers' organizations, biodiversity conservation groups, sustainable agriculture networks and public interest oriented scientists are trying to ensure that farmers' rights are protected, and through the protection of farmers' rights, sovereign control over our biological wealth and its sustainable use in agricultural production is ensured. The conflict over PGRs legislation is a conflict between farmers and the seed industry and between the public domain and private profits, between an agriculture that produces and reproduces diversity and one that consumes diversity and produces uniformity.

On January 29, 1996 at an address at the Indian Institute of Agricultural Research, the US Secretary of Agriculture, Daniel Glickman directly addressed the issue of the protection of seed Multinationals (MNCs). He said,

> 'I hope your new legislation will provide a responsible and reasonable protection to private seed companies, which will encourage them to provide the best seeds available for your farmers. There would be very few inventions of anything, particularly in agriculture, without patent protection because it is the fundamental fact of nature that people will not go through the expense of development of new ideas just for the altruistic benefit of the human race.' (*The Hindu*, 1996)

The US IPRs orthodoxy is based on a fallacious idea that people do not innovate or generate knowledge unless they can derive private profits. However, greed is not a 'fundamental fact of human nature' but a dominant tendency in societies that reward it. In the area of seeds and plant genetic resources, innovation of both the 'formal' and 'informal' systems has so far been guided by the larger human good. Norman Borlaug the scientist behind the Green Revolution and the recipient of the Nobel Peace Prize, made this clear in his statement at a Press Conference at the Indian Agricultural Research Institute, New Delhi on February 8, 1996. He expressed concern against private companies and TNCs gaining control of plant genetic resources and seeds and patenting plants. Professor Borlaug said,

> 'We battled against patenting. I and late Glen Anderson (of International Wheat and Maize Research Institutes) went on record in India as well as other fora against patenting and always stood for free exchange of germplasm.'

He saw IPRs in PGRs as a prescription for famine. Commenting on the US demand for patents he said, 'God help us if that were to happen, we would all starve.' (*The Hindu*, 1996)

It is often stated that IPRs will not stop traditional farmers using native seeds. However, when it is recognized that IPRs are an essential part of a package of agribusiness controlled agriculture in which farmers no longer grow native seeds but seeds supplied by the TNC seed industry, IPRs become a means of monopoly that wipe out farmers' rights to save and exchange seed. This leads to TNC totalitarianism in agriculture. TNCs will decide what is grown by farmers, what they use as inputs, and when they sell their produce, to whom and at what price. They will also decide what is eaten by consumers, at what price, with what content and how much information is made available to them about the nature of food commodities.

The TRIPs Agreement of WTO

The Trade Related Intellectual Property Rights (TRIPs) agreement of GATT provides a framework for the implementation of IPR laws. The section of TRIPs that most directly affects farmers' rights and agricultural biodiversity is Article 27.5.3(b) which states:

> 'Parties may exclude from patentability plants and animals other than microorganisms, and essentially biological processes for the production of plants or animals other than non-biological and microbiological processes. However, parties shall provide for the protection of plant varieties either by patents or by an effective sui generis system or by any combination thereof. This provision shall be reviewed four years after the entry into force of the Agreement establishing the WTO (the World Trade Organization).'

The article thus allows two forms of IPRs in plants, patents and a *sui generis* system.

a) Plant Patents

The first part of this article requires that parties allow patenting of plants and animals produced through 'non-biological' and 'microbiological processes'. The reference is quite evidently to the new biotechnologics of genetic engineering. However, while the moving of species across species barriers through genetic engineering techniques can be defined as 'non-biological' in the sense that such mixing of genetic material would not happen in nature, the 'production' of plants and animals with genes introduced from other species takes place through an essentially biological process of reproduction.

While patenting of plants and animals has become prevalent in the US and UK, the article in TRIPs governing patenting of plants and animals creates major problems that will need to be reviewed in 1999, when the provisions of the Article come up for review. This period is shorter than the transitional period available to developing countries according to Article 65. It would therefore be more useful for developing countries to wait for the TRIPs review before changing their patent regimes.

There are two other articles in TRIPs which allow countries to exclude patenting of plants and animals. These are the article referring to public morality and the article referring to public interest.

The decision in European Parliament against the Patent Directive, and the debates in Indian Parliament on the Patent (Amendment) Act have been strongly guided by the ethical concerns arising from patenting of life.

Patents on plants also raise serious concerns about monopolies over food and agriculture systems. Table 1 shows the trends of corporate conservation of US plants patents.

Broad species patents also raise major concern. In March, 1994, the European Patent Office granted an extraordinarily broad patent to Agracetus Corporation, a subsidiary of W. R. Grace. The patent covers all transgenetic soya bean varieties and seeds, regardless of the genes used and all methods of transformation. This patent has been challenged by RAFI, a public interest group. Dr. Geoffrey Hawtin, Director-General of the International Plant Genetic Resources Institute in Rome, Italy expressed his concern at such patenting:

> 'The granting of patents covering all genetically engineered varieties of a species, irrespective of the genes concerned or how they were transferred, puts in the hands of a single inventor the possibility to control what we grow on our farms and in our gardens. At a stroke of a pen the research of countless farmers and scientists has potentially been negated in a single, legal act of economic high jack.' (RAFI, 1994)

A similar concern has been raised by Dr. Norman Borlaug on the patenting of Plant Genetic Resources.

The concern about plant patents is therefore a deep and widespread concern.

In practical terms, allowing patenting in the field of agriculture will have the following adverse consequences:

- it will encourage monopoly control of plant material by western transnational corporations;
- this in turn will make farmers dependent on corporations for the most critical input in agriculture, i.e. seed. This monopoly control is more far-reaching given the takeover of seed companies by the large chemical and agribusiness corporations which control other inputs into agriculture such as fertilizers, pesticides and herbicides;
- monopoly control on seed linked with corporate control over agriculture will lead to large scale disappearance of farmers' varieties, thus threatening biodiversity conservation as well as farmers' survival. Biodiversity erosion will in turn lead to the erosion of the rich cultural diversity of our country;
- due to royalty payments the prices of seeds will go up;

Table 1 Top 14 Owners of Biotech Plant Patents

Company	Country	No. of Patents
Monsanto	USA	18
DNA Plant Technology	USA	17
PioneerHi-Bred	USA	11
Mycogen	USA	10
ICI/Zeneca	UK	7
Cibat/Geigy	Switzerland	6
Du Pont	USA	6
Rockefeller Univ.	USA	5
Agracetus	USA	5
Agricultural Genetics	USA	4
Amoco	USA	4
Michigan State Univ.	USA	4
Lubrizol	USA	4

Source: RAFIC Communiqué, July/Aug 1995

- the changed economics resulting from IPRs will lead to the displacement of small farmers, who will get into debt and destitution;
- large-scale uprooting of agricultural society, without equivalent absorption in new industrial opportunities will lead to social disintegration, spurt in crime and breakdown of law and order;
- intellectual property protection in the area of agriculture and plant variety will undermine food security since the protected and patented varieties are not linked to food needs, but to the processing and marketing requirements of agribusiness;
- the shift to control of agriculture through the control of seed will also contribute to secondary impacts of other natural resources like land and water into the control of TNCS;
- IPRs in the area of seeds and plants will increase the national debt tenfold;
- the undermining of food security will increase food imports and hence the foreign exchange burden, thus inviting deeper conditionalities from institutions like IMF and World Bank;
- the erosion of food security will create food dependency, turning food into a weapon in the hands of industrialized countries, thus leading to total slavery and recolonization.

Farmers' rights therefore need to be built into patent legislation not just Plant Variety Legislation.

b) A *sui generis* regime

However, countries can create alternatives to the UPOV Convention under a *sui generis* regime, especially in the context of the Convention on Biological Diversity which creates obligations for states to protect biodiversity and indigenous knowledge and practices. The term effective can be defined *vis-a-vis* CBD. This would enable countries to base their *sui generis* legislation on the protection of community rights and farmers' rights and the conservation of agricultural biodiversity as required by CBD.

Farmers' rights need to be the pillar of such a *sui generis* system.

Farmers' rights are an ecological, economic, cultural and political imperative. Without community rights, agricultural communities cannot protect agricultural biodiversity. This biodiversity is necessary not just for the ecological insurance of agriculture. Rights to agricultural biodiversity is also an economic imperative because without it our farmers and our country will lose their freedom and options for survival. Since biodiversity and cultural diversity are intimately linked, conservation of agricultural biodiversity is a cultural imperative also. Finally, without farmers' rights, there is no political mechanism to limit monopolies in agriculture and the inevitable consequence of displacement, hunger and famine that will follow total monopoly control over food production and consumption through the monopoly ownership over seed, the first link in the food chain.

Table 2 illustrates the new concentration of the seed industry and the new convergence of the chemical industry with seed industry.

Until recent times it was the local communities that have used, developed and conserved biological diversity who have been custodians of the biological wealth of this planet. It is their control, their knowledge and their rights that need to be strengthened if the foundations of biodiversity conservation are to be strong and deep.

Tearing down the walls: seed wars at Leipzig

The Fourth Technical Conference on Plant Genetic Resources at Leipzig was supposed to have operationalized 'farmers' rights'. However, the eventual 'success' of the Leipzig Conference was merely that farmers' rights were not fully extinguished by the US and other delegations. The US had put the term 'farmers' rights' in square brackets throughout the Plan of Action. On the concluding day, the negotiations were held up from noon to midnight. The US insisted it would only recognize the individual rights of farmers.

The South insisted that farmers' rights were collective rights of farming communities. The final language that survived states

> 'confirming the needs and individual rights of farmers and collectively, where recognised by national law, to have nondiscriminatory access to germplasm, information, technologies, financial resources and research and marketing systems necessary for them to continue to manage and improve genetic resources'.

While this is not the strongest formulation of 'farmers' rights', by keeping farmers' rights alive, the Leipzig process has kept open the space for conservation of agricultural biodiversity for the future.

Table 2 Countries Which Protect Plant Varieties with Breeders' Rights Systems, as of January 1994

UPOV Members	UPOV non-member states: Laws in accordance with UPOV	UPOV non-member states: Laws substantively in accordance with UPOV	UPOV non-member states: Hybrid System of laws
Australia	Argentina	Chile	Mexico
Belgium	Austria	Kenya	Repub. of Korea
Canada	Norway	Zimbabwe	Rumania
Czech Republic	Portugal		
Denmark	Uruguay		
Finland	Andean Group		
France			
Germany			
Hungary			
Ireland			
Israel			
Italy			
Japan			
Netherlands			
New Zealand			
Poland			
Slovakia			
South Africa			
Spain			
Sweden			
Switzerland			
United Kingdom			
USA			

Source: Carlos M. Correa, Sovereign and property Rights Over Plant Genetic Resources, University of Buenos Aires, Argentina.

The NGOs were committed to keeping the biological and intellectual commons open for local farming communities. This is necessary not merely for the farmers of the South but also for the ecological farmers of the North. IPRs on seeds are interfering in the access of ecological farmers to diverse, organic seed, thus denying consumers access to safe and nutritious foods. Farmers' rights are thus relevant for both North and South. In global solidarity, NGOS made the Leipzig Commitment to Agricultural Biodiversity which opens with the words:

> 'Tearing down the wall. We have gathered in this historic city of Leipzig, which has been at the heart of trade, learning and culture over centuries. Over the last few decades Leipzig had lost its vibrancy and vitality because of a totalitarian regime against which the people started a silent revolution to reclaim freedom. The tearing down of the Berlin Wall is a symbol of peoples' power against totalitarianism. We commit ourselves to tearing down these new Berlin walls.'

Vandana Shiva is Director Research Foundation for Scientific Technology and Natural Resource Policy, New Delhi, India.

References

Bennett, E.,'Threats to Crop Plant Genetic Resources', in J. G. Howkes, *Conservation and Agriculture,* London, Duckworty, (1978).

FAO,'The Realisation of Farmers' Rights', paper presented at Technical Consultation on an implementation framework for Farmers' Rights, Madras, India, (January 15–18, 1995).

Moore Lappé, F. and Collins, J., *Food First,* London, Abacus, (1982).

'Protect private sector in farm research: Glickman', *The Hindu,* (January 30, 1996).

RAFI Occasional Paper, Vol. I, No. 5, (Dec. 1994).

Shiva, V.,'The Violence of Green Revolution: Third World Agriculture, Ecology and Politics/, *Third World Network,* Penang, Malaysia, (1991).

Trygve B., 'Devolution of Plant Breeding', paper presented at the seminar on 'Using Diversity, Enhancing and Maintaining Genetic Resources on Farm', sponsored and organized by IDRC, New Delhi, June 19–21, 1995.

PART 4

Fishing for Food and as a Livelihood

Deforestation, soil erosion and degradation, and the loss of biodiversity have received widespread recognition as major environmental problems faced by contemporary human populations. The loss of food and reduction of biodiversity as a consequence of decline in fish stocks has received less attention, although that pattern may be changing. Of the world's seventeen major fishing areas, those areas richest in abundance of fish caught and consumed by humans, thirteen are considered to be overfished, and several are in serious decline. The loss of protein for coastal populations of humans, in particular, is serious, but the loss of food is only one part of the overall loss. The first article in this section, by Yva Momatiuk and John Eastcott, document the types of losses that occur and the environmental consequences for the northeastern Atlantic off the shores of Newfoundland. That region was once one of the world's richest fisheries, especially the cod fishery, but in a relatively brief period of a few years, the fishery collapsed. Exactly why is not entirely clear, but overfishing likely played a large role. Overestimation of population sizes by fish biologists also contributed to the suddenly acute problem in the early 1990s, but there were many other factors. In 1993, after almost 400 years of successful cod fishing, the fishery was closed, putting communities of fishermen out of work. The collapse of a biological resource led to collapse of a livelihood, with major socioeconomic and biological consequences.

The fate of the Newfoundland cod fishers has been and continues to be played out in many other areas of the world. The second article in this section, by Gregor Hodgson and John Dixon, looks at the decline in coastal fisheries in the Philippines as a consequence of pollution of marine waters. In this case, economic over-exploitation of the fish was not as significant a factor as the vastly increased sedimentation of the water from runoff water in the watershed. The runoff water increasingly is heavily sedimented from road building and logging operations. Here is a case of clear conflict between two types of economic activities, logging for large-scale commercialism and fishing for local food production. The utilization of natural resources at the local level has become caught up in global economics.

The third and fourth articles report on activities taken at the community level for control and restoration of damaged fishing areas. John Kurien shows how a community in India has constructed artificial reefs to provide protection for fish, allowing the fish to reproduce, survive, and thus be restored as an important food source. The community also engaged in collection action to manage the reefs for common property resources. The fourth article, by Denise Heywood, is on an initiative in an earlier stage of development in Cambodia. A freshwater lake there is replenished each year by a periodic backward flow of a river into it. For a few days out of each year, the river flows into the lake rather than out of it, and that replenishes the lake and the biodiversity that it supports, including fish as an important food source. In recent years, however, increased rates of sedimentation have been present in the river, principally from deforestation and consequent erosion, and that sedimentation poses a severe threat to the lake. The local residents and authorities have appealed to UNESCO for assistance in controlling the sedimentation and its impact on the lake. This case study shows, again, the intricate interconnections between local food sources and activities, more regional or national commercial operations, and the linkage across local and international efforts to arrive at some solution to the resultant problems.

The final article in this section also reflects local, national, and global activities concerning commercial and environmental issues. Les Kaufman reports on radical changes that have occurred in Lake Victoria in eastern Africa. Lake Victoria was home to a large variety of fish species, several of which were major food sources for the human populations living around the Lake, more than 30 million of them. A number of anthropogenic, i.e., human-induced, changes occurred beginning in the 1960s that led to a catastrophic decline in fish variety and collapse of several of the fisheries. A number of new species of fish were introduced into the lake, some by design. The Nile perch was one of those, a fish that can grow to several hundred pounds and that is highly predatory. As the Nile perch spread, other species declined, leading eventually to a radical shift in species type and abundance. A loss of an important food source, a loss of livelihood for many fishers, and a loss of biodiversity were among the more serious consequences.

A FISHY TALE

YVA MOMATIUK AND JOHN EASTCOTT

Newfoundlanders have depended on the annual influx of cod for their livelihood for 400 years. But in the last two years, cod stocks have collapsed leaving fishing communities high and dry. Words and pictures by Yva Momatiuk and John Eastcott

Since he began to fish with his father more than 30 years ago, Brian Walsh—a one time mayor of Bay de Verde—has been setting his clock by the seasons. As soon as the snow melts, he and other fishermen come down to the harbour to repair the winter's damages to boats and sheds. Some men put out a herring net or lobster pots as if testing the sea. Come June or early July, they watch capelin arrive in their millions to spawn on the sandy beaches. The smelt-like fish, seldom longer than six inches, are followed by whales, seals and large schools of cod.

In the good times, men would sail out every morning to set cod traps—carefully constructed net boxes which took months to build and thousands of dollars to replace. Some crews used hook and line. But whatever the means, the 150 fishermen of Bay de Verde would bring in cod all summer. In late autumn, after weeks of backbreaking toil, they would settle in for a long winter, building and mending equipment, gathering wood and collecting unemployment benefit. Brian sees no shame in that. As recently as the late 1980s, he and his brother David were catching some 50,000 kilograms of codfish each summer, which sold in the US for about $100,000. He would get just six per cent of that: "If I was working at a salaried job for 12 months to bring $100,000 of new money into the country, I'd be considered a great guy," he jokes.

However, that was in better times. Brian's wife Doreen was working in the fish processing plant and his sons, Patrick and Tommy, sometimes accompanied him on fishing trips. Some time ago, he noticed that the cod were getting smaller and he had to work harder to catch enough. "Imagine planting a certain amount of land to get so much grain, and every year having to plant more land to harvest the same amount," he says. "Eventually, there are not enough hours to plant, and that's what happened to fishing. At first, we were putting in two nets, then four, eight, then 40 nets and two cod traps, just to keep up."

Other fishermen also noticed the alarming decrease in fish size and numbers towards the end of the 1980s. Finally, after a summer of the most appalling cod fishing anyone could remember, Brian called in the federal scientists to ask how they had arrived at the optimistic estimate of cod stocks and the high fishing quotas. He was invited aboard a research vessel sailing along the northeast coast of Newfoundland and on to Labrador, looking for cod. "They were expecting to find plenty of fish," Brian recalls. "But of course, they didn't find any."

On 2 July 1992, John Crosbie, minister of Canada's Department of Fisheries and Oceans, declared a two-year moratorium on the Canadian inshore exploitation of cod, one of the world's greatest sustainable protein resources. In September 1993, the ban was extended to the whole of Eastern Canada, and in spring 1994 all recreational fishing for cod was forbidden. It is clear that the moratorium will have to stay in place for the foreseeable future and perhaps indefinitely. Even though it has turned his life upside down, Brian sees no alternative: "If somebody said to me tomorrow, 'OK, we leave it to you to solve it,' I wouldn't know what to do. There is nothing anyone can do now because there are no fish out there."

When John Cabot landed in Newfoundland in 1497, he observed the sea "swarming with fish which can be taken in baskets let down with a stone." Old Newfoundlanders tell fanciful stories about salmon returning from the sea in such numbers one could cross rivers just by stepping on their backs. St John's, the oldest city in North America, was founded to help funnel thousands of tons of split, salted and dried cod from small fishing boats onto large vessels bound for Europe. Out at sea, squadrons of gannets bombarded the waves, feasting on these rich fishing grounds. The Atlantic seemed inexhaustible, and the fish were there for all the birds to see.

For centuries, cod has been the economic mainstay of fishing villages in the province where money was always scarce and farming virtually non-existent. 'The fish', for this is how Newfoundlanders refer to cod (*Gadus morhua*), is an excellent source of food: big-boned, white-fleshed and delicate in flavour. Attesting to its culinary merits, fishermen who used to work on large boats stipulated that they should be fed cod rather than salmon. It is also popular abroad: about 70 per cent of the annual Newfoundland harvest used to be exported to the US, and 12 per cent to Japan and Europe.

A female northern cod, who may eventually grow to almost two metres, and live up to 50 years, matures in its seventh year and deposits more than one million eggs in the spawning grounds every year. Yet very few of them, fertilised by floating sperm, survive, for the vagaries of the marine environment allow only a small fraction to mature. Little is known about the early life cycle of northern cod.

Although better assessments of young cod numbers is crucial to estimate the size of stocks, any aquasystem is difficult to study: it is just too vast and inaccessible. In addition, cod and other ground fish feeding close to the bottom are hard to count and even find. Consequently, the margins of error in calculations of fish populations tend to be enormous.

We do know that as cod get bigger, they settle to the bottom and feed on capelin, and to some extent on shrimp. They form large schools and, until recently, chased the capelin inshore every year with such regularity that hundreds of fishing communities came to depend on their arrival for survival. But their migration pathways require much greater study. Today's methods include echo sounding and hydroacoustics which allow efficient fish population surveys, but even these elaborate devices, allow only a small area to be investigated. But if the fish were important enough to warrant extensive research, why was one of the most productive fisheries in the world allowed to disintegrate so rapidly?

No one knows for sure what caused the collapse, so everyone is pointing the finger at everyone else. Fishermen, who work close to shore in small boats, blame floating fish factories the size of football fields and trawlers that rake the fragile spawning grounds. Offshore skippers and many fishermen blame seal herds. Harp seal herds alone are now estimated at 3,500,000. Protected from Newfoundland sealers by the recent European ban on all seal products, they are thriving. Fishing plant workers bemoan the number of foreign vessels brazenly harvesting groundfish species, including cod. Provincial authorities tend to blame federal authorities' mismanagement of the stock. Some fishing methods undoubtedly harmed the stock: traps and trawls caught fish indiscriminately, while gill nets, when lost, sank to the bottom and continued catching.

There were other factors. Echo-sounder fish tracking devices and electric net winches used by bigger boats, as well as the growing number of fishermen, have put unprecedented pressure on the stock. It has been a circular process: each innovation designed to improve fishing techniques brought in its wake increased pressure to harvest more fish, in order to pay for the equipment.

Yet the fishery managers disregarded the warnings of disaster made by fishermen years ago."Our greed and shortsightedness have finally burst all the bridges and caught up with us," says Tony Williamson, director of international development in St. John's."The Government reports regarding the size of stocks have been optimistic, and it blinded people. And those who knew what was going on often didn't want to say anything because it was too political." Brian Walsh agrees."I think many people in the scientific community knew how bad it was"he says."And the politicians who wanted to offer something attractive to the people may have said to these scientists:'Go back and see if you can return with some better figures,' and they did."

A look at historical landings of cod reveals a curious pattern. Between 1804 and 1947, Newfoundland catches of cod increased from 75,000 to 200,000 metric tons due to the increase in the number of fishermen, and reached a maximum of 709,000 metrics tons in 1968, only to plummet thereafter. These huge catches may have damaged the stock irretrievably.

So total was everyone's faith in the rosy predictions that there were even concerns about finding new markets to absorb larger catches in the future. But as the catches began to fall, hundreds of desperate Newfoundland fishermen found themselves heavily in debt: reassured that the cod stocks were growing, they ignored the evidence in front of them and invested in new fishing boats and gear.

The 1992 moratorium devastated the economy. In Newfoundland, the poorest and the youngest Canadian province, with a population of half a million, the average per capita income is the lowest in the country and unemployment at 24 per cent, the highest. Just recently, a former Premier of the province declared that Newfoundland has 250,000 too many people to support itself. Today, the fishermen sit idle, plant workers have been laid off, transportation companies and repair shops are closed and unemployment in some fishing communities has reached 80 per cent. And many villages may die out. The job-driven exodus could wipe out the island's culture, where isolation and a singular way of life have preserved traditions and dialects that originated on the west coast of England and Ireland 200 years ago.

The fishermen predict that the new brave talk about 'professionalising' the industry means that even if the stocks rebound, the harvesting of the sea will be turned over to the owners of big boats with sophisticated electronic equipment. This fear is shared by all who have lost their jobs because modern technology has made their skills obsolete.

In the meantime, virtually all cod fishing grounds in Atlantic Canada have closed. The new minister of Fisheries and Oceans promised to send in gunboats if any foreign fishing vessel enters the Grand Banks. But new estimates show that in spite of the moratorium, stocks have fallen 75 per cent since 1992, and continue to decline.

During the last few years, ornithologist Dr Bill Montevecchi, has noticed that many black-legged kittiwakes are failing to reproduce and that the diet among other seabirds has changed. He points out that capelin is the key element in the life cycle of many marine creatures in this region. But the cold water in recent years has meant that the capelin's spawn has been occurring later than usual, and that the fish are undersized. As a result, seabirds are nesting a month later, caught between the lean days of spring and the autumn gales. Some adults are starving and many gulls, whose populations exploded due to the abundance of fish offal that was discarded by boats and processing plants, now fail to reproduce and pray on other birds chicks more frequently. The 'late' pattern, according to Montevecchi, may have affected not only the seabirds, but also the cod. During his research on the misty Funk Island last summer, he was struck by the eerie emptiness of the sea. "In my 10 days there. I heard only one boat moving out in the fog," he says. "I got him on my radio. This guy was going 250 kilometres from Musgrave Harbour to catch some fish at the edge of the continental shelf, taking 36 hours just to get there and set his nets in 500 fathoms of water. He told me, "That's the last of it, boy. When we get those, that will be it."

In Bay de Verde, Brian Walsh tends to the tree in his garden. He says that a lot of people still get up early in the morning, go down to the waterfront where they always went, mill around and go home again. "The only thing we are here for is fishing," he explains. "And if there is no fish, what are we doing here?"

Yva Momatiuk and John Eastcott are a wife and husband team who photograph and write about social and ecological matters.

LOGGING VERSUS FISHERIES IN THE PHILIPPINES

GREGOR HODGSON AND JOHN A. DIXON

The pollution of rivers, lakes, and sea by sedimentation is increasing on a worldwide scale. Sedimentation pollution of costal marine areas is an especially serious problem in Southeast Asia, where fish harvested from coastal waters serve both as a major source of protein for human consumption and a significant source of foreign exchange through exports. One major cause of sedimentation is logging, and in particular the building of roads and other infrastructure needed to get the logs onto the international market.

In 1985, a logging operation was begun in the watershed bordering Bacuit Bay near the village of El Nido, Palawan, Philippines, an important area for two other foreign exchange earning industries—tourism and marine fisheries. The effects of logging-induced sedimentation on the bay's previously pristine marine environment was the subject of a one year study. By the end of the study, only eleven percent of the available commercial forest had been logged, but high rates of accelerated erosion due to logging had already resulted in dramatic increases of sediment transport and discharge into the bay. Sedimentation damage to coral reefs and associated fisheries was rapid and severe.

The detrimental side-effects of logging on watersheds have been well documented throughout the world[1] and include damage to young trees through unplanned felling, soil degradation and soil loss. Although tree-cutting exposes underlying soil to the direct effects of wind and rain by removing protective layers of leaf canopy,[2] the major cause of erosion due to logging operations has been shown to be the construction of the extensive road and skid-trail network,[3] which is necessary to allow for log removal. This erosion in turn is the primary cause of high levels of suspended sediment load in streams and rivers which leads to reduced fish and invertebrate biomass and diversity.[4] In addition, silt deposited in dams and reservoirs results in reduced economic returns.[5]

Although the connections linking watershed erosion, silt-laden rivers, and siltation of the coastal marine environment appear obvious, they have received relatively little attention despite the high economic value of coastal marine life. Terrestrial ecologists have often failed to look beyond the freshwater systems affected by erosion. In the few cases where marine scientists have studied siltation damage to tropical coastal marine species, the origin of the silt has rarely been documented.

Siltation of coastal marine areas occurs throughout the world and may result from logging, agriculture, dredging, construction, and other development activities that expose previously protected soils to the erosive action of wind and rain.[6] Coastal marine pollution is especially serious in developing countries such as the Philippines where the rapidly expanding population depends heavily on marine fisheries to meet protein requirements.

Observed reductions of marine fish stocks in Southeast Asia are partially due to over-fishing,[7] although marine pollution such as siltation probably accounts for a significant percentage of the reported reduction in fish catches in many areas. The negative effects of siltation on fish stocks are often indirect, for example, increased silt load could act indirectly on a tuna fishery by killing planktonic larvae of coral reef fish upon which tuna feed.[8] In this case, the resulting decrease in tuna catch would not be reported until long after the sedimentation event, when the affected year-class of tuna matured and fishermen noticed a reduced catch. Unfortunately, the intricate nature of marine food webs make it difficult to demonstrate conclusively a specific cause-and-effect relationship.[9]

Bacuit Bay

Bacuit Bay is located on the west coast near the northern tip of Palawan Island in the southwest Philippines (Map 1). Palawan is a thin, 425 kilometre-long island bisected by a central mountain range, leaving only a narrow coastal margin of cultivable land that accounts for about 20 percent of the total land area.[10] The weather pattern in Palawan is monsoonal with northern Palawan receiving from 2,000 to 4,000 millimetres of rain each year.[11] It is not however located on a high frequency typhoon track[12] and El Nido has not been hit by a typhoon during the past 30 years.

Ten years ago, Palawan was considered one of the last unspoiled regions in the Philippines, with virgin timber stands and plentiful marine resources, numerous endemic species of plants and animals, and a relatively low population density. More recently, rapid population growth, combined with industrial expansion in mining and logging, has drastically reduced the size of Palawan's remaining wilderness areas. Part of the reason for this expansion is that by the late 1960s, natural resources in the southern Luzon and Visayas regions had been depleted by a rapidly growing population, so that government and private interests turned to Palawan to supply their needs. In addition, migration to Palawan has accelerated due to increases in civil strife in rural areas of other provinces. Half of the nearly five percent annual population increase in Palawan is estimated to result from immigration.[13]

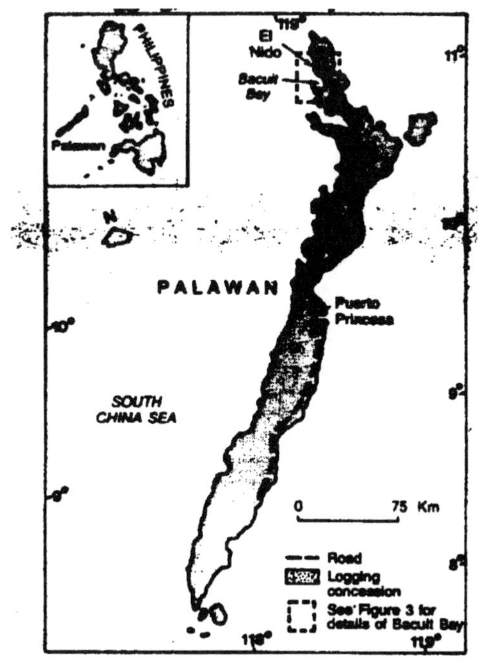

Map 1. The Philipines (inset) aned Palawan Island.

In 1968, almost 92 per cent of Palawan's land area was forested. By 1980, poorly controlled logging and slash-and-burn agriculture had resulted in a decrease in forest area to 70 percent,[14] and by 1987 perhaps only 50 percent of Palawan remained forested. Present forest consumption is estimated to be 200 square kilometres per year, just more than three percent of the 1987 forested area.

Fish are an important resource for Palawan. A high percentage of the total Philippine fish catch is estimated to be taken from the waters surrounding Palawan.[15] The total Philippine catch of many demersal (bottom-dwelling) fish species, however, has been declining in recent years.[16] Previously pristine, the coastal marine environment of Palawan has begun to follow the trend seen in most other areas of the Philippines and is now being subjected to intense fishing pressure, illegal fishing (with dynamite, poison, small mesh nets, and weighted-scareline or *muro ami* fishing), siltation, and heavy metal pollution from mine tailings. Because of its remote location and the lack of roads, port facilities, or other infrastructure, northern Palawan has been one of the last areas to hold-out against this onslaught of intense fishing pressure.

Logging

The Bacuit Bay drainage basin (Map 2)covers 78.3 square kilometres and extends inland to the central Palawan dividing range. Prior to the initiation of logging in 1985, 53 percent of the basin was composed of primary forest. Most of this forest is now included within a large logging concession, encompassing most of northern Palawan. Logging operations in the area bordering Bacuit Bay commenced in

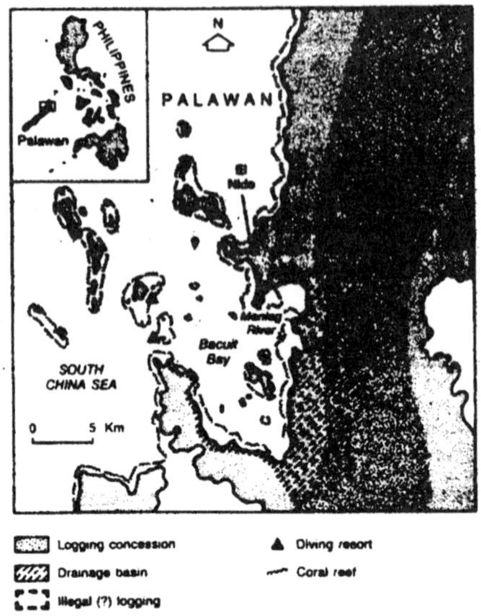

Map 2. The study area showing Bacuit Bay, El Nido village and the overlap of the drainage basin with the logging concession area. The Manlag River drains about 70 percent of the basin.

January 1985, and in January 1986 were temporarily suspended by the logging company for one year.

Philippine law requires that three trees be planted for each tree harvested. This level of planting, combined with diameter limits and selective logging practices that remove approximately half the number of harvestable classes of trees, will theoretically allow a sustainable yield operation with two harvesting cycles during the 85-year timber rotation. Unfortunately, lack of tree planting, overcutting in low-volume timber stands, and excessive felling damage have already been cited as threats to the sustainability of the northern Palawan concession, held by Pagdanan Timber Products Inc. (PTPI).[17]

Transport Infrastructure

Three construction projects connected with the Bacuit Bay logging operations contributed to the serious erosion. First, an earthen pier (known as a 'logpond' in the industry) was constructed. The pier is used to off-load heavy equipment such as trucks and skidders shipped in by barge, as a log storage area and for loading logs onto a barge for transport to the sawmill. The Philippine mahogany (*Dipterocarpus spp.*, locally called *apitong*) logs from this region do not float due to a high resin content and must be transported to a sawmill by barge. In order to construct the pier, approximately 100,000 cubic metres of soil was pushed down into the bay from an adjacent hillside. With only a bare minimum of protective log pilings, the pier eroded rapidly due to wave action and runoff. The second construction project involved clearing 200 square metres for the logging camp, which is used for housing, equipment storage and repair. The camp, located about 3 kilometres inland, was cleaned almost entirely of vegetation and left bare. The third and largest construction project was road-building.

The Bacuit Bay drainage basin terrain is steep (30 percent mean slope), and the land is classified in the "severe erosion hazard" category.[18] Road construction within the hilly drainage basin is generally achieved by cutting into the hillsides, leaving an exposed vertical face on the uphill side. On the downslope side, the extra clay soil is pushed over the edge of the road, spilling down in a wide swathe called a sidecast fill slope. Both the uphill and downhill slopes are destabilized by this type of road construction, increasing the potential for accelerated erosion. During the 1985 and 1986 rainy seasons, the logging roads were frequently blocked by landslides from cuts above them,.

Primary access roads are used daily (for example to transport logs to the log pier) and are usually gravel surfaced. Secondary roads are built to service specific logging zones within the concession and are used only while those areas are productive. Tertiary roads connect secondary roads to the skid trails. Skid trails are the paths cut into hillslopes by bulldozers (Caterpillar tractors) to allow skidders access to specific tree stands. Following tree felling by chainsaw, the skidder transports each log to the log truck onto which it is loaded. Skid trails are normally used for only a few days and then abandoned. In this concession they were designed and built perpendicular to the land contours, thus allowing the greatest possible erosion by maximizing the slope angle for the full length of each skid trail.

Erosion

The results of erosion plot studies show that sheet erosion per kilometre of logged area (roads and cut forest) was about 240 times greater than from uncut forest plots. The sheet erosion for the logging area lying within the Bacuit Bay drainage basin was calculated by extrapolating the erosion plot results (*table 1*). These total sheet erosion values do not include other forms of erosion such as gully or till erosion and mass wasting.

The estimate of logging road area, 0.42 square kilometres, is 83 percent of the cut forest area and is within the range measured in concessions in Malaysia and other tropical regions.[19] When considering erosion from roads, however, the entire area of soil disturbance, not just the width and length of the road surface must be taken into account. Erosion from side-casts along logging roads in

Table 1 Sheet erosion in logging concession area within Bacuit Bay drainage basin, 1986

Area type	Area (km^2)	Erosion (tonnes/km^2)	Total (1000 tonnes)
Roads	0.41 × (3)[a]	14,130	17.4
Cut forest	4.80	260	1.2
Uncut forest	37.00	60	2.1
Total	42.21	14,450	20.7

[a] measured area multiplied by three to account for cut-and-fill slopes.

Malaysia was found to contribute up to 1.5 times the road surface erosion.[20] Field measurements and aerial photograph analysis revealed that cut-and-fill increases effective road width by a factor of three. Using this factor, the total effective road area is 1.23 square kilometres. Roads thus make up three per cent of the drainage basin but account for 84 percent of the surface erosion due to logging (Figure 1). This finding is in agreement with most previous work on the contribution of road-building to total erosion in logging areas.[21]

No attempt was made to measure rill and gully erosion or mass wasting within logged over or undisturbed forest. Therefore, the quantitative contribution of these erosion processes to a total erosion budget for the Bacuit Bay drainage basin is unknown. Sediment from these sources in the cut forest is probably reflected in the high suspended sediment load measured in the major river draining the area. Along logging roads and near the earthen log pier, landslides of various scales were continuously observed. Rill development on most exposed, sloping surfaces, and metre-deep gullies on steep roads made it clear that these processes would contribute a significant amount of soil to a total erosion budget. Colonization and growth of grasses and weeds on level and low slope road surfaces occurred within one year, especially if only a thin strip of land was cleared of vegetation on either side of the road. Steeply sloped roads bordered by 10 metre wide cleared areas showed no plant recolonization two years after abandonment and may continue to erode for decades to come.

Sediment Delivery From the Manlag River

Although accelerated surface erosion was documented within the logging concession, it is important to estimate the sediment delivery ratio—the proportion of eroded soil carried out of the drainage basin by rivers and streams and into Bacuit Bay. From *table 1*, the sheet erosion from the Bacuit Bay drainage basin during 1986 was estimated to be 20,700 tonnes. This figure can be compared with the total 1986 sediment discharge of 35,000 tonnes from the Manlag River, which drains approximately two-thirds of the Bacuit Bay drainage

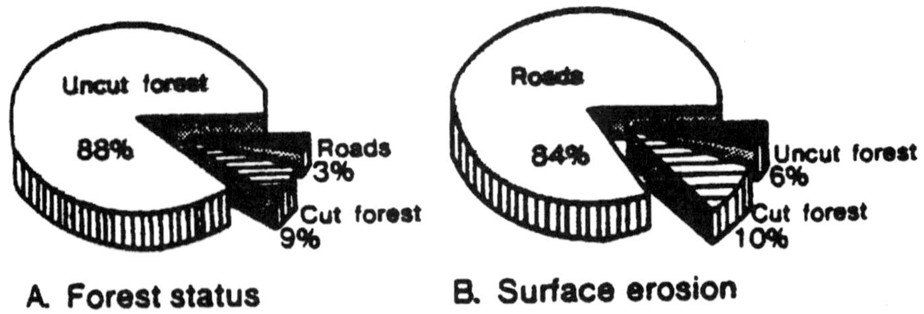

Figure 1. (A) The contribution of roads, cut and uncut (primary) forest areas to the 42.21 square kilometre logging concession area lying within the Bacuit Bay drainage basin in 1986. (B)(The contribution of each of these areas to surface erosion (about 21,000 tonnes) estimated from erosion plot measurements in 1986.

basin. This figure does not include bed load (the heavier material which is washed along the river bottom).

In order to estimate sediment discharge from accelerated erosion due to logging for the entire Bacuit Bay drainage basin, the measured discharge from the Manlag River needs to be adjusted to account for the concession area not draining into the Manlag River and also for the surface erosion component contributed by natural erosion from the uncut forest. These adjustments yield an estimate of about 41,000 tonnes, which is more than twice the sheet erosion due to logging. With a logged area (plus roads) of 5.2 square kilometres, this equals nearly 8,000 tonnes per square kilometre in annual sediment yield. In comparison, annual sediment yields reported from various logging operations in the United States range from about 300 to 20,000 tonnes per square kilometre.[22]

Total Sediment Deposition

Vast amounts of soil eroded due to logging operations in the drainage basin are thus transported to the river and carried to Bacuit Bay. The sum of sediment deposition in the entire bay during the May to December 1986 rainy season was about 128,000 tonnes. An estimate of pristine sedimentation conditions can be made by multiplying the control station sedimentation rate (82.5 tonnes/square kilometre/month) times the entire bay area (120 square kilometres). The result is 9,900 tonnes per month or about 79,000 tonnes for the 1986 rainy season. Subtracting this figure for 'pristine' conditions from the measured total leaves 49,000 tonnes of sediment deposition that could be associated with the logging operations.

Effects On Coral

As a result of this high sedimentation rate, a significant reduction in coral cover and diversity was measured on Bacuit Bay reefs. The control reef with a low (pristine) sedimentation rate showed no significant change in these ecological parameters. The greatest change occurred on the reef closest to the river mouth, which lost nearly 50 percent of its living coral cover. Most of this loss occurred following a storm that resulted in high sediment discharge from the Manlag River for several days.[23]

Although the increase in sediment deposition attributable to logging is only about 60 percent of the 'pristine' rate, any increase in sediment deposition is additive and may stress a living organism, such as a coral, beyond its tolerance limits. Since corals are sessile animals (attached to the bottom), they are unable to escape from temporarily high concentrations of suspended sediment. Instead, they must rely on cleaning mechanisms such as polyp movement, ciliary action, and mucus production to remove the deposited sediment from their surfaces.[24] This clearing process requires energy, and each species has a limit to the rate, intensity, and duration of sediment deposition it can counteract.[25] If the sediment deposition rate exceeds the species-specific sediment cleaning limits

(its biological threshold for this stress), the corals will die. Experiments have also demonstrated that corals held in aquaria and exposed to sediment deposition are more susceptible to lethal bacterial infections than corals held in control aquaria.[26] The corals may also be killed by oxygen or nutrient starvation, or poisoning by coral waste products or bacterial toxins.

Effects of Sediment on Fish

In contrast to corals, fish can swim away from localized high turbidity areas. Some estuarine fish species may even prefer slightly turbid water over clear water during the early stages of their life cycle.[27] It is likely that very high suspended sediment concentrations would be required to kill coral reef fish directly. Laboratory experiments have shown that the white perch (*Morone americana*), a North American estuarine species, is killed by exposure to a suspended sediment concentration of 19,000 milligrams per litre. Mortality is due to gill clogging[28] and oxygen starvation. This extreme level of suspended sediment concentration is probably rarely found in nature. The highest turbidity recorded from the Manlag River was 3,000 milligrams per litre and from Bacuit Bay, 1,000 milligrams per litre. Therefore changes in abundance and diversity of coral reef fish and pelagic (open sea) fish dependent on the bay's food chain are not expected to be caused by fish mortality due to the direct effects of sedimentation. However, little attention has so far been given to the effect of high turbidity on coral reef fish behaviour. For example, intricate mating and territorial behaviour patterns, which are reported to be highly dependent on visual cues,[29] might be disrupted by turbid water conditions. This could result in a reduced reproductive rate that would eventually reduce biomass.

Numerous studies have documented direct and indirect dependence of coral reef fish on the coral reef community.[30] When the coral reef community is damaged, large decreases in fish diversity and abundance result. In one case, the complete destruction of a coral reef at Iriomote Island, Japan, resulted in the loss of 90 percent of fish abundance within two years.[31] Hourigan et al.[32] cite siltation due to dredging as a cause of local extinction of 12 species of butterflyfish at Johnston Atoll in the Central Pacific. In this case, local extinction was correlated with widespread coral mortality resulting in a severe food limitation for those species that feed directly on corals. In general, major shifts in fish diversity and abundance are expected to result from changes in coral reef community structure, which reef fish depend on for food, shelter, reproduction, and juvenile fish recruitment.[33]

Although some coral reef fish feed on coral directly,[34] most depend for their food supply on prey organisms (for example, crustaceans, other fish, or algae living in association with coral reefs). In addition, reef fish use the reef structure for shelter.

Analysis of changes measured in Bacuit Bay coral and fish populations during 1986 produced the following relationships:

- Annual decrease in coral cover of one percent for every additional 400 tonnes per square kilometre of annual sediment deposition in Bacuit Bay
- Annual decrease of one coral species (extinction) per increase of 100 tonnes per square kilometre of annual sediment deposition in Bacuit Bay
- For each one percent annual decrease in coral cover, fish biomass is decreased by 2.4 percent
- For each annual decrease of one coral species associated with coral cover loss, fish biomass decreased by 0.8 percent.

Long-Term Effects of Logging

If a forest is logged and then left undisturbed, it has been generally accepted that surface erosion will eventually decrease with time as plant cover and soil stability increase.[35] The rate of this decrease of total erosion following logging will depend on local conditions such as the variation in annual rainfall.

Tree roots provide soil stability, especially on steep slopes. Following logging, the dead roots of cut trees decay and eventually lose their stabilizing capability, frequently leading to slope failures. Large-scale landslides could cause an increase in total soil erosion and sediment discharge several years after logging has stopped. A study in California found that sediment loads were still significantly higher in streams draining logged-over forest than in control streams six to ten years after initial logging.[36]

In tropical forests, ground cover may colonize and grow more quickly than in mid-latitude forests, thus protecting soil exposed by tree cutting, but the typically shallow layer of tropical forest topsoil may be washed away before recolonization by dipterocarpus tree species can begin. The soil surface of logging roads will generally consist of low nutrient level subsoil and often gravel. When this surface is highly compacted, it will provide a poor environment that only a few specialized plant species will be able to colonize.[37] In addition, gully erosion on steep road surfaces will tend to prevent a rapid decrease in the rate of erosion and sediment discharge from logged-over tropical forest. Observations of forest areas south of El Nido that had been logged several years before, indicated that erosion, especially of the gully and rill type, was still severe on road surfaces.

Resource Use Conflicts

The results of the ecological and economic analysis of the development alternatives for the Bacuit Bay area in the Philippines have implications for coastal development in many other countries. Specifically, the results indicate that sedimentation pollution can seriously degrade coastal marine fisheries in the tropics. Many countries are now at risk from sedimentation pollution due to logging that may already be creating significant resource use conflicts.

The conclusion that much of the erosion is from road-building is of considerable significance. 'Sustainable logging' as proposed by the International Timber Trade Organization and the international development institutions will have to take account of this factor, and ensure that logging which may be 'sustainable' in terms of timber yield, does not destroy local fisheries and therefore the major source of human protein in many tropical countries.

Gregor Hodgson and John A. Dixon carried out this study for the East-West Centre, a public, nonprofit educational institution established in Hawaii in 1960 by the United States Congress.

Endnotes:

1. R. S. Suparto (ed.) *Proc. Symposium on the Long-Term Effects of Logging in Southeast Asia.* BIOTROP, SEAMEO Regional Center for Tropical Biology, Bogor, Indonesia, 1978; U.S. Department of State. *The World's Tropical Forests: A Policy, Strategy and Program for the United States,* Pub. 9117, U.S. Government Printing Office, Washington, 1980; L. Hamilton and P. N. King, *Tropical Forested Watersheds,* Westview Press, Boulder, CO, 1983; P. P. Tucker and J. F. Richards, *Global Deforestation and the Nineteenth-Century World Economy,* Duke University Press, Durham, NC, 1983.
2. S. R. Herwitz. "Raindrop Impact and Water Flow on Vegetative Surfaces of Trees and the Effects of Stemflow and Throughfall Generation," *Earth Surface Processes and Landforms* 12, 1987, 425–432.
3. C. O'Loughlin. "The Influence of Forest Roads on Erosion and Stream Sedimentation—Comparisons Between Temperate and Tropical Forests." Working Paper, East-West Environment and Policy Institute, Honolulu, Hawaii, 1985.
4. A. J. Cordone, *Effects of Logging on Fish Production,* Inland Fish Administration Report, California Department of Fish and Game, 1956; E. Graynoth, "Effects of Logging on Stream Environments and Faunas in Nelson, New Zealand," *J. Mar. Freshwater Research* 13, 1979, 79–109; J. D. Newbold, D. C. Erman, and K. B. Roby, "Effects of Logging on Macroinvertebrates in Streams With and Without Buffer Strips," *Can. J. Fish. Aquat. Sci.* 37, 1980, 1076–1085; D. C. Erman and D. Mahoney, *Recovery after Logging in Streams With and Without Bufferstrips in Northern California,* California Water Resources Center Contribution, University of California at Davis, 1983.
5. L. Hamilton and P. N. King, op. cit. supra 1; R. Goodland and G. Ledec, "Environmental Management in Sustainable Economic Development," *Impact Assessment Bulletin* 5(2), 1986, 50–81.
6. E. D. Gomez, A. C. Alcala, and A. C. San Diego, "Status of Philippine Coral Reefs—1981," in E. D. Gomez (ed.), *Proc. 4th International Coral Reef Symposium, 1981,* Manila, Philippines, 1982; A. T. White, "Effects of Construction Activity on Coral Reef Lagoon Systems," in B. Salvat (ed.) *Human Impact on Coral Reefs: Facts and Recommendations,* Antenne Museum, E.P.H.E., French Polynesia, 1987; B. Salvat, "Dredging in Coral Reefs" in B. Salvat (ed.), *Human Impact on Coral Reefs: Facts and Recommendations,* Antenne Museum, E.P.H.E., French Polynesia, 1987, 165–184.
7. FAO, *Review of the State of the World Fishing Resources,* Circular No. 710, Revision 5, Marine Resources Service, Fishery Resources and Environment Division, Rome, 1987.
8. R. E. Brock. "Preliminary Study of the Feeding Habits of Pelagi Fish Around Hawaiian Fish Aggregation Devices or Can Fish Aggregation Devices Enhance Local Fisheries Productivity?", *Bull. Mar. Sci.* 37(1), 1985, 40–49.

9. R. Jones, "Ecosystems, Food Chains and Fish Yields," in D. Pauly and G. I. Murphy (eds.), *Theory and Management of Tropical Fisheries*, ICLARM Conference Proceedings 9, International Center for Living Aquatic Resources Management, Manila, 1982, 195–236.
10. Bureau of Soils. *Physiography and Soils of the Palawan Island.* Ministry of Agriculture, Manila, 1980.
11. PIADP. "Logging Methods Review." Integrated Environmental Program, National Council on Integrated Area Development, Position paper, Philippines, 1985.
12. PAGASA. *Annual Tropical Cyclone Report,* Philippines Atmospheric and Astronomical Services Administration, Manila, 1957–1987.
13. PIADP, op. cit., supra 11.
14. M. P. Pido. "Palawan: On Natural Resources Conservation." *Canopy Int.* 12(2), 1986, 1–5.
15. ibid.
16. FAO, op. cit., supra 7.
17. PIADP, op. cit., supra 11.
18. Bureau of Soils, op. cit., supra 10.
19. D. A. Gilmour, "Logging and the Environment, with Particular Reference to Soil and Stream Protection in Tropical Rainforest Situations." *Conservation Guide I: Guidelines for Watershed Management.* FAO, Rome, 1977; C. O'Loughlin, op. cit., supra 3.
20. C. O'Loughlin, op. cit., supra 3.
21. G. W. Brown and J. T. Krygier, "Clear-Cut Logging and Sediment Production in the Oregon Coast Range." *Water Resources Research* 7(5), 1971, 1189–1198.
22. reviewed in G. W. Brown and J. T. Krygier, op. cit., supra 21.
23. G. Hodgson. "The Effects of Sedimentation on Indo-Pacific Reef Corals." Ph.D. dissertation, Zoology Department, University of Hawaii, Honolulu, in prep.
24. J. A. E. B. Hubbard and Y. P. Pocock. "Sediment Rejection by Recent Scleractinian Corals: A Key to Paleoenvironmental Reconstruction." *Geol. Rundsch.* 61, 1972, 598–626.
25. G. Hodgson, op. cit., supra 23.
26. ibid.
27. D. P. Cyrus and S. J. M. Blaber. "The Influence of Turbidity on Juvenile Marine Fishes in Estuaries, Part 1, Field Studies at Lake St. Lucia on the Southeastern Coast of Africa," *J. Exp. Mar. Biol. Ecol.* 109, 1987a, 71–91; D. P. Cyrus and S. J. M. Blaber, "The Influence of Turbidity on Juvenile Marine Fishes in Estuaries, Part 2, Laboratory Studies, Comparisons with Field Data and Conclusions," *J. Exp. Mar. Biol. Ecol.* 109, 1987b, 71–91.
28. J. M. O'Connor, D. A. Neuman and J. A. Sherk, "Lethal Effects of Suspended Sediments on Estuarine Fish." *U.S. Coast. Eng. Res. Tech. Pap.* 77(3), 1976, 1–90.
29. R. E. Thresher, *Reproduction in Reef Fishes.* T. F. H. Publications, Neptune City, 1984.
30. K. E. Carpenter, R. I. Miclat, V. D. Albaledejo, and V. T. Corpuz. "The Influence of Substrate Structure on the Local Abundance and Diversity of Philippine Reef Fishes," in E. D. Gomez (ed.), *Proc. 4th International Coral Reef Symposium, 1981,* Manila, 1982; See M. Sano, M. Shimizu and Y. Nose, "Long-Term Effects of Destruction of Hermatypic Corals by *Acanthaster planci* Infestation on Reef Fish Communities at Iriomote Island, Japan," *Mar. Ecol. Prog. Ser.* 37, 1987, 191–199.
31. M. Sano et al., op. cit., supra 30.
32. T. F. Hourigan, T. C. Tricas and E. S. Reese, "Coral Reef Fishes as Indicators of Environmental Stress in Coral Reefs," in *Marine Organisms as Indicators,* New York, Springer-Verlag, in press.

33. J. E. Randall."The Effects of Fishes on Coral Reefs."*Proc. 2nd International Coral Reef Symposium*, Brisbane, 1974, 159–166.

34. E. S. Reese, "Predation of Corals by Fishes of the Family *Chaetodontidae*: Implications for Conservation Management of Coral Reef Ecosystems,"*Bull. Mar. Sci.* 31, 1981, 594–604; E. F. Cox, "Aspects of Corallivory by *Chaetodon unimaculatus* in Kaneche Bay, Oahu," Masters thesis, Zoology Department, University of Hawaii, 1983; T. C. Tricas, "Life History, Foraging Ecology, and Territorial Behavior of Hawaiian Butterflyfish *Chaetodon multicinctus*," Ph.D. dissertation, University of Hawaii, Honolulu, 1986.

35. C. O'Loughlin, op. cit., supra 3.

36. D. C. Erman and E. Mahoney, op. cit., supra 3.

37. Z. Hamzah. "Some Observations on the Effects of Mechanical Logging on Regeneration, Soil and Hydrological Conditions in East Kalimantan," in R. S. Suparto (ed.) *Proc. Symposium on the Long-Term Effects of Logging in Southeast Asia, 1975*, BIOTROP, SEAMEO Regional Center for Tropical Biology, Bogor, Indonesia, 73–78, 1978.

COLLECTIVE ACTION FOR COMMON PROPERTY RESOURCE REJUVENATION:

The Case of People's Artificial Reefs in Kerala State, India

JOHN KURIEN

This paper is about the efforts being made by communities of coastal fisherfolk in South India to build artificial reefs as a means of rejuvenating the ecosystem of their coastal waters damaged by indiscriminate trawling. These initiatives provide the basis for questioning the now influential opinions that in the case of resources in the realm of the commons, precious little will be done to save them from ruin, particularly by those individuals who enjoy access to them. It hopes to add to the accumulating evidence that collective action by the laboring masses in the developing world—peasants, fisherfolk and forest dwellers, to mention a few—to revive and rejuvenate their common pool resources calls to question the undiscriminating policy prescriptions which continue to see only "market or state" interventions to solve issues relating to environmental degradation.

Keywords: Fishing, India, artificial reefs, common property resources

There is now an impressive body of literature which documents the role played by community-level collective action in the management of common property resources. To cite a few examples we have the successful anti-soil erosion and afforestation measures by the farmers of Sukhomajri in the lower Shivalik range of the Himalayas in India (Chopra *et al.* 1990); the peasants of Trobel, Switzerland, who have for over seven centuries managed their communally-owned grazing lands (Netting 1981); the communally-managed irrigation systems of Ilocos Norte, Philippines (Siy 1982); the innovative self-regulation of the coastal waters by the fishermen of Alanya and Aristofersen in Turkey (Berkes 1986); and those in Lofoten, Norway (Jentoff and Kristofersen 1989). Some of this collective action has been endogenously structured by those who use the common property resource. Some others have taken root after appropriate intervention from "outside" led the participants to perceive some gains from collective action within the realm of their use of common property.

The resulting social institutions—some of which have existed over centuries—have evolved sets of rules which the participants understand, agree upon, and are willing to follow. It is the evolving nature of relationships between the participants, within the context of their common property resource, that creates new values and fresh norms, revives cultural knowledge and trust, and facilitates action towards such a desired end. James Coleman refers to such intangible institutional design as a form of social capital. (Coleman 1988).

The impressive evidence of creation of such social capital by community participation in common property resource management from around the world does not seem to have influenced mainstream policy prescriptions in the developed or developing countries: large sections of the academy and many in the business of state craft still emphasize "state" or "private" hegemony as the best means of governing the natural resources used in common.

Such policy prescriptions, particularly the intervention of the state, are also premised on the contention that the victims of a ruined common property resource are the least capable of initiating action to conserve it. An eloquent critique of the foundations of such policy prescriptions together with empirical examples of collective action in governing and conserving the commons has been provided recently by Ostrom (Ostrom 1990).

This article proposes to take the discussion a step further. It provides a case study of "appropriators of the common property resource"—artisanal fishermen of India, fishing in the coastal marine waters—who have taken collective action not only to conserve their fishery resource but *also to rejuvenate it.* Using the medium of artificial reefs, they not only evolved new material designs but also crafted new institutional forms in the process of their cooperative initiatives to resuscitate the coastal marine ecosystem damaged by the use of state-sponsored and market-oriented use of inappropriate technology.

The Fishery and the Fishing Community

India has an illustrious maritime and fishery history and features today among the world's top ten fishing nations. It has a marine coastline length of over 6,000 kms. and a vast network of rivers, lakes, and backwaters that yield a fish catch of over 3.6 million tonnes (Metric ton = 1000 kgs. or approx 2200 lbs). Two-thirds of this harvest is from the marine coastal waters and is netted by the half million active fishermen who belong to specific, traditional caste-bound, artisanal fishing communities who live on the geographic, economic, and social fringes of Indian society.

The most productive coastal marine fishery zone of India is its lower south-west coast region, comprising Kerala State and Kanyakumari District of Tamilnadu (hereinafter referred to as "the region"). It is also one of the world's most important sources of marine prawns. The annual sustainable fish yield from one km^2 of these coastal waters is estimated at 35 tonnes compared to the all-India average of 13 tonnes. On this basis the maximum sustainable yield (MSY) for this region is estimated to be 380,000 tonnes per annum. The major economic species are oil sardines, mackerels, and prawns.

This resource plenitude has led to the larger carrying capacity of this coastal zone and accounts for the greater concentration of traditional fishing communities along this coast than elsewhere in India. Though the region constitutes only a little over one-tenth of the country's coastline, it is home to about a third of the active marine fishermen in India.

The 130 km long coastline of the Trivandrum district of Kerala State and the Kanyakumari district of Tamilnadu is considered to be the most densely populated tract of the south-west coast both with regard to the population of fishermen and number of fishing craft. About 52,000 active fishermen reside in 90 villages along this stretch, and operate about 24,000 small traditional crafts—the majority *kattumarams*—made by tying four logs of soft wood together (see Map 1).

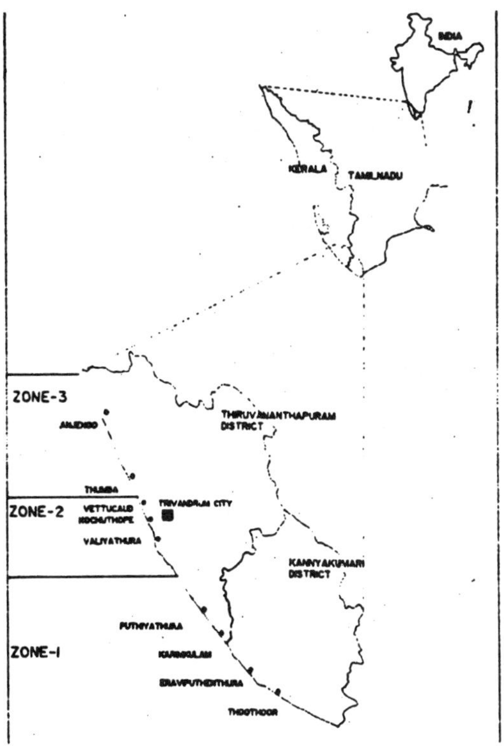

Map 1

This area is not only famous for its dense settlement pattern, but also for the immense diversity of fish resources in its coastal waters. The assortment of passive and selective fishing gear used by the fishermen to harvest these resources is remarkable: specializied small-meshed gill nets; large multi-meshed drift nets; trammel nets; bottom-set nets and traps; boat seines; shore seines and also a variety of hook and line sets.

The fishermen of the area migrate seasonally to other fishing centers in India, with our without their fishing equipment. They have thus come to be known for their skill and daring, their intricate knowledge of the coastal waters, their navigational acumen, and their ability to fish even at the margins of the continental shelf with relatively simple technology. Their knowledge is based on a long cultural continuum of habituated practice stored in the memory and passed on by those who work. It is practical knowledge conditioned into cultural practices. At its core is an elaborate understanding of the nuances of the aquatic milieu and the behavior pattern of living marine organisms.

Early Origins of the Artificial Reef Idea

The idea that certain types of objects when dropped in the coastal waters tend to attract fish has undergone considerable evolution among the fishermen of the region. They refer to all such sunken objects as *kritrima paar* (artificial reefs).

Artificial reefs are man-made or natural objects specifically placed in the coastal waters to attract fish, provide or improve fish or shellfish habitat, and increase fish biomass locally. Their design can range from haphazardly lowered local scrap materials to modern Japanese-style structures in the form of highly sophisticated modules built of concrete, fiberglass, or steel.[1]

Reconstructing from the oral tradition of the older fishermen we provide a brief history of the artificial reefs in the area. It is said that there was an age old practice, among the fishermen of the area who operated the shore-seine, of dumping rocks fastened with coconut fronds into the near shore sea within depths of 5 to 10 meters. Fish tended to aggregate over and around these rocks and were more easily netted by the shore-seines.

Large sunken structures also proved useful in serving the same purpose. It was known that a ship had sunk off the village of Anjengo during the Second World War because the local fishermen had rescued some of its crew. In 1949 a hook fisherman discovered the wreck at a depth of 50 meters. He was amazed by the high hooking rate over the limited area.[2] News of this new fishing spot spread quickly by word of mouth. Expert hook fishermen from the more densely populated southern villages of the area swamped the area and achieved greater success in fishing over the wreck by using artificial bait. The use of artificial bait was new to the fishermen of Anjengo and initially they opposed it.[3] But within a brief period the southerners settled down in Anjengo by marrying into the community.[4] Soon, the use of artificial bait to fish over the *kappal paar* (ship reef) gained widespread acceptance in the village.

During the early forties, a ship berthed at the Valiyathura Pier near Trivandrum city lost its anchors in a storm. Only one was immediately recovered from the sea bottom. The other anchor was found only a decade later when a hook fishermen got his line entangled in it. His loss was temporary because it was compensated by his simultaneous discovery of a rich fishing spot around the *nanguram paar* (anchor reef).

The early practice of dumping rocks with coconut fronds in the shore-seine fishing grounds and the discovery of the rich fishing spots atop sunken structures in deeper waters provided the basis for conscious attempts to erect artificial reefs in deeper waters.

The first examples of such artificial reefs date back to the early 1950s in the villages where natural reefs existed and where there was a tradition of hook and line fishing. The best example of such a reef was the one erected in 1955 in Puthiyathura. The hook and line fishermen of Puthiyathura had been fishing over the natural reefs found close to the shore (2.5 km. away) at a depth of 40 m. Over time they observed that fishermen from the neighboring village of Karimkulam— also fishing off a natural reef at about the same depth—got better harvests. The only difference that they observed between the reefs was that the one off Karimkulam was higher than theirs. This led them to try ways and means of raising the height of their own reef. They dumped granite stones piled on the beach to form anti-erosion sea-walls on the existing natural reef thus raising its height by about a meter. The fishermen observed that six months later their efforts paid off in the form of higher hooking rates.

Another early attempt in 1957 was made in Eraviputhenthura. The hook fishermen of this village had a favorite fishing spot about 1.5 km. off the coast at a depth of 22 m. The sea bottom here had a clay substratum and the larger concentration of fish observed there was attributed mainly to this. In their village the *panchayat* (local administration) authorities had used large concrete rings (3 m. in diameter and 0.5 m. high) to build a community well. One of the unused concrete rings remained on the beach. The fishermen got together, transported this ring on their *kattumarams* to their favorite fishing spot and dropped it there. This was the first known example of an external pre-fabricated structure being used in a known fishing spot as an artificial reef. The new reef was christened *Vatta paar* (ring reef).

Decline of Artificial Reef Construction

For almost a quarter of a century after the late 1950s little was heard of more attempts to create reefs. Several reefs however, were formed by accident and these continued to provide occasional fishing spots.[5] For a while, they generally proved to be good fishing grounds but then ceased to be productive.

The sudden loss of interest in reefs can be largely attributed to the new fishery development strategies of the 1960s. The importance given to the fishery sector changed rather dramatically. India was facing an acute foreign exchange crisis in the 1960s and every effort had to be made to increase exports. New harvesting technologies like bottom trawling were promoted in a big way by the government to cash in on the prawn boom following the marked increase in prawn prices in the international markets.[6] The coastal waters—once the exclusive preserve of traditional fishing communities who viewed the sea as their community asset—became a virtual open access resource to anyone who could afford to make the necessary investments in craft and gear.

The new harvesting technologies and the prawn export processing industry became the domain of a new class of capitalist entrepreneurs. They employed a small section of male workers from traditional fishing communities on the trawlers and women workers in the processing plants. The vast majority of the active fishermen continued their artisanal pursuits with the only major changes being the use of nylon for fabricating their nets and higher prices for their fish.

Fishing effort and fish harvests increased substantially. In Kerala State, from a level of 286,000 tonnes in 1961, the fish harvest increased to 337,000 in 1965 and peaked at 445,000 tonnes (significantly above MSY) in 1971. Since prawns were mainly found in the in-shore coastal waters there was virtually no increase in the fishing area following the introduction of the trawlers. A greater and more powerful fishing fleet was continuously harvesting the same stock of fish in a small area of the coastal sea. This resulted in a conflict over space and product with the large number of artisanal fishermen. From 1975, despite rising investments (or rather because of it) the harvests began to fall below the MSY level indicating that the resource *system* as a whole was adversely affected. The significant drop in productivity (in physical and value terms) resulted in lower income levels despite rising fish prices. A crisis was in the making (See Table 1).

Table 1 Select Data on Kerala State's Marine Fishery Sector

	1961	1965	1971	1975	1981
Fish Harvest (000 tonnes)	286	339	445	420	274
Harvest as % of MSY (MSY = 380,000 t)	7	89	117	110	72
Price of Fish (Rs/tn)	120	200	660	1760	2000
Value of Harvest (Rs.mn)					
Current Prices	35	68	293	740	548
Constant Prices (60–61)	35	48	138	197	144
Active Fishermen (000) 3	80.7	88.6	101.9	111.6	127.9
Fishing Crafts					
Non-Mechanized	20667	20900	21718	25100	26271
Mechanized	152	501	1780	2105	3038
Productivity per annum					
Physical (Kg/worker)	3540	3820	4370	3760	2140
Value (Rs/wkr-current)	430	770	2900	6600	4300
(Rs/wkr-const)	430	550	1370	1760	1134

Source: Kurien, 1987

The year 1979 witnessed the beginning of a socio-ecological movement spearheaded by the artisanal fishermen of the region, with the assistance of several social activists, formed a trade union. The union presented three major demands to the state authorities: (a) a total ban of trawl fishing during the monsoon period of June, July and August which was the breeding season for many species of fish; (b) reinstating the in-shore coastal commons to the artisanal fishermen in which trawlers would be totally prohibited; and (c) a greater share of assistance to the artisanal fishermen in the fishery development budget of the state (Kurien 1992).

By 1983, this movement had attained the dimensions of a major political force in the region. In a parliamentary democracy with numerous coastal constituencies, a restive fishing community was no solace for politicians, irrespective of their political color. Regardless of whether they were in or out of power, they began to take the fishermen's demands seriously.

In 1984, the second and third demands of the fishermen were conceded by the government. Legislation to demarcate and regulate the coastal waters was enacted and a police force to enforce these measures was set up. The emphasis on promoting prawn exports using state finances was relegated and the funds were earmarked for investment which would directly benefit the artisanal fishermen. A big push was given to upgrading their harvesting technology through the supply of outboard engines, beach landing crafts and new fishing gear. Welfare measures for artisanal fishermen and their families were also increased significantly.

The demand for a total monsoon trawl ban was conceded only in 1989 after three scientific commissions had studied the matter following the fishermen's refusal to compromise on this issue.

The importance of this socio-ecological movement was that it was the fishermen's first collective macro-initiative to re-establish their historical rights of exclusive access to the in-shore coastal waters. In the words of an artisanal fishermen's union leader, on his sixth day of an indefinite fast before the government secretariat demanding greater regulation of the coastal waters: "Our struggle is to ensure a future—for us and the fish."

Resurgence of the Artificial Reef Idea

The conflicts and impasse in the current pattern of fisheries development, the ecological crisis and the macro-level political mobilization around it, brought renewed vigor to the search for alternatives.

The gains of the fishermen's movement in pressuring the state to legally recognize their traditional rights over the coastal commons was no small achievement. However, neither the state nor the unions had the enforcement structure necessary to protect the claims to the coastal commons. The costs of such an enterprise would have been prohibitive. An important and practical way to overcome this, on the part of the fishermen, was to initiate steps to define and protect the coastal zone within limits of individual fishing communities. The village-level measures taken by artisanal fishermen to achieve this led to violent clashes with the trawlers, litigation by the trawler lobby, and accusations from the state that the fishermen were taking the law into their own hands.

The search for alternatives to rejuvenate their coastal commons was, however, on a different plane. At the village level, one important manifestation of this search was in the form *gramakuttoms* (village gatherings) at which fishermen and social activists discussed the need to revive, strengthen, and enhance their cultural beliefs and knowledge about fish and the nature-processes in the sea.

These discussions highlighted that fishermen had a strong conservationist ethic towards the fishery resource. They had a keen awareness and knowledge of the totality of the aquatic eco-system and viewed *Kadalamma* (mother ocean) as their community asset and a life-giving system rather than a hunting ground. Bringing ruin to her, individually or collectively, was something they could not comprehend.

Hook and line fishermen from the Trivandrum District were in the forefront of the village movement to help *Kadalamma* rejuvenate herself. Their success in fishing had depended exclusively on their encyclopedic knowledge of the natural reefs, sand banks in the sea and other fishing spots. They were the first to be affected by the damage to the demersal ecosystem caused by incessant trawling. Though they were victims of a tragedy of the coastal commons, they did not become strait-jacketed prisoners when faced with this dilemma.

FISHERMEN'S KNOWLEDGE ABOUT REEFS. Discussions with the younger fishermen illustrated very clearly that although two decades had elapsed since artificial reefs were constructed, knowledge about the reef-sea-fish interactions had been passed down to them from the older generation and kept alive by their own practice of fishing over them. There was a holism about their understanding which stemmed from their concern with the whole resource system rather than just the fish in it.

Fishermen consider reefs as an important basis for ecosystem rejuvenation. This association is premised on their understanding that underwater structures in the sea cater to the *adisthana avasyangal* (basic needs) of fish: their need to feed; their strong desire for protection, rest, and shade; and their urge to breed.

Consequently, for an artificial reef to be a source of food to fish, the kind of materials used to build it gained importance. The materials used should be those on which benthic vegetation would aggregate quickly thereby ensuring adequate food supplies. The artificial reef needs to be erected in areas where the sea bottom is naturally productive.

To serve the needs of protection, rest and shade for fish, the structure and the position of the artificial reef are determining factors. Only an artificial reef of sufficient height will provide shade. Solid structures are not conductive for protection and rest as they do not provide hiding places from predators. If fish are to make artificial reefs their breeding grounds, then the prerequisites of food and protection become imperative.

For fishermen to be able to catch the fish which use the artificial reefs in such a wide variety of ways, the reefs should be aligned on the seabed in the east-west direction. Given the north-south direction of littoral current in the region, this is the best alignment to ensure that the maximum number of fishermen can fish over an artificial reef at any given time without getting their hooks entangled. They have also learned that artificial reefs should be located in the "fish channel"—a path which was identified as being between 25 and 50 m. depth in the in-shore sea. Referring to the way this total understanding of fish behavior helped to induce fish into the artificial reefs, one fisherman remarked: "The fish teaches us and then we teach the fish (a lesson!)."

Sharing Knowledge. The process of sharing the knowledge which they had accumulated over the years was achieved by creating unique possibilities for inter-village discussions. The forum for this was provided by the Program for Community Organization (PCO), a voluntary organization—initiated by this author and a few other social activists—which had been in close contact with these fishing communities for over two decades since the early 1960s.

In the initial years, the work of the PCO was based on its association with groups of fishermen and their families in individual villages in the area. Land based issues like cooperative marketing of fish as well as employment and training for the youth were the focus of activities.

The crisis in the fishery in the late 1970s changed all this. The village specific "land-based" approach gave way to a more generalized "sea-based" strategy for mobilization and awareness creation among the active fishermen of the area as a whole. The focus shifted towards analyzing the various facets of the relationship between fishworkers and the sea, the ways and means by which these bonds were disrupted as a result of state-sponsored fisheries development, and the manner in which the links could be strengthened.

The first important outcome of this change of orientation was the cadre-formation training and movement-oriented research backup provided to the newly formed fishermen's union referred to above. The second, close on the heels of the first, was the support in the search for micro-level alternatives for "greening" the coastal commons.

The exchange of information regarding natural and artificial reefs was part of the second strategy. The initial attempt by the PCO was to encourage fishermen to articulate the slowly disappearing oral traditions of their knowledge system so that it could be documented, and, where possible, refined and complemented with the knowledge of modern science.

The pedagogy used for this process including arranging for the older and more experienced fishermen from villages with pre-1960 experience of fishing over reefs to share their knowledge with others. One revelation during these sharing sessions was the difficulty of fishermen in expressing physical reality on a two-dimensional scale using the blackboard. This difficulty was overcome by their own suggestion that it would be easier for them to explain the science of reef building using a small glass fish-tank or pond using stones and leaves.

These animated sessions, attended by fishermen from several villages in the area, laid the bare foundations for a people's science movement. It also served to foster a new group of leaders vastly different from those who spearheaded the unions. Charisma, public-speaking talents, and political acumen got low priority; fishing skills and hands-on grasp of technical details and community organization skills became more important.

The need for a better understanding of natural processes that take place underwater around an artificial reef highlighted the scope for a far greater degree of collaboration between coastal fishermen, marine biologists, oceanographers, and other scientific personnel. It revealed the exciting possibilities of undertaking what Norgaard referred to as a "coevolutionary development process" of the two knowledge systems. (See Norgaard 1984, Kurien 1987).

Such discussions, arranged by the PCO, gave a big impetus to the learning process. They also revealed the sharp differences in the approach of scientists and experienced fishermen towards the marine ecosystem. While the former were preoccupied with their ability to make precise descriptions about fish behavior and the nature of the sea, the latter were more adept at isolating the crucial interrelations which aided their ability to catch fish.

Promoting true dialogue was a slow and not too successful venture. For the fishermen, these sessions provided an opportunity to widen their horizons and provided the opportunity of getting to know what went on around them and also in other parts of the world vis-à-vis artificial reefs.

Towards People's Artificial Reefs (PARs)

The renewed initiatives among fishermen to erect artificial reefs needs to be situated against this background. Several groups of fishermen from as many as 22 villages in the area decided to take creative, collective action to erect reefs in the coastal waters off their villages. These actions need to be viewed as the micro-level expressions of their historic macro-level socio-ecological movement. We have termed the reefs erected in this process as "people's artificial reefs" or PARs.

During the years 1979–1989, the pace at which PARs were erected increased substantially (See Table 2). The stretch of coastline where PARs have been erected can be further divided into three zones with the city of Trivandrum, the capital of Kerala State, as the reference point. (See Map 1 and Table 3).

The PARs erected during this period witnessed changes both in relation to the materials used and the institutional processes associated with the erection of and access to the PARs.

CHANGING MATERIALS. The "first generation" reefs erected before 1960 used materials available at the village which the fishermen could obtain without money. This "whatever-you-can-get" approach to collecting materials to erect reefs continued until the 1980s. The fishermen of Eraviputhenthura who were the first to use a pre-fabricated structure (cement well ring) in 1957 erected their first PAR in 1980 with the large unused iron wheel of a damaged roller that was abandoned in the village. In Valiyathura an unused telephone post was used in 1983. However, with more organized and widespread efforts to build PARs, it was unlikely that such an approach could continue for long.

Table 2 People's Artificial Reefs Erected in Trivandrum and Kanyakumari

	Before 1960	1979–1983	1984–1989
No. of PAR's Erected	2	9	21

Source: Adapted from Kadappuram, 1989

Table 3 Percentage Distribution of PAR's Along the Coast

Area		Pre-1960	Post-1979
Zone 1	Villages south of Trivandrum City	100	40
Zone 2	Villages Around Trivandrum City	—	50
Zone 3	Villages north of Trivandrum City	—	10

Source: Adapted from Kadappuram, 1989

In the "second generation" PARs erected between 1984–86, it was noticed that while granite stones wrapped in coconut fronds remained the main component, three more important additional items were widely used: large rocks packed in monofilament net bags; the screw-pine plant; and cement well rings. All these items had to be specially ordered or collected specifically for the purpose and paid.

After 1986 there was a further spurt of changes. The use of old automobile tires tied to cement rings was tried out by the fishermen in Vilyathura. In Thoothoor, the granite stones were painted red before they were transported to the sea. Along with this came the measures taken to protect the PARs from being covered by fishing nets. Rings with built-in hooks, anchors and concrete pillars placed at the four corners of the PAR structure were examples of the measures taken by the fishermen.

Changing Institutional Forms. The discussions about materials and designs was accompanied by the concerns about the micro-level institutional processes and forms to be adopted in erecting reefs and having access to them.

The information which we have of the pre-1960 reefs (in Zone 1) indicates that the erection was the result of the initiative of a small group of fishermen in the village, but that the access to the reef was open to the whole village community. There were no perceived "private costs" to the group: the materials used were obtained free of cost and the labor involved in erecting the reef considered *shramadan* (voluntary labor). As a result the "promoters" of the reef were not overtly preoccupied with questions about property rights or private returns.

In the post-1980 period we observe a quiet accretion of social capital in the form of institutional designs associated with the PARs. For a more analytical appreciation of these changes it may be useful to consider two facets of the activity: (1) the initiatives for erecting PARs and (2) the claims for access to the PARs.

Initiatives for erecting reefs came from four quarters: individuals in the village, groups of fishermen within a village, the whole village community, and external agencies. These were not mutually exclusive type of initiatives. For example, certain leading individuals may play a crucial role in the initiatives taken by a specific group of fishermen to erect a PAR. Alternatively, an external agency—e.g., an NGO, the church/temple,—may financially or otherwise assist the whole village community in erecting a reef. By and large, these efforts were motivated by the concern to rejuvenate the "resource system" of the damaged coastal commons and thus rejuvenate the fish stocks.

The claims for access to the PARs came from two sources: a group in the village community or the whole village community. These access claims were motivated by differing understandings about the rights of appropriating "resource units" of fish from the PARs.

We shall enumerate below five "erection-access" combinations which have been observed in the post-1980 period. At present, some combinations co-exist and some of them are likely to emerge as dominant forms.

Individual Erection—Group Access. One of the first post-1980 reefs was erected in 1983 by an ex-serviceman from the fishing community in the village of Valiyathura (Zone 2). He organized the erection of a reef partly using material freely available on the beach (stumps of coconuts etc.) and also by arranging for the purchase of granite, cement rings etc. This reef came to be known after him as *Ouseph Parr* (Ouseph's reef). The total cash investment was estimated to be about Rs. 1000/ which Ouseph claims to have raised by pledging his wife's gold ornaments. He hoped to recover this investment by collecting a rent from the fishermen who were accorded access to the reef by him. The reef became productive in about three months and became known as a good fishing spot. Due to the difficulty of monitoring and preventing free access to the reef. Ouseph's hopes of profit were only partially realized.

Group Erection—Group Access. Ouseph Paar became the inspiration for fishermen to consider taking their own collective initiatives to erect reefs as the means of rejuvenating the coastal commons and enhancing their fish harvests. Small groups in the villages joined together, collected equal shares, and erected reefs.

Initially, these small group initiatives were restricted to the villages around Trivandrum city (Zone 2), where hook fishermen were in a minority (See Map 1). In 1984 the fishermen of Kochuthope, a village just north ofValiyathura erected a PAR which cost them Rs. 6000. Initially only the 100 hook and line fishermen who contributed to the erection had access rights to the PAR. Subsequently, when the PAR was built-up again, the membership was raised to 300. Their collective experience of restricting the use of the PAR to the members who contributed funds for its erection was fairly successful.

There was a spurt in the number of such "group erection—group access" PARs in the Zone 2 villages during the years 1984–85. The PARs were located just outside the range of the shore-seine operations (at about 15m depth) which was about 1.5 to 2.0 kms from the shore and just within keen eye-shot.

The fishermen of Zone 2 enumerated five distinct advantages of this location: (a) it prevented conflict with shore-seine operators; (b) it allowed easier monitoring of the PAR for "poachers," mainly from the neighboring villages; (c) it was within rowing distance, thus dispensing with the need for mechanical propulsion to reach the PAR and thereby reducing the costs of fishing; (d) proximity also made the PAR accessible to the older and the very young fishermen; (e) since the reef fish were fresh and brought quicker to the shore they fetched a higher shore price.

At no stage during these years (1984–87) were PARs considered to be providing a major source of the fish harvests. However, the positive experience surrounding the erection of and access to PARs resulted in a favorable disposition towards enhancing the scope and dimensions of such group initiatives.

Group and External Agency Erection—Group Access. In early 1988, the fishermen of Valiyathura, who were members of a village-level cooperative part of the apex organization called the South Indian Federation of Fishermen Societies (SIFFS), decided to erect a reef. They formed a group of 100 members with each one contributing Rs. 100 towards the cost of reef material. They approached SIFFS for a grant of Rs. 10,000. The amount was granted and the SIFFS also took the initiative to arrange for fishery scientists to be associated with the erection as well as the monitoring of the biological aspects of this PAR.

The nature of access rights did not materially affect the activities of the scientists who were monitoring the biological changes that were taking place underwater. However, to assess the economic viability of PARs—a task which SIFFS intended to do—restricted access rights would make the task much easier. The concept of "group erection-group access" was therefore supported by SIFFS on the grounds that it was easier to monitor such reefs to establish their economic viability since the fish harvest is restricted to a well defined group.

Group Erection—Community Access . Our inquiries reveal that at about the same time as the fishermen in the villages around Trivandrum City were erecting their "group erection-group access" PARs, fishermen in the villages south of Trivandrum (Zone 1) where hood fishing predominated and where natural reefs were once in abundance, were erecting PARs adopting what they termed the *utsava shylee* (festival approach).

The erection of and access to these PARs was analogous to the funding and celebration of a village festival. For village festivals, funds are collected from all the households in the village on the basis of a "whatever-each-one-can-give-happily" approach. Contributions should be in cash or kind. The actual initiative and hard work of organizing the festival is undertaken by a core group (a festival committee) whose primary gain is social recognition. Any number of volunteers are accommodated. While the grandeur of the festival is proportionate to the total funds raised, access to the fun of the festival day is open to all in the community irrespective of their contribution.

The analogy can be fully extended to the PARs erected by fishermen in Zone 1. These PARs are assets financed by voluntary contributions of the entire households in the village; erected by a group of able-bodied, young fishermen; with the access to the PAR being open to the whole village community.

Community Erection—Community Access. It was during discussions at the PCO that fishermen from the southern villages (Zone 1) heard about the system of "group access claims" to PARs erected by groups of fishermen in the villages around Trivandrum (Zone 2). They were astonished and remarked: "You don't show your narrowness at a village festival or at sea. Both are for all." A very animated discussion followed in which the implications of the philosophy of stewardship of natural research was raised. They were of the opinion, that as children of the sea and as those who know the secrets of the sea, such a narrow and partisan approach was not warranted *among themselves.*

The fishermen from the southern villages (Zone 1) suggested that while access to the PARs should be open to all, certain community agreements need to be evolved to restrict fishing effort by individuals. For example, a limit should be placed on the number and size of hooks. The use of lights to fish over the reef at night should be prohibited. Only one member of a household should fish at the PAR. These would ensure a more equitable distribution of the catch. Community sanctions for those who violated these norms were appropriate.

These inter-village discussions resulted in a self-critical review of the other approaches. Despite some initial resistance and skepticism—backed also by reservations to this approach from SIFFS—the fishermen in Zone 2 made a quiet transition of community access to PARs. The discussions transformed the situation from one where village communities took isolated decisions to one where the pros and cons of seemingly appropriate decisions could be tested against the merits of the more rewarding experiences of others.

The most recent initiative (1991) to create a PAR by the fishermen of the village of Thumba (for the first time in Zone 3) bears witness to the effectiveness of committed information exchange about meaningful experiences. In true festival style the whole community of Thumba was motivated to raise funds to erect a PAR which they claimed would help to rejuvenate a natural reef which was recently destroyed by the indiscriminate fishing of a large fleet of trawlers. They formed a *sahodara samajam* (brotherhood fraternity) in which one member from each of the fishermen households in the village was given membership. This was a means of formalizing total community participation in the complete venture. The *sahodara samajam* elected a "works committee" which would be responsible for erecting the PAR, deciding on the norms/restrictions on access, and for settling conflicts should they arise. Every household made a financial contribution. This fund was matched by an equal grant from their church.[7] Some technical assistance was obtained from the PCO on the appropriate shapes of the specially designed concrete structures to be used.

From the above description of the changing institutional forms we can discern a concerted movement towards great community involvement in the erection of and access to PARs. Sharing of experiences through a committed exchange of information facilitated largely by an NGO led to a steady process of accretion of institutional social capital. Implicit in this has been both a collective learning and self-transformation process for large sections of the community of fishermen. The result has been the growing support for an institutional-choice which spreads both the costs and the benefits more evenly within the community.

These changing institutional forms challenge the influential predictions that only state or market solutions can allocate and protect common resources. They also call to question the assumption that those who are caught in a "commons dilemma" would rarely invest time and money in the design and supply of institutions to conserve it.

The initiatives of the fishermen of the region also illustrate that people who have a very intimate association with natural resources as a source of livelihood, given the appropriate circumstances, can empower themselves to go beyond macro-level collective action aimed at *conserving* resources to micro-level initiatives for *improving* and *rejuvenating* them.

Enter the State

Policy makers, government bureaucrats, and scientists were unaware of the growing importance and numbers of PARs until this fact was brought to their notice by the Program for Community Organization in 1987. When first appraised of the matter, they viewed these initiatives as quaint since they did not fit into the usual patterns of development and management strategies initiated to meet the crisis of overfishing of the coastal commons.

Having been in the past decade preoccupied with ways and means of getting the maximum amount of fish out of the sea with a minimum of social conflict and law and order problems, it was not surprising that they initially considered the "throwing of granite into the sea" to be a waste of time, materials, and money. Measures such as providing outboard engines or better craft to artisanal fishermen to fish in deeper waters and putting into action a marine police force to regulate fishing in the coastal waters were more in line with their concerns.

With the spread of the idea of PARs among the fishermen and the scientists, policy makers have begun to take cognizance of this matter. Initiatives of the fishermen along the lower east coast of India (Tamilnadu) to build reefs and the role of NGOs specializing in rural technology in assisting their efforts has prompted the Department of Science and Technology of the Government of India to initiate a national program on artificial reefs. The purview of the program has now been extended to include the lower south-west coast region.

As a pilot project of this national program a prefabricated reef was placed in the coastal waters of the district of Alleppey in Kerala State. This was done with much official fanfare but no participation from the local fishing community. The initiative was publicized as a resource conservation measure. There were to be no restrictions on access to the reef.

Unfortunately, the last one heard of the reef was also on the day of the official function. Fishermen of this district who have for decades used encircling nets to harvest large shoals of pelagic fish (which have no need for a reef habitat) have never known the use of reefs. Consequently the estimated Rs. 50,000 invested in this state-sponsored project was really a misplaced investment in every sense of the term.

The Future of PARs: New Meanings, New Directions

The changes in material, designs, and institutional forms have resulted in PARs attaining a wide range of "meanings" to the artisanal fishermen.

PARs are no longer just the collective effort of "throwing granite into the sea." They have become the artifact around which fishermen can make creative use of their accumulated, transgenerational knowledge about the aquatic milieu and the behavior of fish.

PARs are also seen as the appropriate physical structures for a fencing of their exclusive fishing zone against the incursion of trawlers. PARs are the rallying point for collective action in evolving institutional processes and institutional forms for coastal resource rejuvenation.

While our description of the changing institutional forms highlights a noticeable convergence towards community *access* to PARs, the initiatives to *erect* the PARs are likely to receive the support of external agencies in the future. If the present level of investments can be maintained (on average, about Rs. 10,000–15,000 per PAR), community initiatives to erect PARs are likely to continue. Since community erection is closely associated with community access, an open question still remains: Will the coastal commons in the long run become "privatized" by village communities with differing economic endowments and fishing skills?

On the other hand, the success of PARs is creating a growing appreciation among the fishermen of the need for bigger and better designed physical structures for PARs—e.g., specially designed concrete structures. This is creating a latent demand for a greater level of financial investment in artificial reefs beyond the capacity of local village communities. The opportunity for more organized state involvement becomes an imminent possibility.

State investment will foster the negation of community access and favor *open access* for all active fishermen. It will also tend to "standardize" artificial reef construction taking out from it the "people's science" element and the popular, decentralized participation in creating this social asset. People would lose their autonomy and become passive recipients of the benevolence of a central authority. The values associated with PARs as concrete expressions of the people's will and resolve to nurture a damaged ecosystem will tend to get lost in the euphoria of reefs as a means of enhancing fish harvests and quick profits.[8]

This likely divergence in the future initiatives in erection of and access to PARs point to "higher order" dilemmas and choices of the future, (Ostrom 1990). The course that emerges will depend on a multiplicity of factors. Most important in this will be the manner in which these micro-level initiatives mesh into the larger macro-level actions which attempt at pressuring the state to redefine its power equations with the fishing community.

Lessons in Collective Action

The actions of the fishermen of the area in erecting PARs questions the conclusion that people confronted with a"Hardinian tragedy"of a ruined commons, will in accordance to Olson, *not* join hands for collective action unless coerced to do so. It also raises the issue about the assumption of the prisoner's dilemma game; that there cannot be any change in the structure of the game.

By examining the brief history of PARs we see that once those caught in a dilemma meet, discuss, exchange views, and learn from each other, it is unlikely that they will retain the status quo structures. They evolve new learning, new institutions and new trust among themselves—attributes which contemporary theories of collective action are only beginning to address.

We have observed how the resurgence of reef building, in the wake of the marine resource crisis, has also brought with it the reaffirmation of the community's responsibility to nurture the resource-system which is the basis of their survival. It has also provided a new, collective motivation for them to articulate, sharpen, and expand their knowledge base by greater interaction between themselves and with scientists and social activists. The resolve to build PARs has also provided fresh foundations for the spontaneous growth of new village leadership and widespread community participation. All of these are essential ingredients for sustainable collective action.

There is no claim that PAR building will make any substantial contribution to healing the wounds inflicted on the coastal ecosystem of the area. Nor can it be said that the future of these small-scale fishing communities and the common property resource which they consider to be their inheritance depend overwhelmingly on the success or failure of PARs as technological artifacts.

If an alternative development paradigm is to emerge, with sustainability of the coastal resource system at its heart, then the key to the future will be cultural, socio-economic and political empowerment, and even more importantly the social capital of institutions crafted during the collective action.

Endnotes

1. The extent to which artificial reefs increase fish biomass or redistribute existing stocks of fish is not clear. However, even if they do not substantially increase fish production, they can be used as effective fisheries management tools. The increased standing fish crop around artificial reefs reduces fishing effort and, therefore, saves time and fuel. Fishermen in developing countries often must limit their efforts because of high fuel costs. Furthermore, artificial reefs can be used to create fishing grounds for artisanal fishermen who use traps and hook and line gear. (BOSTID 1988).
2. Using their depth gauging plumb-lines the fishermen have estimated the wreck to measure 50 meters in length and 45 meters in width. The top of the wreck was just 15 meters below the surface. This spot was until recently (until 1985) considered to be one of the major fishing grounds of the region. Fishermen say that too many drift and gill-nets which got entangled to it have wrapped around the top of the wreck and reduced its productivity.
3. In the oral tradition of the village there is a couplet which alludes to this discovery and the use of artificial bait: "Sukkurappan kandupidicha kappal paar/Irayillathe meen pidichu

thekken maar" *(Translation):* Sukkurappan discovered the shipwreck (but the) Southerners fished over it without live bait.

4. Fishing communities of this region follow a matrilocal tradition.
5. The most interesting examples are that of the spent rocket heads (fired from the equatorial rocket launching station, Trivandrum) that dropped into the sea. Initially the authorities provided a compensation to fishermen who "discovered" these heads—when their nets got entangled on them and partially destroyed in the process. Soon the fishermen observed that the rocket heads became good fish aggregating devices.
6. International prawn prices began to increase with the stoppage of Chinese exports after 1949 and the rising per capita incomes in Japan. Prawns were not a preferred seafood in this region. It commanded a price much lower than that of other species of fish until this export boom.
7. The funds at the command of the church are completely raised by contributions from the daily earnings of the fishermen of the community. Such funds are controlled by the priest and a committee comprised largely of non-fishermen from the community. Church funds are normally utilized for building/extending church structures and for religious festivities. Against this background, the contribution for an occupational purpose attains great significance because of its potential demonstration effect along the coast.
8. This is likely to be an imminent possibility given the keen observation by fishermen that cuttle fish *(sepia pharonis)* which has a big export demand, spawn in the PARs. Whether fish other than specific "reef fishes" breed on artificial reefs still seems an unsettled issue among American and Japanese scientists. The former continue to debate whether artificial reefs actually increase productivity or merely attract and concentrate organisms from surrounding areas. Japanese scientists generally have little doubt that artificial reefs, *when properly designed, sited and placed* can be used to increase the productivity of desired species. (Sheeby 1982).

John Kurien is Associate Fellow at the Center for Development Studies, Trivandrum—695 011, Kerala State, India. He is an activist-researcher whose main interests are the political economy of fisheries development and issues relating to environment, development and collective action. *This is a revised version of a paper presented at the Indo-Pacific Fishery Commission's Symposium on "Artificial Reefs and Fish Aggregation Devices as Resource Enhancement and Fisheries Management Tools" held in Colombo, Sri Lanka in May 1990. The author acknowledges here the efforts of John Fernandez Kadappuram, and Titto D'Cruz, who through the Program for Community Organization motivated the fishermen of the region to articulate their knowledge and strive towards community-erected-community-used PAR's. He was also personally associated with some of the efforts of the fishermen in their erection of PAR's and also in their collective discussions. Both proved an invaluable learning experience. The author also wishes to thank Raj Kamini Mahadevan, Dilip Menon, and the anonymous reviewers of Human Organization for their helpful comments.*

References Cited

Berkes, F. 1986. "Local-Level Management and the Commons Problem: A Comparative Study of the Turkish Coastal Fisheries." Marine Policy 10:215–229.

BOSTID. 1988. Fisheries Technologies for Developing Countries. National Washington DC: Academy Press

Chopra, K. *et al.* 1990. Participatory Development: People and Common Property Resources. New Delhi: Sage Publications.

Coleman, J. 1988. Social Capital in the Creation of Human Capital. *American Journal of Sociology 94* (Supplement): S95–S210.

Jentoff, S. and T. Kristofersen. 1989. Fishermen's Co-management: The Case of the Lofoten Fishery. Human Organization 48(4):355–364.

Kadappuram, J. F. 1989. People's Technology in Fisheries, India. (Paper presented at the ITDG International Seminar, London)

Kurien, John. 1987."Knowledge Systems and Fishery Resource Decline: A Historical Perspective"*In* Ocean Sciences: Their History and Relation to Man. (Proceedings of the IV History of Oceanography Congress). Lenz, W. and Deacon, M. eds. Deutsche Hydrographische Hamburg: Zeitshrift.

Kurin, John. 1992."Ruining the Commons and Responses of the Commoners: Coastal Overfishing and Fishworkers' Actions in Kerala State, India" *In* Grassroots Environmental Action: People's Participation in Sustainable Development. Ghai, D. and Vivian, J. M., eds., London: Routledge.

Netting, R. McC. 1981. Balancing on an Alp. Cambridge: Cambridge University Press.

Norgaard, R. B. 1984. Coevolutionary Development Potential. Land Economics 60(2): 160–173.

Ostrom, E. 1990. Governing the Commons: The Evolution of Institutions for Collective Action. Cambridge: Cambridge University Press.

Sheeby, D. J. 1982. New Approaches in Artificial Reef Design and Applications. Annapolis: Aquabia Inc.

Siy, R. J. 1982. Community Resource Management: Lessons from the Zanjera. Quezon City: University of Philippines Press.

REVERSAL OF FORTUNE

DENISE HEYWOOD

As a lake it is a geographical phenomenon, but as Cambodia's economic lifeline the Tonle Sap is now under threat.

Cambodia's Tonle Sap, or Great Lake, is one of the geographical wonders of the world. Each year, during the monsoon season, a strange phenomenon occurs: the flow of water in a river running from the lake reverses and increases the lake's size more than fivefold. But the survival of this ecologically important area—on which Cambodia's largely agrarian and fishing society depends—is increasingly under threat.

Recognizing the importance of this natural legacy, the Cambodian Government has decided to seek special protection by nominating part of the lake—at its confluence with the Tonle Sap river, where migratory and resident species of birds and fish congregate—as a World Heritage Site. No other natural sites in mainland Southeast Asia, apart from a national park and wildlife sanctuary in Thailand, comes under its aegis. UNESCO's World Heritage Committee will make its decision in December 1995.

Situated in the centre of Cambodia, the Tonle Sap is connected to the Mekong—one of the three largest rivers in the world—by the Tonle Sap river. It is during the monsoon that the Tonle Sap river reverses direction and the waters from the Mekong drain into the lake, flooding an area of 13,000 square kilometres—nearly a seventh of Cambodia's total land surface—creating the largest freshwater lake in Southeast Asia. The villages of stilted houses on the lake's shores are forced to retreat two or three kilometres. In the dry season, from November to May, the lake reverts to less than a fifth of this size—an area of just 2,500 square kilometres.

This unique reversal phenomenon is thought to have started about 5,000 years ago, when the Mekong 'captured' the small Fleuve Bassac and Tonle Sap rivers which drain the lake. During the floods, the Mekong, which is at a higher altitude than the lake, rises so much that instead of draining into the South

China Sea, it flows back into the lake. The process is like the movement of a tide in an estuary: the water surges towards the lake, like water into a hole. When the lake has filled, and the river and lake levels have balanced, the flow ceases. After the rainy season, when the Mekong is low, the waters once again drain back south and away to the sea.

The annual flooding of the flat central lowlands ensures that the countryside is covered with nutrient-rich mud for growing rice, Cambodia's principal food staple. And, apart from attracting thousands of wetland birds, the lake's abundant supply of fish is an important food and income source for the local people.

But the Tonle Sap's role as the country's economic and dietary lifeline is now in jeopardy. For since the 1960s, sedimentation and uncontrolled fishing practices, have been threatening the lake's ecosystem. Khamliene Nhouyvanisvong, UNESCO's special representative to Cambodia, is so concerned that he says UNESCO must help whether or not it gains a heritage listing. "It's a must for the survival of the Khmer," he says. He proposes a zoning system to help protect certain areas of the lake from tourism, an area the government hopes to develop in the future. "We will divide the lake into zones, but we won't stop activities that have been going on for centuries."

Concern for the Tonle Sap was first raised in 1963 when French biologists J. Carbonnel and P. Guiscafre noticed that logging had depleted the forest surrounding the lake. Before the civil war of the early 1970s, Cambodia had 70 percent forest coverage; today, less than 30 percent remains. Cambodia's timber export industry earns the country between $40 and $50 million annually. The logic of a forestry policy that puts logging above all else, however, is now being publicly questioned. A few months ago, King Norodom Sihanouk blamed massive deforestation for the country's recent floods and drought. "Cambodia could become a country of wilderness like some countries in Africa," he warned.

Deforestation, and the resulting erosion of soil, has also increased the rate of sedimentation into Tonle Sap. The situation has been further aggravated by the gem-mining activities in Pailin, a Khmer Rouge stronghold in the west of the country. The blasting of river-deposited unconsolidated soil adds large quantities of sediment where the Pailin river drains into the lake.

Heng L. Thung, a geographical information systems' advisor to the Cambodia Environmental Advisory Team, a UN development programme, believes that the annual reversal of water has been the Tonle Sap's saviour. "Sedimentation is the natural death of a lake," he explains. "It enters the lake, weeds grow, interlaying occurs, and then, in a temperate climate it becomes a bog. In the desert, there is an alkaline deposit and it becomes like the Dead Sea.

"If this 'capture' by the Mekong had not taken place, the lake would slowly fill in and die."

Nevertheless, a recent report prepared by the French Institute for Scientific Research and Cooperation for Development, reveals that sedimentation has reduced the lake's depth—which varies seasonally from between two to 10 metres—resulting in smaller fish catches and the extinction of some species. In the last 15 years, Nhouyvanisvong estimates that more than 10 species have disappeared. The elephant fish, a species that is unique to the Mekong, is now endangered. "It only eats the grey shrimp. If the shrimp survives, then the elephant fish will survive. The real problem is the food chain."

Heng Thung does not believe, however, that fish numbers have declined. "How do they know?" he asks. "Who has counted the fish in the lake? The Department of Fishery is currently making a statistical inventory. But it is a complicated procedure. If the fishermen don't catch a particular species for six months, it doesn't mean that the fish doesn't exist. Maybe," he adds, "the fish have got smarter."

Until recently, the plentiful supply of fish has attracted thousands of birds and fish-eating water fowl to this important wetland. Today, certain species of birds risk extinction. Ornithological records are scarce, but extensive studies carried out by British ornithologist Peter Carr reveal 105 different bird species, many rarely seen elsewhere. These include the spot-billed pelican, great cormorants, Indian shags, Chinese pond herons, great egrets, Brahminy kites, yellow bitterns and whiskered terns. But Carr has observed a dramatic decline in large water birds, especially white pelicans, black-necked storks, black-shouldered ibis, Indian skimmers and Sarus crane.

"The Sarus crane," says David Ashwell of the International Union for Conservation of Nature, "is a globally threatened species. So is the greater adjutant stork which breeds only in Assam and here." As the largest relatively undisturbed wetland in Asia, Carr adds, its diversity makes it especially important for birds. "It requires protection to safeguard its riches for the future." says Carr.

Meanwhile, Cambodians this month celebrate the reversal of the river's flow in a three-day festival that dates back to the 12th century. Thousands of spectators will throng the riverbank in Phnom Penh as 200 brilliantly painted longboats complete a 1,000-metre course. The festival represents the importance water has played in Cambodia's past, and its vital role in the future. Whatever UNESCO's decision, it is certain that Tonle Sap's future will depend on how committed the government and the people of Cambodia are to protect and sustainably manage this natural legacy.

Denise Heywood is a free-lance journalist based in Phnom Penh.

CATASTROPHIC CHANGE IN SPECIES-RICH FRESHWATER ECOSYSTEMS:

The Lessons of Lake Victoria

LES KAUFMAN

There is at least an even chance of maintaining a diverse, productive, and healthy tropical ecosystem in Lake Victoria

Living communities change frequently. Hurricanes, volcanoes, predator outbreaks, and disease destroy old patches of forest and reef, making way for new species. Shifting continents redistribute the earth's great floras and faunas. Collisions with extraterrestrial debris may bring passage from one era of life to the next. Humans, too, are a force for ecological change, perhaps now the dominant force. Conversion of tropical forest to cropland or human habitation rapidly degrades species richness.

In the marine realm, the plight of the whales, sea turtles, and seabirds is widely known. Marine fishes and invertebrates are thought to be buffered against extinction by high fecundity and widely dispersed pelagic larvae (Robins 1991), but when more is known about marine animals more may be found to be endangered (Earle 1991, Kaufman and Strysky 1991, Tudge 1990). As for freshwater fishes and invertebrates, the evidence is more telling. Freshwater communities are exposed to the full brunt of deforestation, mining, poor land-use practices, exotic species introductions, and development. Therefore, it is not surprising that more than one in five freshwater fishes may be threatened or endangered (Kaufman 1989, Williams and Nowak 1986).

Entire freshwater fish faunas are disappearing (Avise 1990, Kaufman 1989, Tudge 1990, Williams and Nowak 1986, Williams et al. 1989). The infamous snail darter is but one representative of the vanishing mollusc and stream fish fauna of the Mississippi Basin and Appalachia. The desert pupfish is one of dozens of endangered desert fishes and insects in the US southwest and Mexico (Williams et al. 1989). Aquarium fishes bred domestically and available in pet stores are almost gone from their native haunts in Madagascar, Sri Lanka, and Borneo (Andrews and Kaufman in press). The endemic species flock of Lake Lanao, in the Philippines, is virtually extinct (Kornfield and Carpenter 1984). Though usually with fewer species to lose, high-latitude freshwater ecosystems are showing similar signs of stress and disruption (Taylor et al. 1984, Wells and McLain 1972, Williams et al. 1989).

The recent history of the fishes of Lake Victoria, East Africa (Figure 1), exemplifies a pace and magnitude of change that is alarming. In 1988, the World Conservation Union *Red Book of Endangered Species* listed the hundreds of endemic fishes of Lake Victoria under a single heading: "Endangered." The most exuberant expression of vertebrate adaptive radiation in the world (Table 1) is now in the midst of the first mass extinction of vertebrates that scientists have ever had the opportunity to observe, an event as exciting as it is depressing.

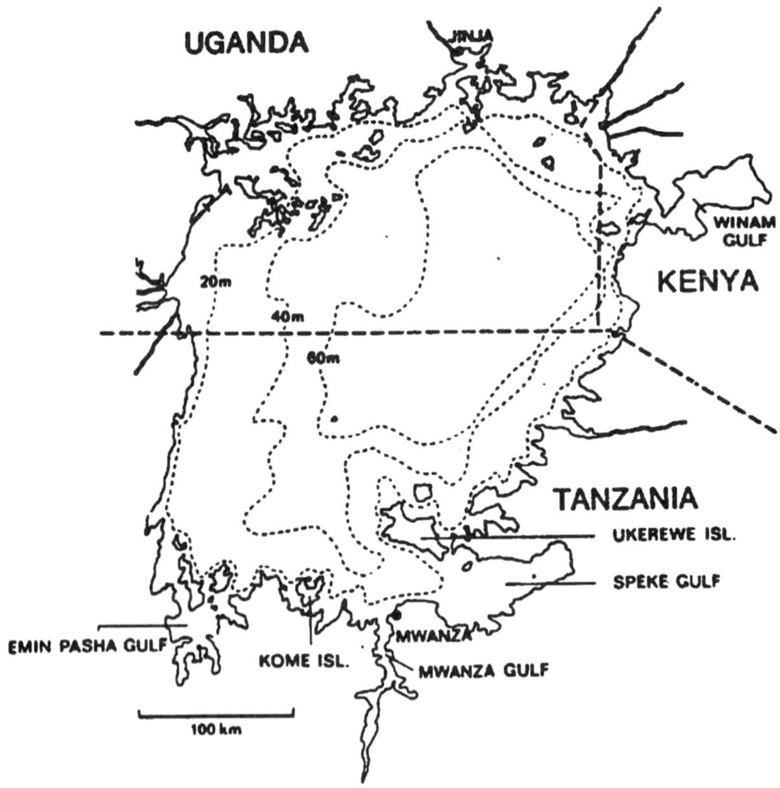

Figure 1. Lake Victoria, showing bathymetry. Note Winam Gulf in Kenya. The cluster of islands in the northwest sector is the Sesse Islands. Adapted from Witte et al. 1992.

Table 1 Examples of trophic diversification in Lake Victoria. Nomenclature follows Greenwood 1979, except for undescribed species, which are listed, by convention, as "Haplochromis" with nickname in quotes to avoid confusion with described taxa.

Trophic groups	Taxa
Plant eaters	
Pelagic phytoplanktivore	"Haplochromis""kribensis"
Detritivore 1 (soft algae)	*Enterochromis* spp.
Detritivore 2 (diatomaceous ooze)	*Oreochromis esculentus*
Rock scraper	*Neochromis* spp.
Plant scraper	*Haplochromis lividus*
Algal browser	"Haplochromis""kruising"
Leaf chopper	*Xystichromis phytophagus*
Algal gardener	*Neochromis nigricans* (?)
Arthropod eaters	
Crab eater	"Haplochromis""smoke"
Prawn eater	*Psammochromis* spp.
Generalized insectivore	*Astatotilapia* spp.
Rock picker	"Haplochromis""rockpicker"
Fat-lipped sucker	*Paralabidochromis chilotes*
Forceps-toothed picker	*Paralabidochromis victoriae*
Rock-reef, low-foraging zooplanktivore	"Haplochromis" *nyereri*
Rock-reef, high-foraging zooplanktivore	"Haplochromis""double stripe"
Soft-bottom zooplanktivore	*Astatotilapia piceata*
Sand/rock, pit-hunting zooplanktivore	"Haplochromis""deep black"
Pelagic zooplanktivore	"Haplochromis""argens"
Sand sifter	*Psammochromis* spp.
Parasite picker	"Haplochromis" *teunisrasi*
Mollusc eaters	
Winkler	*Ptyochromis sauvagei*
Oral snail crusher	*Macropleurodus bicolor*
Oral clam crusher	*Hoplotilapia retrodens*
Pharyngeal crusher	*Labrochromis* spp.
Fish eaters	
Ambush piscivore	*Pyxichromis orthostoma*
Pursuit piscivore	*Prognathochromis dentex*
Snout-engulfing paedophage	*Lipochromis maxillaris*
Egg snatcher	*Astatotilapia barbarae*
Scale eater	*Allochromis welcommei*

More than 30 million people who depend on the lake are feeling the consequences of roller-coaster changes in the fauna and lake environment. The most important freshwater food fishes in East Africa have already vanished from the marketplaces and very nearly from the planet. A fishery that once drew on hundreds of species, mostly endemic, now rests on three: a native pelagic minnow called the omena (*Rastrineobola argentea*) or dagaa in Tanzania; the introduced Nile perch (*Lates niloticus*), known as mbuta; and the introduced Nile tilapia *Orechromis niloticus* (Figure 2).

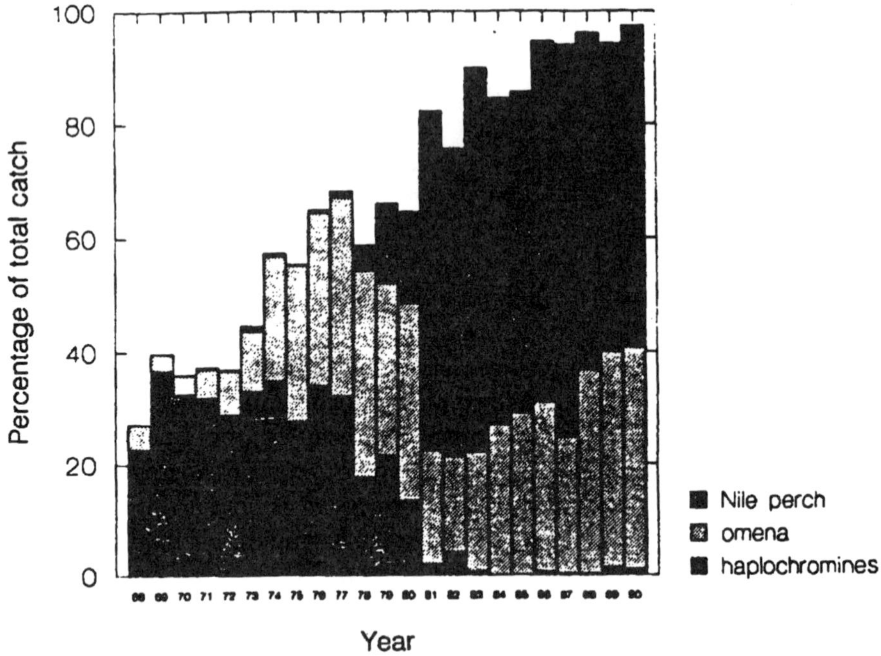

Figure 2. Percentage of the commercial catch in Kenya made up of haplochromine cichlids, omena, and Nile perch between 1968 and 1990. The remaining 40% to 70% of the catch before 1981 was dominated by native lungfish (*Protopterus*), catfishes (*Synodontis* and *Clarias*), carps (*Labeo* and *Barbus*), elephant-nose fishes (*Mormyrus*), and both native and introduced tilapiine cichlids (*Oreochromis* and *Tilapia*). (From Getabu 1987, FAO Report #388.)

Scientists, fishermen, and environmentalists have decried the loss of Lake Victoria's native species, and others have praised the introduction of Nile perch. Some have even referred to that fish as a savior. Now the savior threatens to destroy itself, the lake ecosystem, and a major source of protein in the midst of the world's fastest-growing human population. In Lake Victoria, as elsewhere, human welfare is intimately linked to concern for species conservation and ecosystem integrity.

The African great lakes and their fishes

To understand Lake Victoria, one must see both the lake and its fauna in the context of its giant East African neighbors, Lake Malawi and Lake Tanganyika (Table 2). The faunas of all three lakes exhibit the products of rapid speciation from very few ancestors. Best known is the fish family Cichlidae, of which more than 90% of the species in each lake are endemic (Greenwood 1984). Catfishes, mormyrids (elephant-nosed electric fishes), carps, gastropod and bivalve mollusks, insects, and crustaceans also have produced clusters of endemic species, but, with the possible exception of the Lake Tanganyika gastropods, these radiations are much less diverse and morphologically varied than those of the cichlids. Closely related but less species-rich cichlid flocks also occur in the

Table 2 Physical characteristics and biodiversity profiles for the three larger lakes of East Africa. Lakes Victoria and Malawi excel in species and trophic diversity. Lake Tanganyikan cichlids exhibit the greatest reproductive, form, and phyletic diversity.

	Lake		
Physical characteristics	Victoria	Malawi	Tanganyika
Surface area (sq km)	62,000	28,000	32,600
Maximal depth (m)	100	700	1470
Effective depth (m) (depth at which oxygen levels fall below 1 ppm)	100 (before 1986) 40 (after 1987)	200	150
Biological characteristics			
Species diversity (Cichlidae)	Medium (350+)	High (500+)	Low (170+)
Trophic diversity	High	High	Medium
Reproductive diversity	Low	Low	High
Form diversity	Low	Medium	High
Phyletic diversity	Low	Medium	High
Age of radiation (thousands of years)	14–225	<1500	3000+

nearby, smaller lakes Edward, George, Kivu, Kyoga, and Nabugabo (Greenwood 1974, 1981).

The cichlid faunas in the three great lakes in East Africa are strikingly similar and often cited as examples of evolutionary parallelism (Eccles and Trewavas 1989, Fryer and Iles 1972, Witte 1984). Three apparent lineages are present in the lakes: haplochromines, tilapiines, and lamprologines. The haplochromines are a species-rich and geographically widespread lineage. The tilapiines are species poor but also widespread. The lamprologines are species rich, but in the great lakes occur only in Lake Tanganyika.

As a group, the cichlids stand in contrast to the Nile perch and omena, both of which invest heavily in fecundity and little in parental care. The life history profile of cichlids in general entails small broods and extended parental care. Most East African haplochromine and tilapiine cichlids breed relatively few (5 to 100) large eggs and develop their young in their mouth, but the substratum-spawning lamprologines and tilapiines place large clutches (hundreds to thousands) of small eggs in nests on the lake bottom. All defend their young until they are self-sufficient. They have limited dispersal (except for seasonal inshore movements to spawn or release young) and strong site attachment—characteristics that should in theory make them highly vulnerable to extinction (Jablonski 1986, Gaston and Lawton 1990).

Curiously, this view is sharply at odds with certain other of the cichlids' attributes. They are aggressive, behaviorally and physiologically adaptable, phenotypically plastic, and prone to extraordinary evolutionary diversification (Avise 1990, Fryer and Iles 1972, Liem 1974). Much of this versatility has been attributed to a fundamental reorganization of the pharyngeal jaws that cichlids share with several closely allied fish families (Kaufman and Liem 1992).

Many cichlids can individually alter tooth and skull morphology in response to a change in diet (Greenwood 1965, Kaufman 1989, Meyer 1990, Sackley 1991, Witte et al. 1990). One cichlid, the Victorian snail-crusher *Astatoreochromis alluaudi*, was introduced to West Africa to help control the snail vectors for bilharzia, but these cichlids fed on insects instead of snails and stopped producing the massive dentition and musculature necessary to crush snail shells (Slootweg 1987).[1]

Cichlids have escaped from tropical fish hatcheries into the canals and everglades of south Florida, where they reproduce more successfully than do native sunfishes (Centrarchidae; Courtenay and Robins 1973, Hogg 1976, Taylor et al. 1984). Introduced all over the world as a ready source of homegrown protein for developing nations, tilapiine cichlids may have affected hundreds of native fish communities.

In short, cichlids have adapted to an incredibly wide range of conditions. One might expect such adaptability and ecological versatility to offer some measure of protection against extinction. That it has not done so in Lake Victoria may be one indication of the magnitude of change that has taken place.

Recent changes in the Lake Victoria ecosystem

The recent history of Lake Victoria is one of dramatic change in limnological parameters and native fishery stocks from the late 1960s to the present (Ogutu-Ohwayo 1990). Overfishing, exotic species introductions, deleterious land-use practices, and pollution from various sources all contributed to the oxygen depletion and mass extinction of indigenous fishes now taking place.

The earliest known anthropogenic changes in the fish assemblages were the result of overfishing. Gill nets and other modern fishing methods, introduced by the British in the early part of the century, led to the rapid depletion of important anadromous fishes that once had spectacular spawning runs up the rivers that flow into Lake Victoria, and of the ngege, *Oreochromis esculentus*, which originally supported the lake's most important fishery (Graham 1929, Ogutu-Ohwayo 1990).

Several exotic species of tilapia were introduced from other lakes and rivers during the late 1950s to replace the devastated stocks of native fishes. At first these introductions were not very successful, but the Nile tilapia, *O. niloticus*, eventually took hold and prospered.

High rainfall during the 1960s submerged and destroyed extensive beds of littoral vegetation, eliminating important spawning and nursery grounds for native species. This natural occurrence must have contributed to the decline of at least those native species dependent on nearshore habitats.

Beginning in 1934, Nile perch (*L. niloticus*) were introduced to the lake (Fryer 1960). Nile perch persisted as a minor component of the fauna for two decades, until the early 1980s. Then the species underwent a staggered series of population explosions in the eastern, northern, and southern waters of the lake. The trigger for the Nile perch irruptions is not known; it is interesting and mysterious that the fish should have persisted for so long and at such low densities before the explosion.

The leading hypothesis for the native fishes' demise is that they were consumed by the introduced Nile perch: the decline in native fishes and the increase in Nile perch population are almost perfectly reciprocal (Figure 2), Ogutu-Ohwayo 1990, Witte et al. 1992). Indeed, Nile perch seem to have regulated fish diversity in the past. Those East African lakes naturally inhabited by Nile perch have few haplochromine species, whereas those without the predator host species-rich communities (Greenwood 1981).

The dramatic disappearance of native fishes from Lake Victoria took place within a remarkably short time period, mostly between 1973 and 1982 (Witte et al. 1992). As recently as 1978, the haplochromine fauna was intact. Haplochromines contributed approximately 80% of the biomass and Nile perch less than 2%, with the remainder consisting of the introduced Nile tilapia and native noncichlids.

By 1983 in Kenya (Figure 3) and by 1986 in Tanzania, the native fish community had been virtually destroyed, and Nile perch comprised better than 80% of the catch. The remaining 20% consisted of Nile tilapia, the tiny native omena, and a small remnant of other native fishes. A community of more than 400 fish species collapsed to just three co-dominants: the Nile perch, the Nile tilapia, and a single

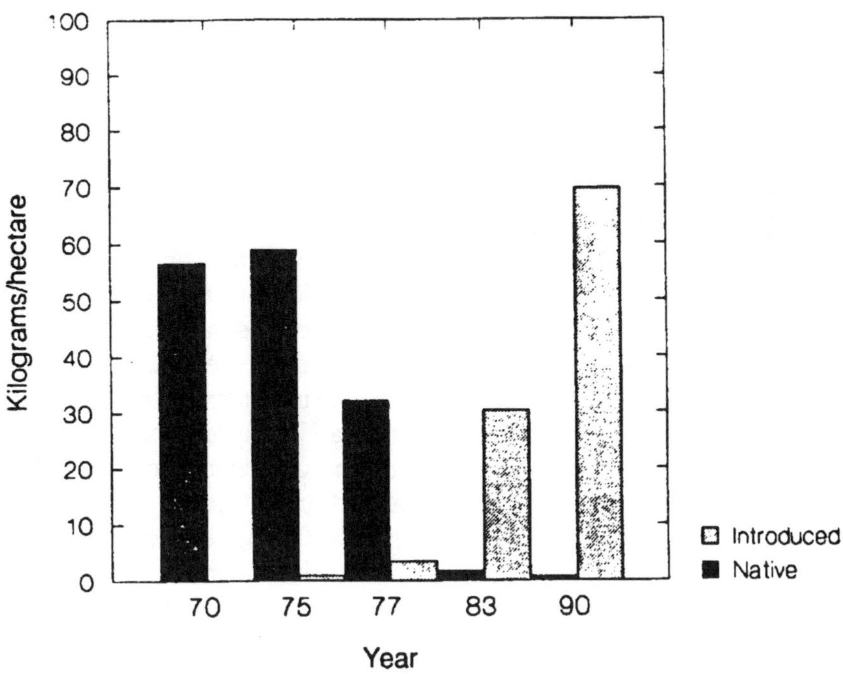

Figure 3. Demise of the native fishes of Lake Victoria, as illustrated by fisheries surveys in Kenyan water by the Kenyan Marine and Fisheries Research Institute. Standing stock estimates derived from the following numbers of hauls: 1969–1970 = 19; 1975 = 69; 1977 = 167; 1982–1983 = 54; 1989–1990 = 41. (Data courtesy P. Ochumba, A. Asila, and J. Ogari of KMFRI.)

indigenous species, the omena (Hughes 1986). Meanwhile, as its own supply of forage fish dwindled, the Nile perch exhibited an astonishing ecological shift. In essence, it has now become a whale, subsisting largely by scooping up great heaps of a tiny native shrimp, *Cardina nilotica.* The next most important item on the Nile perch menu is Nile perch; cannibalism is now a major link in the food chain.

At first, Lake Victoria was cited as a classic example of what can go wrong with alien fish introductions in tropical freshwaters (Barel et al. 1985, Balon and Bruton 1986). Local fisherman complained bitterly about Nile perch. The huge, clumsy fish destroyed the fishmen's nets and equipment, which had been designed for smaller quarry. Nile perch flesh is oily, and without refrigeration quickly rots. Sun-drying, a method of preservation that worked for the cichlids, does not work for large Nile perch. Nile perch could be smoked, but this method requires wood. Trees are in short supply around most of Lake Victoria, a decline that some think may have been accelerated by this new demand.

Such was the story of disaster told in the foreign press. But during the late 1980s the situation changed, as foreign aid and investors helped bring refrigerator trucks and processing plants to the lake shore. The fatty, flaky white flesh of Nile perch became popular on African hotel menus and a successful export to Europe and Israel. Scientists deplore the loss of haplochromine biodiversity (Barel et al. 1983, Witte et al. 1992), but the importance of biodiversity is readily questioned when a commodity of immediate value appears in its place.

The African view of the Lake Victoria's Nile perch is complicated, and largely misunderstood by outsiders who have caught only bits and pieces of this ecological soap opera (e.g., Avise 1990, Dawes 1986, Harrison et al. 1989, *New Scientist* 1988, Tudge 1990). The small-scale subsistence fishery that focused on the native cornucopia has taken a severe beating, with most species simply gone from the marketplace. My conversations with villagers in Kenya and Uganda revealed that, regardless of how they may feel about the Nile perch, they miss the native fishery with its great variety in taste and texture.

The revenues generated by the Nile perch fishery are much greater than those ever realized from the lake's native species, but the relationship between changes in the lake's fauna and changes in revenues generated by the fishery is not a simple one. Landings data, such as those presented in Figure 4 Kenya, suggest that the usable productivity of the lake increased by at least half an order of magnitude over 15 years. Landings data can be misleading, however. Because the traditional nearshore fishery was less mechanized, less efficient, and less wide- ranging than the new fisheries aimed at Nile perch and omena, even if productivity were unchanged, landings would be expected to soar. Some Ugandan and Tanzanian fishermen may be selling their catches in Kenya, where prices tend to be higher, thus skewing estimates of landings by country.

Experimental fishery surveys are less vulnerable to such bias and tell a strikingly different story. Experimental surveys are carried out on a routine basis by the Kenyan, Ugandan, and Tanzanian fishery research organizations; estimates of standing stocks for introduced and native fishes in Kenya's Winam Gulf region are presented in Figure 3. The sharp switchover from native to introduced species

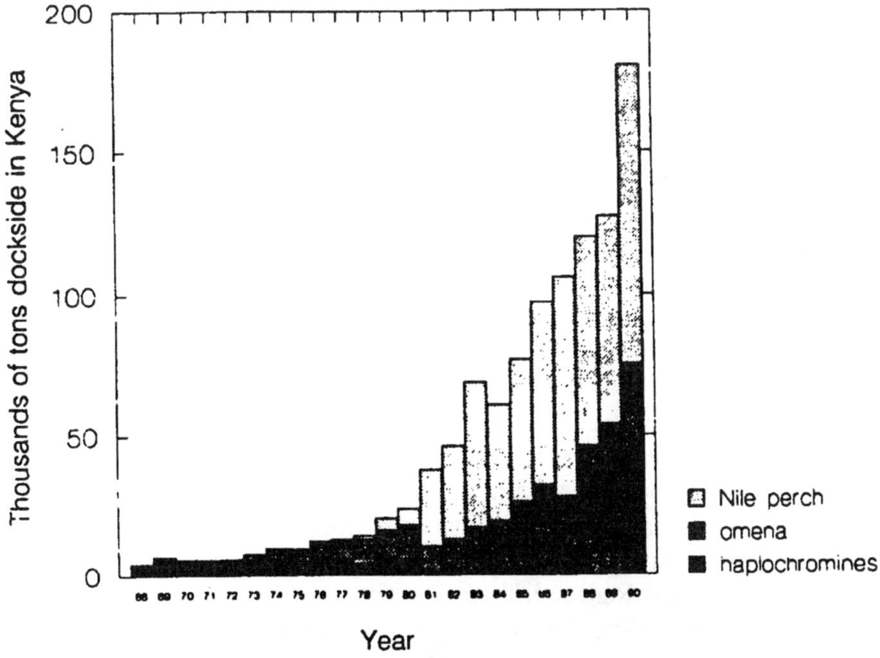

Figure 4. Total landings for haplochromines, omena, and Nile perch in the Kenyan waters of Lake Victoria, 1968–1990. (From Getabu 1987, FAO Report #388.)

is still there, but by these data, standing stocks of introduced species were only marginally higher in 1990, if at all, than were the stocks of native species 20 years earlier.

Although more tonnage is being landed, and more valuable tonnage at that, the total amount of fish biomass produced annually by the lake may not have changed as much as first thought. From a marketing standpoint, the new product may be easier to handle because it lacks the native species' bewildering variety. This change is analogous to the common practice of clearcutting rainforest and replacing it with one or a few fast-growing, easily processed, exotic timbers.

The distribution of wealth resulting from the Nile perch fishery is also different from that of the original, artisanal fishery. Some local fishermen may actually be worse off despite the apparent plenty. Large-scale operations that exploit the introduced species for foreign currency are doing well. The small-scale fishermen and fishmongers, who never went hungry and who relied for their livelihoods on the traditional tastes and interesting diversity of the native species, are a vanishing breed. The specter of protein malnutrition in the lake basin has been raised by socioeconomists, an incredible irony in a place exporting nearly 200,000 tons of fish protein annually.

Mysterious die-offs of Nile perch, and evidence that the stocks are being overfished, have raised additional concerns. There is serious worry about the future of Nile perch, a concern much stronger than the lingering nostalgia for native food fishes and the biodiversity of haplochromines. But most people do not realize that there is now an even more serious problem: the entire lake has been placed in jeopardy by profound changes in the structure and dynamics of the ecosystem.

Along with irruptions of Nile perch, the early 1980s brought another surprise: the regular appearance of dense algal blooms and associated low oxygen levels in the lake's shallow waters (Ochumba and Kibara 1989). Anoxia brought stinking masses of dead Nile perch to the surface. Fish kills had probably occurred on Lake Victoria before,[2] but not with such frequency or intensity. In March of 1987, the US National Undersea Research Program sent a remote-operated vehicle and team of investigators to Lake Victoria to meet with local scientists as an overture to a planned Large Lakes of the World Program.

During a series of demonstration dives on the lake, the group discovered that below 30 meters there was no fish life, and the bottom was littered with dead and dying organisms. This observation was the first hint that the anoxia described by Ochumba (Ochumba and Kibara 1989, based on data collected in the early to mid-1980s) might in fact be extensive in the lake's deeper waters and thus worthy of grave concern.

The Lake Victoria Research Team was formed to spur multinational research in the northern waters of the lake, as part of a larger Lake Victoria Research and Conservation Program coordinated through the New England Aquarium in Boston. The National Undersea Research Program provided seed support for hydrological and remote-operated-vehicle survey work, and the National Science Foundation funded preliminary biotic survey activities. The US Agency for International Development, the Pew Foundation, the Canadian International Development Research Council, and the Conservation Food and Health Foundation also assisted.

The results of the first phase of work by Lake Victoria Research Team members are alarming. It appears that coincident with the Nile perch explosion was a violent change in the physical environment of the lake. Before 1978, aerobic life penetrated into the lake's deepest waters. The lake was well-mixed, although oxygen was lower in the deeper waters (more than 60 meters) for brief periods during the long (approximately February through April) and short (December) rainy seasons (Talling 1966). Fish biomass was high even in deep water, though the rains may have initiated inshore migration by fishes requiring higher oxygen levels (Hoogerhoud 1983).

In contrast, now Lake Victoria appears to be stratified for the entire year. Seasonality is still apparent, but mixing is restricted to periods during June and July when the oxycline (the depth at which oxygen falls precipitously), though still persistent, sinks to approximately 50 meters[3,4] (Figure 5a). The region between 50 and 25 m is subjected year-round to frequent, severe deoxygenation. We have observed local upwellings of hypoxic water, along with extensive fish kills, at the surface near rocky islets. Thus, most of the lake's volume is incapable of supporting any but the most rugged species. Adult Nile perch themselves are probably not tolerant of prolonged hypoxia.

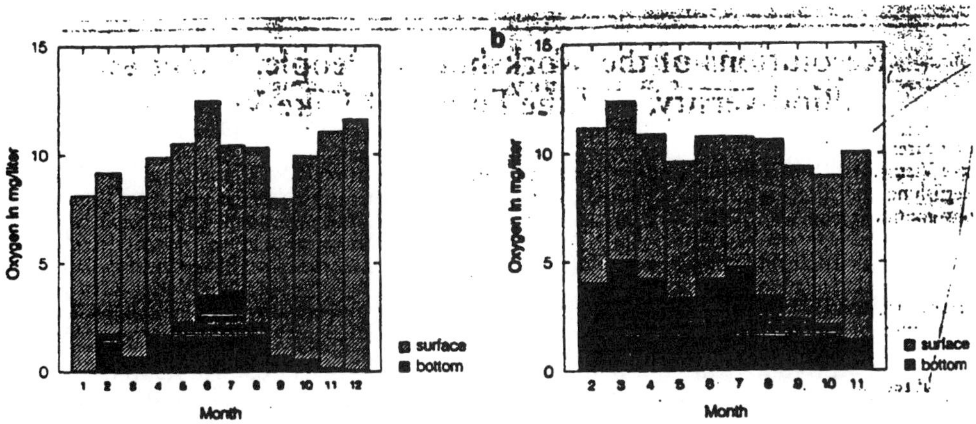

Figure 5. Mean monthly oxygen concentrations at the surface and bottom (depth 40–60 m) of Lake Victoria (measurements were made with a Hydrolab multiprobe). **a.** Open lake, based on six stations in the open waters of northeastern Lake Victoria (Kenya and Uganda). **b.** Rusinga Channel, Kenya, based on two stations located near the mouth of Winam Gulf. Note higher oxygen levels in Winam Gulf during those months for which comparative data are available. (From Lake Victoria Research Team; 22 missions conducted during 1989–1991.)

One interesting sidelight was the discovery that in the deepest portions of Rusinga Channel at the mouth of Kenya's Winam Gulf oxygen levels are consistently higher than at equivalent depths in the main lake (Figure 5b), from which the gulf is partially isolated by a shallow sill. This protected region is inhabited by several tiny, undescribed deepwater haplochromines.

Perhaps the greatest surprise from the work using the remote-operated vehicle was the observation that the lake's small, native detritivorous shrimp, *C. nilotica,* was abundant and active in the midst of massive kills of omena at and below the point where oxygen levels declined toward zero. This observation is important because it suggests that the shrimp might have a refuge from Nile perch. The predator, which as an adult is believed to require a level of at least 5 ppm oxygen in its water, would probably be unable to destroy this food base the same way that it wiped out the old one.

Myriad other aspects of the lake's ecology also appear to have changed: productivity and turbidity have both increased, papyrus swamps seem to be on the decline, snails seem greatly increased in abundance, locals say lake fly (mostly *Chaeoborus* swarms) are larger and more prevalent than before, and water hyacinth, a noxious floating plant introduced from South America, has invaded Lake Victoria from Rwanda via the Kagera River; now it is spreading and increasing in the lake. It will take a while to assess the validity, the timing, and the relative importance of such changes.

The collapse of the species-rich Lake Victoria ecosystem and its replacement by a highly simplified, largely exotic community raises a plethora of varied and important questions. The answers must address the dynamics of speciation and extinction, the wisdom of alien introductions, and the practices of fisheries management in the tropics.

What happened to Lake Victoria?

The recent changes in Lake Victoria are astonishing, but they are not without precedent: this lake has a tumultuous past. The Victorian Basin as a whole was formed approximately 750,000 years ago when crustal uplifting and doming reversed the westward flow of the Kagera and Katonga rivers into the new lake basin (Kendall 1969). The basin's lakes have varied in depth, shape, and number several times since its formation (Greenwood 1974, 1981, Schulz et al. 1990). On reflection, there must already have been several cycles of explosive speciation and mass extinction, with the recent faunas of lakes Victoria, Kyoga, Edward, George, Nabugabo, Kanyaboli, and Kivu being only the latest bursts in a series of massive speciation events (Greenwood 1981). Lake Victoria itself may have completely dried up only 14,000 years ago (Scholz et al. 1990.) The speciation rate that this history would imply (more than 350 species arising in 14,000 years) is overwhelming.

The current crisis probably began with the eutrophication of the lake basin. Algal productivity in Lake Victoria has increased fourfold over several decades, with a concomitant drop in water transparency (Ogutu-Ohwayo 1990). This change could be due to an increase in anthropogenic nutrient loads and/or changes in nutrient dynamics within the lake. Even before the Nile perch irruptions, the Lake Victoria ecosystem probably had already been altered by eutrophication. The huge stocks of detritivorous haplochromines observed before the collapse of the 1980s might have been at least partly a reflection of the increase in primary productivity.

It is remarkable that the Lake Victoria ecosystem was able to cope with increased productivity over several decades without an earlier collapse. This resilience may have been related to the high proportion of haplochromine cichlids, which eat virtually everything—especially each other, with more than half the haplochromine species once preying on other haplochromines (Greenwood 1981, Witte et al. 1992). In addition to the tight internal recycling that they provided, the haplochromines helped to move biomass and nutrients in the course of their vertical and horizontal migrations (e.g., Goldschmidt 1989).

This biological mixing could have been as important in lake dynamics as is the physical mixing of the water column. First, it could have retarded the accumulation of rotting biomass in the deeper portions of the lake. The importance of this effect should not be underestimated: 80% of the haplochromine biomass, or approximately 60% of the lake's total biomass, consisted of cichlids that fed in the lake's thick layer of detritus. The second effect of biological mixing was to expose much of the lake's biomass to potential export through predation on fishes by birds (Wannink 1990), snakes, crocodiles, otters, and other amphibious organisms, including, of course, the human population.

In sum, Nile perch may have decoupled the recycling capacity in Lake Victoria by converting the recycling machinery itself, (i.e., the haplochromines) into Nile perch flesh. A primary question is whether the present, still presumptive links from detritus to shrimp to Nile perch are indeed less efficient than those of the original food web. If so, then the impact of this cessation of recycling and biological mixing, combined with a long-term increase in anthropogenic nutrient loading on the lake and resulting increase in algal growth, could have caused a build-up of biomass in the lake's deeper waters. Biological oxygen demand would surge, and the deeper water column would be rapidly depleted of oxygen during periods of thermal stratification.

Native fishes, seeking reliable refugia in their seasonal moves between shallow and deep water habitats, may have faced a choice of death by asphyxiation in deeper water or death by Nile perch predation in the more oxygen-rich shallows. Each force—deoxygenation and predation—selected for a mutually exclusive refugium. It is most likely this combination of complementary impacts, rather than any one factor by itself, that brought about the rapid disintegration of the haplochromine assemblage.

Alternative hypotheses can be constructed, but their assessment is hindered by a lack of comprehensive data. For example, little is known about the potential for the new food web to functionally replace the old one. The shrimp *Caridina* is both a detritivore and, as seen from the remote operated vehicle, a facultative anaerobe, so it could transport organic material from the benthos into the lake's upper waters. We do not know, however, if the shrimps actually extend over much of the lake bottom, or if a significant portion of the population is available for consumption by Nile perch.

It is clear from temperate lakes, however, that changes in shrimp abundance can have profound impacts on aquatic food webs (Spencer et al. 1991). Lake fly larvae (*Chaeoborus*) such as those in Lake Victoria are well-known for their vertical migrations and tolerance of low oxygen levels, and when they metamorphose into winged adults, they could export benthic nutrients to the birds that feed on them. We do not know, however, if there are trophic links that could funnel benthic-derived nutrients into the fly larvae. One of the most interesting questions concerns the extent to which Nile perch juveniles can replace haplochromines as a manifold for nutrients to higher trophic levels (mostly adult Nile perch). Small Nile perch differ from large ones only in scaling, whereas 60% of the haplochromine species were piscivores that varied considerably in form and behavior as well as size.

Could it happen again?

The story of Lake Victoria, told and retold at scientific meetings, has become a kind of ballad, warning of the dangers of meddling with nature. Are other tropical lakes equally vulnerable to this kind of disturbance? Strong superficial similarities among the cichlid faunas of Lake Victoria, Tanganyika, and Malawi suggests that they are (Witte 1984). On the other hand, comparison of different facets of biodiversity in the East African lakes reveal properties unique to each fauna (Table 2). These differences are worth exploring in terms of their implications for conservation biology.

Lake Malawi has the highest species diversity of the three great lakes. Lake Tanganyika's cichlids exhibit the greatest variation in body form, as well as the greatest diversity in reproductive behavior. Lake Tanganyika also has the highest phyletic diversity—the greatest number of distinct, ancient lineages, each of which gave rise to independent radiations within the lake basin (Nishida 1991). Lake Tanganyika's great age and complex history probably explain the high phyletic diversity, which in turn provided the broad morphological substratum from which the lake's high form and reproductive diversity are derived. In contrast, the haplochromine flock in Lake Victoria exhibits four characteristics that clearly set it apart from the cichlid faunas of Lake Malawi and Lake Tanganyika:

- Species diversity out of proportion to age. The cichlid fauna of Lake Tanganyika could be anywhere from 3 to 100 times older than that of the Lake Victoria, yet Lake Victoria has (or had) two to five times as many haplochromine species as Lake Tanganyika. The implied rates of speciation in Lake Victoria are phenomenal regardless of whether one estimates age on the basis of tectonic, sedimentary, or molecular clues.
- High functional diversity. The Victorian haplochromines exhibit behavioral and dietary specializations comparable to those of the Malawian and Tanganyikan cichlids. In fact, Lake Victoria shares guilds with Lake Malawi that are absent in Lake Tanganyika. The best example is the functional group of piscivores known as baby suckers, or paedophages (Wilhelm 1980). Paedophages are predators with large mouths, thick rubbery lips, and odd, stout, outward-pointed anterior teeth. Paeodophages take the head end of brooding females into the mouth, suck out the eggs or young, and then eject the spent female to eventually produce another meal.
- Low phyletic diversity. The entire Victorian haplochromine flock is descended from a single ancestral stock, no more than 225,000 years ago (Meyer et al. 1990). Lake Malawi may be monophyletic, but it has had time to produce several distinct descendent lineages. Lake Tanganyika's cichlids are definitely derived from a diverse ancestry (Nishida 1991).
- Low form diversity. The high species and functional diversity of Victorian haplochromines is out of proportion to their limited variation in body form. The low form diversity greatly complicates taxonomic work with these organisms, though data on both behavior and morphology support their validity as biological species (Hoogerhoud et al. 1983).

These attributes make the Victorian radiation especially important to evolutionary biologists, but they also raise a red flag. Species that share so much in genetics and outward appearance may also share vulnerabilities.

No matter how adaptable they may be individually, all of the Victorian haplochromines are likely to share many of the same weaknesses. Just as genetic diversity among individuals may be critical to the long-term survival of a species, the genotypic and phenotypic diversity among species could provide a community with some hedge against mass extinction and ecological collapse. If so, Lake Victoria may have been unusually vulnerable to the dramatic events that have overtaken it.

Resolutions of the Workshop on People, Fisheries, Biodiversity, and the Future of Lake Victoria

After careful consideration of available evidence, based on research results, we conclude that the environment of Lake Victoria is changing rapidly and the fishery is unsustainable at its present composition and yield. Due to overfishing and the development of an export market, the amount of fish protein now available for local consumption is inadequate. The changing condition of the fishery will affect the social and economic welfare of millions of people in the Lake Victoria basin. Prospects for the future are cause for grave concern and reason for immediate action designed to resolve key unknowns and establish research and management policies that can be developed and applied expeditiously. The resolutions presented below outline the urgent need to understand current changes in limnology and the environment, to understand how fish biology affects the sustainability and management of the fishery, and to understand and mitigate the social and economic impacts of anticipated changes and varied management options.

General recommendation

It is necessary to form a Lake Victoria Fisheries Commission to harmonize research and management strategies for the Lake Victoria Basin. An important scientific function of the commission should be to foster close international cooperation, standardization of research and management methods, cross-calibration of scientific instruments, and continuity of monitoring.

Limnology and environment

- There is a need to develop general ecosystems model of Lake Victoria that includes the physical, chemical, biological, and human factors required to understand and predict lake productivity.
- Due to increased oxygen depletion, loss of fish habitat and fish kills are extensive. There is urgent need to understand the controls on oxygen distribution and levels in Lake Victoria, including the influence of low levels on fish stocks.
- Tremendous alterations in the food web have occurred in the last 30 years. Understanding the effects of these changes on water quality and lake productivity requires determination of the flows of nitrogen, phosphorus, sulfur, silicon, oxygen, and carbon into the lake ecosystem.
- There is need to determine the energy flow through the two major trophic pathways (grazing and detritus) that couple fish productivity with primary production.

- A research and management program for the wetland and forest habitat in the riparian zone of the lake and its waterways should be undertaken, with special attention to the water hyacinth.

Fish biology

- We must quantify interactions among food-web structures, water quality, and life-history characteristics of fishes in order to understand the relationship between fish production and the causes of eutrophication and anoxia in Lake Victoria.
- There is continuing need for stock assessment such as that now underway with support from the European Community. Such assessments are necessary to guide ecosystems modelers, policy makers, fisheries investors, and fisheries managers.
- Refuge areas (fish parks) should be established to protect the diversity of the native fish species and maintain spawning stocks of commercially important species.
- The possible contribution of aquaculture and management of alternative fisheries to the fishery industry should be fully explored, including their potential for restoring and rehabilitating indigenous stocks.
- Proposals for the introduction of exotic flora and fauna to the Lake Victoria ecosystem should be submitted to the Lake Victoria Commission for evaluation. Such proposals must be treated with great caution and rigorously evaluated with respect to biological, economic, and social impacts. Introductions must meet with the approval of all three riparian nations. Management of species already introduced is essential to the welfare of existing fisheries.

Policy, management, and socioeconomics

- The quality of runoff water from the land is deteriorating and may affect fish habitat. Fishery scientists should define water quality requirements of the fishery and work closely with limnologists in drafting overall guidelines for the water quality of Lake Victoria.
- Proposals for fisheries planning and development should be submitted to the Lake Victoria Commission for detailed social and environmental impact assessment before implementation. All proposals must be cognizant of the changing patterns of ownership and the control of harvesting and processing facilities. They should also analyze the differential incomes received by the various participants in the fishery industry.
- The fisheries planning process needs to be integrated with the development of information systems encompassing regional, social,

economic, and biological data. Programs should be developed to assist those groups most likely to suffer the effects of predicted declines in the current fishery.

- Research should identify the factors that facilitate or impede local participation in fisheries management, and fisheries officials should be familiarized with programs in other countries that involve local fishers and fisheries management.
- The supply of fish within the lake basin available to local populations is insufficient to meet their needs. Studies should be conducted to determine if the expansion of the Nile Perch fishery and export marketing have contributed to reduced dietary protein and malnutrition in local people.
- An economic assessment of the value of the native fauna is needed, taking into account their value for food, aquarium trade, medicine, and tourism.
- The feasibility of levying a tax on fish exported from the basin to support the research, management, and monitoring of the fisheries should be determined. Proceeds should be directed initially toward the effective implementation of regulations on fishing gear, methods, and allowable catch.

Workshop was held in Jinja, Uganda, 17–29 August 1992. It included scientists from Uganda, Tanzania, Europe, and North America. The meeting was sponsored by the US National Science Foundation.

What can be done?

Talk of large-scale manipulation to improve or "rescue" Lake Victoria never fails to fuel heated debate, because not everyone regards management of severely damaged ecosystems as a practical enterprise. At the rate that tropical ecosystems are deteriorating, however, it would behoove us to begin to think seriously about management.

There is some room for optimism. Huge Lake Erie, declared "dead" only 30 years ago, is once again an important recreational and fishery resource. The success in Lake Erie is empowering, so long as it is understood that Lake Erie was rehabilitated, not restored; the present biotic assemblage is dominated by introduced species. In addition, the story of Lake Erie is not finished. The introduction and spread of the exotic zebra mussel (Roberts 1990) is a new challenge.

To do nothing at all in Lake Victoria would clearly be foolish, because there are genuine threats facing the people of the lake basin. Most immediate is the threat to the fishery posed by the lake's hypoxic bottom waters and physical instability. There have been many local fish kills caused by mixing events that have trapped fish populations in oxygen-depleted water. Exploration with remote-operated vehicles has shown that the fish kills seen on the surface are even more extensive on the lake bottom at 30 m and deeper. Thus, a large-scale fishery collapse from a major mixing event is a real possibility. Another threat may come from sharp reductions in the abundance of snail-eating fishes, which will probably affect the prevalence of bilharzia. This effect can be complex, either raising or lowering the incidence of disease according to the age distribution of the snails.

Wholesale species extinctions have made it impossible to restore Lake Victoria, but that observation does not rule out an attempt to rehabilitate it. Cautious management of the watershed, the lake ecosystem, and fishery could have positive effects. Much more must be known about the lake ecosystem before any sophisticated manipulations can be attempted, but meanwhile there are numerous actions that can bring definite benefits to the people of the region.

The identification and reduction of anthropogenic nutrient inputs is one potential action. This challenge could be combined with afforestation projects with indigenous species to rebuild some of the native forests and their biodiversity, reduce materials and contaminants in runoff, and help restore the dwindling supply of firewood. Careful management of the Nile perch population could aid both the Nile perch fishery and the survival of native species. The establishment of separate management areas for native and introduced species could be a useful approach, because there do appear to be distinct refugia, or at least geographically circumscribed remnants of the native fauna. Native food fishes, if kept from extinction, can certainly be restored to the marketplace through aquaculture.

Many of the native species are great value, and their remnant populations are potentially significant economic assets. For example, the anadromous ningu (*Labeo victorianus*) lends itself well to hatchery-style aquaculture and in situ restoration efforts, much like those employed for North American trout and salmon species. The ngege, though smaller and slower growing than the introduced Nile tilapia, is reputedly the most delicious tilapia species in the world. Its genome is a prize for the tilapia aquaculture industry, whose products are derived from selective breeding and hybridization of wild gene pools. The ngege is also one of the most specialized phytoplanktivores among the East African lake fishes; it could someday assist in cropping algal blooms in Lake Victoria.

Similar arguments can be made for the lake's haplochromines. They can be eaten (though never among the most desirable species, they were locally important both for food and medicinal use), help control disease-carrying snails and insects, serve as forage for larger edible fishes, and offer intriguing possibilities for polyculture because of their functional diversity.

Perhaps the greatest opportunities for positive change come from development projects now slated for the lake basin. Though traditionally, thought of as harbingers of further devastation, plans to expand industry or tourism can, if properly planned, and executed, bring great benefit to the local people, to the investors, and to the long-term security of biodiversity and the environment.

An excellent example of this potential, as well as the potential for disaster, is a project planned for Uganda's Sesse Islands. More than 80 small islands and islets located in the lake's northwest corner, the Sesse Islands, are still largely forested and are an important conservation area for flora and fauna, both terrestrial and aquatic. They would be an ideal location for experiments in sustainable development, habitat conservation, and species restoration. Whether the investors can muster the wisdom, the foresight, and the morals to conduct this project in an exemplary manner remains to be seen. If they fail, Uganda will lose one of the last shreds of the Lake Victoria Basin still in a reasonably intact state.

There is at least an even chance of maintaining a diverse, productive, and healthy tropical ecosystem in Lake Victoria, drawing on both native and introduced species as building blocks. The details of the new system depend on the ingenuity of restoration biologists, multinational cooperation, and socioeconomic vision. Most important, those involved must be prepared to accept the fact that some aspects of the system will be highly nonlinear or totally unpredictable and therefore beyond control.

There is no lack of dissenting views. Colleagues and proposal reviewers have stated that Lake Victoria should be written off so that scarce resources can be focused on conservation in Lakes Malawi and Tanganyika. This idea is not popular around Lake Victoria. It is considered in the same vein as the view that AIDS is beneficial because it will solve the problem of overpopulation in the region. The continued presence of Nile perch in the lake raises obvious doubts about reintroduction, but data from Tanzania, Kenya, and Uganda indicate that at least 10% of the original fauna in the well-sampled nearshore region is persistent and sympatric with high Nile perch densities. It would pay to learn which taxa are resistant to extinction in the lake as it is today, why they are resistant, and what other native species might with some assistance also have a chance to survive.

To the scientist, the haplochromines of Lake Victoria have a great deal to teach about both speciation and extinction. To the conservationist, the remnants of the native fauna are the building blocks for attempting to reconstruct the Lake Victoria ecosystem. We echo Aldo Leopold's advice to intelligent tinkerers: do not throw away the pieces. These pieces are the most critical of all resources for the future, provided that a safe, productive context for them can be created in a future ecosystem.

In 1987, the New England Aquarium proposed programs to save the pieces of three disintegrating fish faunas through captive breeding: Lake Victoria, North American desert fishes, and Appalachian stream fishes (Kaufman 1987). Lake Victoria was chosen as the first trial project, and today the program engages the efforts of more than 30 institutions, led by the New England Aquarium and the Columbus Zoo aquarium.

Captive breeding programs only make sense as part of in situ conservation efforts (Kaufman 1987, 1989, Reid 1990). Ironically, their greatest contribution may be in acting as a wedge to build awareness and garner institutional and grant support for programmatic work, such as that of the Lake Victoria Research and Conservation Program. Certainly the fauna rescue and captive breeding program of Lake Victoria has had this effect, even though it had its humble beginnings in an attempt to maintain an array ofVictorian haplochromines as a living archive for research purposes, an effort initiated in the mid-1980s by researchers at the University of Leiden and the British Museum. The Lake Victoria Research and Conservation program picked up where the earlier research left off.

Now it is possible to contemplate using rescued haplochromines in polyculture and reintroduction. Furthermore, the experiences with the Lake Victoria breeding program can inform efforts to conserve freshwater fish species around the world, the focus of a workshop held in Columbus, Ohio, in October, 1992. Hopes are running high for coordinated ecosystems, fisheries, and resource management planning for Lake Victoria as a whole.

The Lake Victoria Research and Conservation Program has focused on three objectives: to understand biological processes in the lake and its watershed well enough to offer management alternatives to decision makers in the three countries; to track and where possible prevent the extinction of natives species; and to restore endangered food fishes to the marketplace through aquaculture. In August 1992, scientists, policy makers, and government administrators met on the Ugandan shoreline of Lake Victoria to discuss the condition of the lake and the further steps that might be taken.

Preservation of virgin natural ecosystems, the world's"last great places"in the words of The Nature Conservancy's celebrated program, is a major preoccupation of conservationists, and deservedly so. But perhaps this approach is too limited to the situation of the moment. Lake Victoria may inspire some attention to salvaging orphaned pieces of ecosystems that can never again be precisely as they were.

Conservationists scan the scars of civilization for those places where the odds and the payoff for biological salvage and reconstruction are the greatest. On land, such a place may be Madagascar, Hawaii, or the dry forest of Costa Rica (Janzen 1986). On water, it is without question Lake Victoria. The value of conservation efforts, even in badly damaged ecosystems, should not be doubted. Anything is better than waiting 20 millennia for a new fauna, or a new source of food, to evolve.

Acknowledgments

Seed funds for the Lake Victoria Research and Conservation Program were provided by the New England Aquarium for coordination and public education, the National Undersea Research Program for hydrology and visual exploration by remote vehicles, the National Science Foundation for preliminary faunal survey and research planning, the Environmental Protection Agency for rescue and propagation of endangered native food fishes, the Institute for Museum Services for haplochromine captive breeding program, US-AID and IDRC of Canada for limnological studies and the Conservation Food and Health Foundation for work by African colleagues. The author is grateful to the Pew Scholars Program for Conservation and the Environment for core support. The data and insights presented are the result of a team effort, led in particular by P. Ochumba, J. Ogari, and E. Okemwa (Kenya) and F. Bugenyi, R. Ogutu-Ohwayo, P. Basasibwaki, and B. Kudhongonia (Uganda), with major contributions by W. Cooper, M. Gophen and R. Hecky, P. Kilham, and R. Robinson. The author thanks P. H. Greenwood, F. Witte, R. Lowe-McConnell, R. Welcomme, and five reviewers for their stimulating comments. We applaud the efforts of our other colleagues and friends of the fisheries research institutes of Kenya, Uganda, and Tanzania, who have worked for so long, with so little, to understand Lake Victoria and brings its benefits to their people and the world. This article is dedicated to the memory of Peter Kilham.

References cited

Andrews, C., and L. S. Kaufman. In press. The role of captive breeding in the conservation of endangered freshwater fishes. *6th Conference on Breeding Endangered Species.*

Avise, J. C. 1990. Flocks of African fishes. *Nature* 347:512–513.

Balon, E. K., and M. N. Burton. 1986. Introduction of alien species or why scientific advice is not heeded. *Environ. Biol. Fish.* 16:225–230.

Barel, C. D. N., R. Dorit, P. H. Greenwood, G. Fryer, N. Hughes, I. B. N. Jackson, H. Kanawabe, R. H. Low-McConnell, F. Witte, and K. Yamaoka. 1985. Destruction of fisheries in Africa's lakes. *Nature* 315:19–20.

Courtenay, W. R. Jr., and C. R. Robins. 1973. Exotic aquatic organisms in Florida with emphasis on fishes: a review and recommendations. *Trans. Am. Fisheries. Soc.* 102:1–12.

Dawes, J. 1986. Lake Victoria cichlids face extinction. *Aquarist and Pondkeeper* July:22.

Earle, S. 1991. Sharks, squids, and horseshoe crabs—the significance of marine biodiversity. *Bioscience* 41:506–509.

Eccles, D. H., and E. Trewavas. 1989. "Malawian Cichlid Fishes: The Classification of some Haplochromine Genera." Lake Fish Movies, Herten, Germany.

Fryer, G. 1960. Concerning the proposed introduction of Nile perch into Lake Victoria. *E. Afr. J. Agric.* 25:267–270.

Fryer, G. and T. D. Iles. 1972. *The Cichlid Fishes of the Great Lakes of Africa: Their Biology and Evolution.* Oliver and Boyd, London.

Gaston, K. J. and J. H. Lawson. 1990. The population ecology of rare species. *J. Fish Biol.* 37(Suppl. A):97–104.

Getabu, A. 1987. Aspects of the Lake Victoria fisheries with emphasis on *Oreochromis niloticus* and *Alestes sadleri* from the Nyanza Gulf. Kenya Marine and Fisheries Research Institute, Kisumu, Kenya.

Goldschmidt, T. 1989. An ecological and morphological field study on the haplochromine cichlid fishes (Pisces: Cichlidae) of Lake Victoria. Ph.D. dissertation, University of Leiden, Leiden, Netherlands.

Graham, M. 1929. The Victoria Nyanza and its fisheries: a report on the fish survey of Lake Victoria 1927–1928 and appendices. Crown Agents for the Colonies, London.

Greenwood, P. H. 1965. Environmental effects on the pharyngeal mill of the cichlid fish *Astatoreochromis alluaudi. Proc. Lin. Soc. Lond.* 176:1–10.

_____. 1974. Cichlid fishes of Lake Victoria, East Africa: the biology and evolution of a species flock. *Bull. British Museum (Natural History) Zoology Suppl.* 6:1–134.

_____. 1981. Species flocks and explosive evolution. Pages 61–74 in P. H. Greenwood and P. L. Forey, eds. *Chance, Change and Challenge: The Evolving Biosphere.* Cambridge University Press and British Museum (Natural History), London.

_____. 1984. African cichlids and evolutionary theories. Pages 141–154 in A. A. Echelle and I. Kornfield, eds. *Evolution of Fish Species Flocks*. University of Maine Press, Orono.

Harrison, K., O. Crimmen, R. Travers, J. Maikweki, and D. Mutoro. 1989. Balancing the scales in Lake Victoria.

Hogg, R. G. 1976. Ecology of fishes of the family Cichlidae introduced into the fresh waters of Dade County, Florida. Ph.D. dissertation, University of Miami, Coral Gables, FL.

Hoogerhaud, R. S. C., F. Witte, and C. D. N. Barel. 1983. The ecological differentiation of two closely resembling *Haplochromis* species from Lake Victoria, *H. iris* and *h. hiatus* (Pisces: Cichliadae). *Neth. J. Zool.* 33:283–305.

Hughes, N. F. 1986. Changes in the feeding biology of the Nile perch, *Lates niloticus* (*i.*) (Pisces: Centropomidae), in Lake Victoria, East Africa since its introduction in 1960, and its impact on the native fish community of the Nyzanza Gulf.

Jablonski, D. 1986. Mass extinctions: new answers, new questions. Pages 43–61 in L. S. Kaufman, and K. G. Mallory, ed. *The Last Extinction*. MIT Press, Cambridge, MA.

Kaufman, L. S. 1987. Caught between a reef and a hard place: why aquariums must invest in the propagation of endangered species. Pages 362–385 in *Proc. Amer. Assoc. Zool. Parks and Aquaria*. Front Royal, VA.

Kaufman, L. S. 1989. Challenges to fish faunal conservation programs as illustrated by the captive biology of Lake Victoria cichlids. Pages 105–120 in B. L. Dresser, R. W. Reece, and J. Maruska, eds. *Fifth World Congress on Breeding Endangered Species in Captivity*.

Kaufman, L. S., and K. F. Liem. 1982. Fishes of the suborder Labroidei (Pisces: Perciformes): phylogeny, ecology, and evolutionary significance. *Breviora* 472:1–19.

Kaufmann, L. S., and A. Strysky. 1991. Designation of marine species for protection under US Endangered Species Act. Report to the Office of Protected Resources, New England Aquarium, Boston.

Kendall, R. L. 1969. An ecological history of the Lake Victoria Basin. *Ecol. Monogr.* 39:121–76.

Kornfield, I., and K. E. Carpenter. 1984. Cyprinids of Lake Lanao, Philippines: taxonomic validity, evolutionary rates and speciation scenarios. Pages 69–84 in A. A. Echelle and I. Kornfield, eds. *Evolution of Fish Species Flocks*. University of Maine Press, Orono.

Liem, K. 1974. Evolutionary strategies and morphological innovations: cichlid pharyngeal jaws. *Syst. Zool.* 22:425–441.

Meyer, A. 1990. Morphometrics and allometry in the trophically polymorphic cichlid fish, *Cichlasoma citrinellum*: alternative adaptions and ontogenetic changes in shape.

Meyer, A. T., D. Kocher, P. Basasibwaki, and A. C. Wilson. 1990. Monophyletic origin of Lake Victoria cichlid fishes suggested by mitochondrial DNA sequences. *Nature* 347:550–553.

New Scientist. 1988. Monster fish may be innocent of ecological crimes. *New Scientist* 1622:34.

Nishida, M. 1991. Lake Tanganyika as an evolutionary reservoir of old lineages of East African cichlid fishes: inferences from allozyme data. *Experentia* 47:974–979.

Ochumba, P. B. O., and D. I. Kibara. 1989. Observations on the blue-green algal blooms in the open waters of Lake Victoria, Kenya. *African J. Ecol.* 27:23–34.

Ogutu-Ohwayo, R. 1990. The decline of the native fishes of Lakes Victoria and Kyoga (East Africa) and the impact of introduced species, especially the Nile perch, *Lates niloticus* and the Nile tilapia, *Oreochromis niloticus Environ. Biol. Fish.* 27:81–90.

Owen, R. B., R. Crossley, T. C. Johnson, D. Tweadle, I. Kornfield, S. Davison, D. H. Eccles, and D. E. Engstrom. Major low levels of Lake Malawi and their implications for speciation rates in cichlid fishes. *Proc. R. Soc. Lond. B* 240:519–553.

Reid, G. M. 1990. Captive breeding for the conservation of cichlid fishes. *J. Fish Biol.* 37 (Suppl. A):157–166.

Robbins, C. R. 1991. Regional diversity among Caribbean fish species. *BioScience* 41:458–459.

Roberts, L. 1990. Zebra mussel invasion threatens U.S. waters. *Science* 249: 1370–1372.

Sackey, P. 1991. Morphological and behavioral plasticity in fishes. Master's dissertation, University of Massachusetts at Boston.

Scholz, C. A. and B. R. Rosendahl. 1988. Low land stands at Lakes Malawi and Tanganyika, East Africa, delineated with multifold seismic data. *Science* 240:1645–1648.

Slootweg, R. 1987. Prey selection by molluscivorous cichlids, foraging on schistosomiasis vector snail, *Biomphalaria glabrata*. *Oecologia* 74:193–202.

Spencer, C. N., B. R. McClelland, and J. A. Stanford. 1991. Shrimp stocking, salmon collapse, and eagle displacement. *BioScience* 41:14–21.

Talling, J. 1966. The annual cycle of stratification and phytoplanktion growth in Lake Victoria (East Africa). *Int. Rev. Ges. Hydrobiol.* 51:545–621.

Taylor, J. F., W. R. Courtenay Jr., and J. A. McCann. 1984. Known impacts of exotic fishes in the continental United States. Pages 322–373 in W. R. Courtenay, Jr. and J. R. Stauffer, eds. *Distribution, Biology, and Management of Exotic Fishes*. Johns Hopkins University Press, Baltimore.

Tudge, C. 1990. Underwater, out of mind. *New Scientist* 3 November: 40–45.

Wannink, J. H. 1990. The pied kingfisher (*Ceryle rudis*) and dagaa (*Rastineobola argentea*): estimating the food intake of a prudent predator. In *Proceedings of the Seventh Pan-African Ornithological Congress*, Nairobi, Kenya.

Wells, L., and A. McLain. 1972. Lake Michigan: effects of exploitation, introductions, and eutrophication of the salmonid community. *J. Fish. Res. Bd. Can.* 29: 889–898.

Wilhelm, W. 1980. The disputed feeding behavior of a paedophangous haplochromine cichlid (Pisces) observed and discussed. *Behavior* 74:310–323.

Williams, J. E., J. E. Johnson, D. A. Hendrickson, S. Contreras-Balderas, J. D. Williams, M. Navarro-Mendoza, D. E. McAllister, and J. E. Deacon. 1989. Fishes of North America endangered, threatened or of special concern: 1989. *Fisheries* 14(6):2–20.

Williams, J. E., and R. M. Nowak. 1986. Vanishing species in our own backyard: extinct fish and wildlife of the United States and Canada. Pages 107–139 in L. S. Kaufman and K. G. Mallory, eds. *The Last Extinction*. MIT Press, Cambridge, MA.

Witte, F. 1984. Ecological differentiation in Lake Victoria haplochromines: comparison of cichlid species flocks in African lakes. Pages 155–167 in A. A. Echelle and I. Kornfield, eds. *Evolution of Fish Species Flocks*. University of Maine Press, Orono.

Witte, F., T. Goldschmidt, J. H. Wanink, M. J. P. van Oijen, P. C. Goudswaard, E. L. M. Witte-Maas, and N. Bouton. 1992. The destruction of an endemic species flock: quantitative data on the decline of the haplochromine species from the Mwanza Gulf of Lake Victoria. *Environ. Biol. Fishes* 34:1–28.

Endnotes

1. Haplochromine Ecology Survey Team, 1992, University of Leiden, personal communication.
2. R. L. Welcomme, 1992, personal communication. FAO, Fisheries Research and Environment Division.
3. L. Kaufman, 1992, manuscript in preparation.
4. R. Hecky, 1992, manuscript in preparation, Freshwater Institute, Canada.

Les Kaufman is chief scientist at the New England Aquarium's Edgerton Research Lab, Central Wharf, Boston, MA 02110.

PART 5

Examining Tourism and National Parks

Tourism is principally a phenomenon of the developed world, at least in its modern form. People have, of course, been travelers for much longer spans of human history than the modern age, and travel in one form or another is probably as old as the human species. What has occurred recently, however, is unprecedented. People travel, go on vacation, visit new sites and places, in effect tour, in numbers that are already huge but that continue to grow. The major destination sites, such as Paris, Rome, and New York receive millions of tourists each year, as does the Caribbean and other recreational areas. If tourism is defined liberally, as spending several nights away from home while engaging in some form of recreational activity, then tourism is the world's largest industry in terms of monetary expenditures, exceeding even the petroleum industry, electronics, and the military complex.

Two recent developments of tourism have led social scientists, especially anthropologists, to become interested in the topic. One is that the sheer volume of tourists to sites such as national parks and to archaeological sites have begun to severely degrade the environment and the monuments that lead people to visit in the first place. At famous archaeological sites such as Chichen Itza, a Mayan site in the Yucatan peninsula, the movement of tourists has to be controlled to keep them from damaging the site, and sensitive areas are sometimes placed off limits altogether. In addition, the magnitude of the revenues generated from entrance fees tends to shift the decisions about restoration of the park from archaeology per se to tourists' interests. As in the cases of logging and fishing, tourism and the local environment come into conflict over the use of resources and the protection and conservation of those resources. The second development in tourism is related to this local-global interrelationship.

As tourism expands farther and farther into areas previously not included in tourist circuits, the impact on local populations and their environments intensifies. That has been the case especially in ecotourism, where an effort is made to have the environment and ecosystems as the principal attraction for tourists. As the volume of tourists increases, their impact on the environment also increases, paradoxically diminishing the value of what served as the attraction in the first place. Hikers in the Annapurna Range of the Himalayas, for example, have contributed significantly to deforestation through their use of firewood for heating and cooking.

National parks are currently caught in the middle of a number of controversies. The principal one concerns the justification of the existence of parks themselves. Valuable land is set aside for recreational use, land that in the view of some could better be put to other commercial uses. That issue has pitted environmentalists against developers, and that conflict has become confrontational and volatile at times. When biodiversity is associated with the land use issue, especially in terms of endangered species, the conflict can become very acute, as in the case of the spotted owl in the northwest coast area of the United States. The first of the articles in this section addresses some of the issues surrounding national parks. Ken Wu addresses an issue that splits the environmental community itself. Some environmentalists argue that national parks represent only a token investment in terms of what is needed for environmental protection, and that educational and activist efforts should be devoted to broader and more fundamental issues. Within that framework, Wu argues for the value of parks, taking the stand that fighting for actual land, the biodiversity that it contains, and the recreational value that it represents is preferable to more abstract and idealistic notions.

Sue Wheat identifies problems and issues about tourism in the Central American country of Belize. Belize has become one of the ecotourist meccas, due largely to the small islands, or cays, that lie offshore in the Caribbean. Tourism is the mainstay of the economy, and the country desperately needs to maintain if not increase the volume of tourists, but as Wheat notes, the tourism industry is in a Catch 22 type of situation. Coral reefs and tropical rain forests represent the two major attractions, and tourists in growing numbers are attracted to those areas. By their very nature, however, those ecosystems are fragile. They can be easily damaged, and large numbers of tourists will inevitably lead to their degradation. Belize is faced with a way of trying to manage the resources. Natural resource management is a growing concern worldwide, but it is particularly acute in ecotourism 'hot spots,' such as Belize. One particular problem that management faces is that the information needed is not available, and what tends to be least available is information about the human populations in the area in terms of their knowledge of and use of the resources. Anthropological research is acutely needed in many of these cases.

The third article, by Antonio Massinga, addresses a similar dilemma in the East African country of Mozambique. As he notes, Mozambique is one of the world's poorest countries, and it desperately needs economic development.

Pursuit of development has led the country to seek private commercial development of its resources, including its beaches on the Indian Ocean. As development proceeds, however, environmental destruction increases, sometimes at alarming rates. Part of a solution is to include environmental protection centrally into plans, something not occurring at the present.

The fourth article, by Manuel Cesario, reports on a situation in a national park in Brazil where efforts have been made to incorporate local communities into park management, showing that essentially everyone gains when there is a sharing or responsibility and accomplishments. He shows that in this particular case, the presence of the park promotes not only ecosystem protection but maintenance of cultural systems of knowledge and of the health status of local human populations. The fifth article addresses the needs for control of tourism and local community involvement and management in the Mediterranean Basin. Andreas Ashiotis documents some of the negative impact on 'uncontrolled' tourism and suggests ways in which benefits would derive from management of resources toward sustainability. The sixth article, by Gordon Matzke and Nontokozo Nabane, presents an in-depth case study of an instance of community-based management conservation and management organized and promoted at a national level in the southern African nation of Zimbabwe. The program, Communal Areas Management Program for Indigenous Resources (CAMPFIRE), has had an extraordinary effect on local communities in terms of educational and economic well-being, all the while protecting the natural environment. It is an illustration of what can be accomplished when there is an articulation of local and national interests that leads to concerted action.

ECO-FORESTRY OR PROTECTED STATUS?

Some Words in Defense of Parks

KEN WU

The philosophical criticisms of parks and protected areas must be most vigorously addressed, for the development of such anti-wilderness environmental arguments is on the rise . . .

In the 1990s, the tremendous growth in the environmental movement has been accompanied by numerous changes within the movement, many for the better but some for the worse. Among the detrimental changes has been an increasing tendency among certain segments of the environmental movement, including some engaged in struggles against wilderness destruction, to downplay the need for parks and protected areas. Many such individuals and groups are either fighting wilderness destruction in a vacuum, that is, with no clear alternative to the destruction, or are calling for "eco-forestry" and other forms of supposedly benign, environmentally-friendly resource extraction. I hope to show here that a call for anything other than protected status in priority wild areas is to the detriment of native biodiversity. I'll examine the primary arguments against park establishment that some environmentalists use and the strategic consequences of not advocating parks.

A Rebuttal to Some Primary Arguments Against Parks

Much of the lack of advocacy for protected areas can be attributed to ignorance. Many activists simply do not have an overview of the status of endangered ecosystems in North America, and are unaware that it is precisely in parks and protected areas that ecosystems are healthiest and most secure from environmental destruction. Hence, they do not understand the importance of directly campaigning for protected status, as opposed to mere moratoria on the destruction which usually get lifted later.

However, it is the philosophical criticisms of parks and protected areas that must be most vigorously addressed, for the development of such anti-wilderness environmental arguments is on the rise, as exemplified by William Cronon's essay, "The Trouble With Wilderness," in the recent anthology *Uncommon Ground: Toward Reinventing Nature.*[1] Dave Foreman, David Johns, George Wuerthner, Mike Matz, and Reed Noss have already responded to many critiques of the wilderness concept and Wilderness Areas in The Wildlands Project anthology, *Place of the Wild,*[2,3,4] as well as in *Wild Earth,*[5,6,7] and there is no need for me to repeat their refutations. Still, I would like to add a few insights of my own, because I think it is crucial that such misguided criticisms of parks (the most common Canadian wilderness designation) are refuted once and for all before they gain a further foothold in the movement. The environmentalist arguments against protected areas, and my rebuttals, are as follows:

1. **The concepts of parks and wilderness separate humans from nature when, in fact, humans are a part of nature. Thus, parks reinforce the man/nature dualism of western civilization.**

Of all the arguments against protected areas, this one takes the cake for being ill-considered and just plain illogical. That human society *should be* in harmony with Nature does not mean it is in harmony—far from it, thus the whole environmental crisis. There's a difference between what *should be* and what *is.* Industrial society with its automobiles, factories, DDT, and shopping malls is certainly not one with Nature, and by using the word *wilderness* we are not somehow creating a dualism; a dualism already exists. There is a world of a difference between a parking lot and a prairie, a clearcut and an old-growth forest. Human civilization has already separated from Nature, from the wilderness; the task is to put humans back into harmony with Nature by developing an environmentally harmonious society *and* by protecting Nature in wilderness parks *while industrial society still exists.* Wilderness advocates didn't create the human/Nature dualism; agriculture, technology, and industrial society did by destroying Nature, thereby creating an obvious distinction between wilderness and human society. We must recognize this wilderness/civilization dichotomy if we are to overcome it. Creating parks, protecting the Nature that people are supposed to be a part of, is the most important step in transcending that dualism.

2. **"Eco-forestry" and environmentally harmonious lifestyles and practices are needed, not more parks. It is not humans per se that are at fault, but rather the ways we live that are destructive.**

Fair enough. Hunting/gathering lifestyles have more or less allowed the ecosystems in which they occur to remain intact. Arguably, such lifestyles are environmentally-harmonious. However, eco-forestry, permaculture, and organic agriculture with the use of today's advanced technologies and with the present human overpopulation are far cries from hunter/gatherer lifestyles.

In eco-forestry, large numbers of trees, up to the annual growth of the forest, are removed and used for lumber. This is in contrast to the small number of trees, if any, removed by hunter-gatherers to make the odd boat or building. True, where a swidden (slash and burn) agricultural system is also practiced along with hunting and gathering, as in many tropical aboriginal cultures, many more trees are taken. This may represent the beginning of a primarily agricultural lifestyle in such peoples, which would certainly be environmentally destructive, as with all agriculture. Agriculture is the destruction of native organisms in an area and their replacement by one or a few species useful for humans. However, in *primarily* hunting/gathering societies, swidden takes only tiny fractions of the forest cover, which are quickly reclaimed when the small patch clearings are abandoned in a couple years. In contrast, through selection logging and commercial thinning, which are much more practical possibilities than eco-forestry in an industrial society, trees far in excess of the annual growth may be removed, to the point where forest interior conditions are lost. Tree removal aside, problems of road building, habitat fragmentation, soil compaction, erosion, stream damage, and the introduction of exotic species arise even with selective forestry practices. Nor should indigenous practices of burning tracts of forests to provide better grazing for ungulate prey be used to justify alternative forestry practices. Increasing numbers of studies are revealing the differences between logged and burned areas,[8] such as changes in soil chemistry, successional species composition, and the presence of gradients of defoliation in burned areas but not logged areas. Clearly, wild Nature and areas used for forestry are not the same. Eco-forestry may be needed in areas not available for protection, but such practices are not appropriate everywhere and are not a replication of natural processes.

Some opponents of protected areas cite the example of indigenous peoples living in harmony with Nature to deny the necessity of protected areas in which human habitation is prohibited. Fine, let's have protected areas that include the protection of native hunting/gathering tribes. Most wilderness advocates would support the continuance of indigenous hunter/gatherers living in protected wilderness areas, as long as the native peoples possess traditional technologies and populations (as with several tribes in tropical Africa, Asia, and South America). Few protected area advocates, however, would support native peoples with industrial technologies and larger populations harvesting resources in protected areas, especially not for commercial purposes. This is where ecocentric environmentalists often differ from more anthropocentric environmentalists, who support native peoples with chainsaws, bulldozers, rifles, steel traps, and snowmobiles exacting resources in proposed protected areas. Support of native hunting/gathering lifestyles does not negate the need for protected areas. Rather, it is a justification for protected areas that include hunter/gatherers.

Proponents of native sovereignty may object to the notion of native people living in parks controlled by colonial governments, whether here in North America or elsewhere in the world. Native sovereignty may be a legitimate right, but in the meantime before the ruling governments are either pressured into accepting native sovereignty or are overthrown, it does neither the environment nor native people any good to have corporations destroy wilderness. Parks are the best means within the present society to prevent this.

3. **Changing society to become environmentally harmonious is the crucial task, not creating more parks that exist parallel to consumer society without challenging its fundamental basis. Industrial society will eventually destroy protected areas anyway through pollution (ozone layer depletion, greenhouse effect, acid rain, etc.) and by opening park borders in times of resource scarcity.**

This is a critique used by both reformists and radicals. Its two main problems are that it confuses the means with the ends and that it is strategically unsound. First, from an ecocentric perspective, the continued existence of Earth's complete natural biodiversity is the most fundamental goal. To achieve this goal, we must advocate *both* the protection of this biodiversity in wilderness parks—a particular means that is also identical to the ends—and the establishment of an environmentally harmonious society so that pollution and population growth don't destroy protected areas and the rest of Nature. Thus, when one pushes for new environmental laws to regulate logging practices or to curb pollution or, more fundamentally, when one works to dismantle industrial society, it is to ensure the long-term security of protected areas and all species, including humans. Yet, the critics of protected areas, believing the primary task is the survival of the human species, do not see any reason to protect wilderness; a world with the basic necessities for survival—clean air, water, soil, and renewable agriculture—is all that is needed to secure the human existence. The existence of the world's vast array of biodiversity in functioning ecosystems (some species may be reserved in genetic banks), is for the most part not a necessity for human survival; the garden vision, as critiqued by Roderick Nash,[9] is seen as sufficient.

To some critics, wilderness protection is simply a means to "save the planet," meaning to secure human existence, while the reform or replacement of industrial society is the most crucial task for ensuring human survival. Such people have confused the means of creating a green society to secure wilderness with the ends.

In addition to being anthropocentric, this critique is strategically unsound. If, as many confused park critics claim, protection of more wilderness would be great but society must be changed first, then it will simply be too late for most wild areas and species by the time The Revolution succeeds. Already, most of the parks and designated Wilderness Areas in the US and in southern Canada are surrounded by agriculture, clearcuts, and urban development. If it weren't for the protective designations, these natural areas would be long since destroyed.

4. **Our parks have failed miserably in halting the loss of biodiversity. Most parks, too small to begin with, are located in high elevation areas of rock and ice or lands otherwise unsuited for human use, while the most productive and diverse low elevation ecosystems have**

been largely left out. In addition, parks have been subject to industrial tourism that has destroyed much of their biotic integrity.

As George Wuerthner points out: "The fact that our present preserve system does not work as well as it should does not mean that it could not work."[10] That our parks are too small to maintain healthy populations of all their species doesn't mean we should not advocate parks; it means we fight to get bigger parks, as in the proposed Northern Rockies Ecosystem Protection Act. That parks are rarely established in old-growth forests or prairie grasslands doesn't mean we stop advocating the creation of parks; it means we work to get old-growth forests and prairie grasslands protected. For example, here in British Columbia, the tremendous push by the public to protect old-growth rainforests has resulted in significant tracts of prime, low-elevation old growth being protected in recent years, such as the Carmanah, Megin, Stein, Khutzeymateen, Boise, Kitlope, Mehatl, Skagit, Clendenning, and Niagara Valleys, as well as South Moresby. These are not lands marginal for human use; they are worth billions of dollars in timber value. That some parks contain ski resorts, grazing, and logging doesn't mean parks are useless; it means we fight against ski resorts, grazing, and logging in parks. Moreover, to say that parks have failed is to accept a very narrow and uninformed view of ecosystem protection. Alpine and subalpine ecosystems, which have their own unique species that are just as important in their own right as old-growth endemics, have been reasonably well protected. All other ecosystems partly protected by parks—including the small and moderate-sized tracts of productive, economically valuable lands—also represent partial victories. Park creation is a process, where all areas protected thus far are victories while still more and larger parks must be created to complete an ecologically viable system of protected areas.

Ultimately, if one believes that Nature has intrinsic value and that humans cannot improve it, then there really is no truly environmental option other than to leave wilderness as is, and to secure it from future human alteration; this is the definition of a protected area, or what is often called a "park." Some people have a problem with the word *park*, because it holds some connotation that wilderness is for human recreation; fine, then let's call them "Ecological Reserves" or "Wilderness Reserves." But to not advocate the protection of an endangered ecosystem because of a name, and thus allow it to be clearcut or strip-mined, is a crime.

5. **Nature needs human management to stay healthy. For example, exotic species must often be controlled, prescribed burns must be set in isolated habitats, predators must sometimes be controlled to allow endangered species to recover their populations, and new individuals must be introduced into small, isolated populations to prevent inbreeding. Thus, because nature must be managed, there is fundamentally nothing wrong with managing a landscape through selection forestry, controlled grazing, or limited agriculture.**

This argument is made by some conservation biologists and land managers who realize that active management of some wild areas is necessary to maintain their natural character. Humans have so disrupted natural populations and processes that human intervention is often needed to correct past mistakes. The difference between correcting and managing *human-induced mistakes* on Nature and managing *Nature itself*, however, is huge. One can still advocate parks even if the areas of concern need such corrective management; their protective status should nonetheless forbid the managing of Nature itself. Unfortunately some people lump both managing human mistakes and managing Nature under the general concept of "management," and support "alternative" forms of commodity extraction in place of protective status, thinking that such activities are fundamentally no different from prescribed burns or the elimination of exotics.

Many of these anti-park arguments are used in William Cronon's influential essay, "The Trouble With Wilderness." I shall leave other authors in this issue to debunk Cronon's claims.

Implications of Failing to Advocate Parks

With the main philosophical arguments against protected areas out of the way, the strategic implications of not calling for full protection can be examined. Environmentalists' failure to call for protected status commonly has one of two "best-case" consequences:

1. A moratorium is placed on the destruction, whether by court injunction (in the US) or by simple government decree (in Canada). Moratoria can always be lifted, so the same fight will be repeated all over again, except that political circumstances may not be as favorable the next time around; new anti-environmental politicians may be in power, the "Wise Use" backlash may have grown, or the environmental movement may be on the downswing. Moratoria are not solutions.
2. A half-baked solution results in which the pristine status of the area is compromised. This may include smaller clearcuts, limited road building, or, very unlikely, implementing the alternative forestry suggested by the environmental group (which, as already discussed is not a replication of Nature). These half-baked solutions are often harder to overturn than the full-scale onslaughts, as such reforms may render complacent much of the more moderate environmental movement. Meanwhile, the wilderness is progressively eaten away at a reduced pace.

Conclusion

Clearly, direct calls for the establishment of protected areas are necessary if wilderness areas are to be saved once and for all. Of course, there is no guarantee that protected areas will not be opened up in the future for development, but there is no guarantee on anything in society; protected status is the most secure way to ensure the survival of native biodiversity.

Sometimes in building a coalition with non-environmental groups that share an opposition to a proposed development, a direct call for protected status may destroy the alliance. Some locals may be against the development of gas wells in their area but still want to continue grazing their cattle, or may oppose logging plans but still want to commercially trap. In such areas, environmentalists must use their judgment in deciding whether the coalition is worth temporarily forfeiting a protected area. In any case, the ultimate goal of the campaign should be complete protection once the immediate threats are defeated. In addition, one must question whether a coalition with other groups is desirable in the context of the overall campaign, especially if such groups oppose all protective designations and will end up becoming the opposition after the common threat is defeated. As a general principle in wilderness campaigns, the sooner one calls for complete protection, the better.

Ken Wu is a recent graduate of the University of British Columbia's ecology program. He is a co-founder of the Radical Ecocentric Network and a correspondent for Wild Earth.

Endnotes

1. William Cronon, "The Trouble With Wilderness," *Uncommon Ground: Toward Reinventing Nature,* William Cronon (ed.) (New York: W.W. Norton and Company, Inc.) Chapter 1.
2. Dave Foreman, "Where Man is a Visitor," *Place of the Wild,* David Clarke Burks (ed.) (Washington, DC: Island Press) 225–235.
3. David Johns, "Wilderness and Human Habitation," *Place of the Wild,* 149–160.
4. George Wuerthner, "A New Vision for the West," *Place of the Wild,* 203–213.
5. Dave Foreman, "Wilderness Areas are Vital," *Wild Earth,* Winter 1994/95, 64–68.
6. Mike Matz, "Lock it Up," *Wild Earth,* Winter 1994/95, 6–8.
7. Reed F. Noss, "Wilderness—Now More Than Ever," *Wild Earth,* Winter 1994/95, 60–63.
8. Reed F. Noss, *The Wildlands Project: Yellowstone to the Yukon,* Canadian Parks and Wilderness Society, 1993, video.
9. Roderick Nash, *Wilderness and the American Mind* (New Haven: Yale University Press, 1982) 379–388.
10. George Wuerthner, "A New Vision for the West," *Place of the Wild,* 204.

TAMING TOURISM

SUE WHEAT

It's known as the "smokeless industry". It's now the world's biggest. And increasingly, tourism claims to offer a sustainable development option to countries short on cash but long on environmental riches. In fact, tourism, even ecotourism, is causing widespread environmental and cultural disruption. In this article, Sue Wheat compares the best of ecotourism with the rest.

"Wanted—an island to buy or rent" reads an advertisement in the *Belize Review,* giving an Arizona contact address. Now that Belize has become a mecca for nature lovers—especially from its neighbour the US—such advertisements are commonplace. With its idyllic tropical island cayes, the second largest barrier reef in the world, abundant flora and fauna in spectacular rainforest and ancient Mayan ruins, this small, Central American state attracts huge interest from both pleasure seekers wanting to get away from it all and keen-eyed investors.

However, Belize is just one of many developing countries which has seen a rapid expansion in its tourist industry. Cheaper flights and increased leisure time have stimulated a dramatic increase in "long-haul" travel, in many cases to parts of the world that other industries have never touched. Thirty years ago, few people could have imagined the ease with which it is now possible to reach the remotest corners of the world. Today, the search for the exotic destination is on with a vengeance. The combination of palm-fringed islands, adventure, intriguingly different cultures and strong Northern currencies, has meant that international tourist arrivals in developing countries have more than doubled since 1976.

According to the World Travel and Tourism Council (WTTC), tourism is already the world's largest industry and is growing at a phenomenal rate—23 per cent faster than the overall world economy. The number of people now traveling for tourism purposes is around 500 million and is expected to rise to 937 million by 2010. For an industry which doesn't have to manufacture anything itself and simply exploits ready-made features, these are astounding figures.

But just because the tourism industry does not have factories does not mean that its impact on the world is negligible. In fact, the "smokeless industry", as it has been described, is increasingly being blamed for massive environmental, cultural and social damage. In a 1992 report entitled "The Tourism Industry and the Environment", the Economist Intelligence Unit (EIU) argues that a growing number of disasters such as polluted beaches, degraded coral reefs, wildlife parks without wildlife and disappearing indigenous cultures, are the result of tourism and are "part of an accelerating pattern".

Growing awareness of these problems and pressure on the tourist industry to become environmentally friendly have prompted a search for alternative forms of tourism. "Eco-tourism", "responsible tourism", "sustainable tourism", and "low-impact tourism" are just some of the terms used to describe these new approaches. Ecotourism is one of the most commonly used, referring to a niche market for environmentally aware tourists who are interested in observing nature. But what starred off as a small-scale, locally-owned initiative is now succumbing to overdevelopment. Eco-tourism has become the fastest expanding sector in the industry.

The Ecotourist Mecca

Although tourism has only become an important factor in Belize's economy in the last decade, it is probably the world's most famous ecotourist destination today and has hosted two major international ecotourism conferences. Tourism revenue now accounts for 26 per cent of its gross national Product (GNP) and international tourists number around 250,000—a massive increase from the average figure of 11,000 in the 1960s.

As has happened in many developing countries, Belize is caught in an economic Catch 22 and is looking to tourism for a way out. A drastic fall in world market prices for its cash crops forced the country to find other sources of foreign exchange, much of it needed to meet obligations on its US$103 million foreign debt.

But like other countries which have taken this route, Belize did not take account of the likely environmental consequences of promoting tourism. These impacts are most graphically illustrated in coastal areas where the ecosystems are particularly fragile. "Coastal development in Belize involves the clearing of mangrove swampland, drainage and infilling, using topsoil literally shaved off the wetland savanna a few miles inland. This involves the destruction of two distinctive ecosystems," explains Erlet Cater, Senior Geography Lecturer at Reading University and an expert on sustainable tourism in Belize.

Ironically, the international standard Biltmore Plaza hotel, the venue for the 1991 Caribbean Ecotourism conference, was built on land reclaimed in this way. And hotels are themselves far from benign in their effect on the environment, argues Cater. In fact, their impact can be disastrous. In catering to western standards, for instance, hotels use up large amounts of energy and generate huge quantities, of waste.

Attempts to protect the environment on which Belize's national income is so dependent are now underway. Three new protected marine areas have been declared and the Government is cooperating with various environmental organizations. Coral Cay Conservation, a non profit-making British marine research group, provides scientific expertise and resources to help develop management initiatives for the protection and sustainable use of the country's coastal resources. And Edinburgh University's geography department is mapping the country's mangroves with satellite imagery, as part of a plan to control and coordinate tourism development, urban expansion, fisheries, coastal protection and timber use.

Coral Cay Conservation works closely with local government departments and trains local people in scientific techniques, so that the project can eventually be Belizean-run. "We are there to protect the biodiversity of Belize, but the last thing we want to do is say tourists can come and look but the fishermen can't fish," explains Samantha Flint, manager at Coral Cay Conservation.

Flint feels that the erosion of the indigenous culture and the exclusion of local people from tourism policy are storing up problems for Belize. With an estimated 90 per cent of coastal building developments now under foreign ownership, Belizeans may already be permanently alienated from their own land.

Ian Munt, who has worked in Belize as an urban planner, and Egbert Higinio, from the University of West Indies in Belize, have investigated the impacts of tourism on the country and maintain that "despite some promising results, much ecotourism in Belize merely replicates the problems characteristic of traditional mass-tourism—foreign exchange leakage, foreign ownership and environmental degradation."

They point to the US$50 million "sustainable tourism" development on the highly developed Ambergris Cay as an example of how ecotourism is simply being used as a marketing label. The Belizean Government bought back two-thirds of the 8,000-hectare cay from its US owner in an attempt to "reinvest in the Belizean people" and proposed "an integrated and ecologically sound resort development" for the caye. However, this includes at least one international hotel, two "all-inclusive spa hotels", three to five upscale lodges, two golf courses, 1,000 luxury homes, polo fields and stables. As Munt and Higinio point out, "it is difficult to see how this development really differs from any other."

Trickle-down Effect

Belize is certainly not alone in suffering these problems. The Western desire for the exotic has been identified by many developing country governments as the route to easy wealth and a panacea for various economic ills. However, the proverbial economic trickle-down effect invariably means just that—a trickle. While tourism employs more than 127 million people worldwide, in developing countries, the local people normally work in menial, seasonal and low-paid positions, with managerial jobs going mainly to expatriates working for multinationals.

Transnational corporations (TNCs), such as airlines, hotel chains and building giants, now dominate the industry and much of the financial benefit. Many of these TNCs have a turnover larger than the gross domestic product of the developing countries in which they operate. According to the World Tourism Organization (WTO), at least 55 per cent of the foreign exchange earned by tourism leaks back to developed countries through spending on imports for the tourist industry, such as fittings, foodstuffs and labour. In some Caribbean countries. this figure may be as much as 70 per cent. Although the developing world increased its share of international arrivals during the 1980s, its share of tourism revenues actually fell, according to an EIU analysis of WTO figures.

Across the world, the hidden costs of tourism are often severe. On the resort islands of Thailand, much of the indigenous population has been forced to migrate, because the price of basic necessities and land is beyond their purchasing power. In many developing countries, children skip school to sell to tourists, and traditional livelihoods are abandoned. Local farmers and landowners in tourist destinations are often harassed by developers wanting to build on their property; fishermen are prevented from using public beaches taken over by hotels.

In Tanzania, the Government forced the Masai from their centuries-old homeland in the Ngorongoro Crater, in order to open up the park for tourism. As a result, posing for photographs is one of the only ways they can avoid poverty. The indigenous Berawan in Sarawak are in a legal battle with the Royal Mulu Resort Hotel, which is planning to expand and construct a golf course on their ancestral land. The Berawan stress that they welcome tourists, but they are concerned that they might lose their Native Customary Rights to land."Although most of Mulu workers are Berawan, they do not hold any position or partake in the development process within Mulu. Is this development for us? Certainly we are left out," says a Berawan statement.

However, many local communities around the developing world have had enough of being exploited by the tourist industry and are trying to garner a better share of the proceeds. The Toledo Ecotourism Association (TEA)—a grassroots organization of Mayan indigenous people in southern Belize—is organizing village stays for tourists in settlements built by the Mayans with local materials. Tourists are looked after by the villagers themselves, ensuring that all the financial benefits are kept within the community. Special emphasis is put on preserving their ancient culture and fragile environment. Aware of the danger of their tourism business becoming a "monoculture", the TEA members are also developing a firm agricultural base. Their main product is cocoa, which they are selling to a UK company for organic chocolate.

Cash for Conservation

Village communities in Zimbabwe are taking a similar approach to a different tourist resource. By joining the Communal Areas Management Programme (CAMPFIRE), they acquire ownership of the wildlife in their area and profit from it by charging tourists who want to take part in sustainable hunting and photo-safaris,"In a country where 95 per cent of the population had been forced into 30 per cent of the land mass during colonial rule and wildlife had been given 15 to 20 per cent of the land, the people had great animosity towards wildlife,"explains Keith Madders, Director of the Zimbabwe Trust, which manages CAMPFIRE. Now that the wildlife has an economic value for the community their attitude has changed. As a result, CAMPFIRE has lifted many people out of poverty, often doubling their income. Many animals once under threat of extinction are also out of danger, as it is in the communities' interests to deter poachers.

The money from tourism is paid directly to the CAMPFIRE communities and used at their discretion. The people of Mahenya have so far funded a school, a clinic and a grinding mill with their profits. In 1992, the community decided to distribute the Z$70,000 (£7,700) made from hunting four elephants and one buffalo between each of its 391 households, giving each one z$180 (£2O) cash after costs were deducted. In rural Zimbabwe, that is a useful sum, even if it's not a fortune. There are now 22 districts involved in CAMPFIRE—A rapid growth from the two that existed in 1987—and it is expected that similar projects will start in other African countries.

While community-based initiatives may be more socially beneficial than mass tourism, they are not without their problems. How long will a low-impact project stay low impact once it is successful? How can such groups compete with the discount-prices of the large tour operators? The EIU concludes that "the future of tourism. therefore, must lie in the way mass tourism is conducted and not in a move to'alternative' tourism."

The WTTC suggests environmental protection is best achieved through voluntary codes of conduct. Its Environment Research Council lists more than 100 agreements and guidelines. They also point to the"International Hotels Environment Initiative", with its manual of good environmental practice for hotels.

However, Tricia Barnett of Tourism Concern maintains that the tourism industry has to go beyond cosmetic environmental changes. It must accept that "the environment" includes both the natural and social environment, she says. "While the tourism industry is patting itself on the back for recycling brochures or reducing energy consumption, the really serious social issues such as displacement of people from their homes, loss of livelihoods and lack of access to public land are being ignored."

Voluntary regulations have certainly not helped those currently fighting to save their land and livelihoods. Some 50,000 Sri Lankans demonstrated against the building of the Kandalama Hotel on an important local reservoir. Campaigners are seeking to deter European tour operators from using the hotel, which is set to open this year. In West Java, farmers have been harassed, imprisoned and their homes burnt down after peacefully demonstrating against their illegal eviction by a company planning a tourist resort and golf course on their land. The case of one farmer imprisoned for 10 months has now been taken up by Amnesty International.

For travelers searching for the exotic, and for the industry itself, it is a psychological hurdle to accept tourism as a human rights issue. But this is what it has become. The ever-increasing number of tourists will continue to cause cultural disintegration, the marginalisation of local people and environmental degradation, unless the tourism industry, governments and local communities work together. Tourism is no different to any other multibillion pound industry. Even in Paradise, profits have priority over people. One can only hope that community-based tourism initiatives will be able to prosper enough to give Paradise back to the people who live there.

Sue Wheat is a freelance journalist specializing in environmental and development issues.

BETWEEN THE DEVIL AND THE DEEP BLUE SEA:

Development Dilemmas in Mozambique

ANTÓNIO MASSINGA

Mozambique is one of the world's poorest countries and highly-dependent on foreign aid. After three decades of devastating warfare, the country has adopted an IMF-inspired structural adjustment programme to attract foreign investment. Mainly South African entrepreneurs are investing in eco-tourism, exploiting the country's beaches and wildlife reserves. Others such as pulp and paper companies are also sizing up the country's land for plantations. The multiple grabs for Mozambique's resources have confronted environmentalists at times with uncomfortable dilemmas. Defeating one potentially damaging development proposal may only be possible by supporting another.

Mozambique's beaches along its 2,500 kilometer coastline and its once abundant wildlife inland have drawn visitors since the colonial era. White elites from Mozambique, Rhodesia and South Africa came to one of the richest game parks in southern Africa, Gorongosa, which was considered to be the "showpiece of colonial Mozambique's conservation effort"[1] and several other hunting preserves *(coutadas). Coutada* 16 in Gaza province, created in 1961, became one of the most commercially-developed hunting reserves in Mozambique.

Tourist infrastructure was built to accommodate the visitors, encouraged by the colonial government which believed that tourism was critical to the country's economic development. Between 1967 and 1971, the number of foreigners visiting Mozambique rose from 292,000 to 548,000 while tourism's contribution to the country's balance of payments increased from 272 million escudos (US $130 million) to 417 million (US $180 million.)[2] Guides recall the "hundreds of hotels, nightclubs and bars" in "cosmopolitan cities" such as Beira in the center of the country and Maputo (known at the time as Lourenço Marques), the capital in the south.[3]

Today, most of these hotels and holiday complexes are in ruins after some 30 years of warfare. From the mid-1960s, the Mozambican Liberation Front (Frente de Libertação de Moçambique or Frelimo) led the resistance against Portugal, which had colonized the country during the "scramble for Africa" in the late nineteenth century. In 1975, Mozambique gained its independence and Frelimo came to power. The white Portuguese elite all but deserted the country, while Frelimo's principled attitude towards its white-ruled neighbors cut the country's official economic links with Rhodesia and South Africa. As a result, Mozambique lost 50 to 60 percent of its foreign visitors.

In response to Frelimo's victory and its support of black opposition movements in Rhodesia, Ian Smith's minority government supported the formation of a guerrilla army, the National Resistance Movement (Resistência Nacional de Moçambique or Renamo) to oppose Frelimo and carry out insurgency within Mozambique. After the 1980 Lancaster House Agreement which granted Rhodesia black majority rule, South Africa became Renamo's principal supporter along with conservative circles in the United States which were keen to stem supposed Soviet influence in Africa.

The guerrilla war between Frelimo and Renamo destroyed most of the country's tourist infrastructure: hotels and bungalow parks became battlefields or were simply looted. The Gorongosa game park became a Renamo stronghold and was stripped of game by soldiers hunting for food and trading in illegal ivory. Even after the 1985 Nkomati Agreement with South Africa under which bilateral relations with Mozambique were officially resumed and the border reopened, tourism was not rekindled; continued Renamo raids made travel by road unsafe.

Warfare devastated the country's economy. Its mainstay, agriculture, and its major export, prawns, were disrupted and transport infrastructure destroyed. Under the pressure of economic collapse, the Frelimo government took a structural adjustment loan in 1987 and went on to implement the various accompanying IMF/World Bank conditions. Government subsidies were cut, the local currency devalued, and market restrictions and regulations lifted. Almost all large enterprises, nationalized soon after independence, were targeted to be privatized again and restrictions on foreign investments were gradually lifted.

Consolidation of the Peace

Mozambique's civil war came to a close with the end of the Cold War and of apartheid in South Africa. In 1992, Frelimo and Renamo signed a peace treaty, and two years later, elections under the auspices of the United Nations resulted in the country's first multi-party parliament.

The ceasefire and economic liberalization have made the country attractive once again to foreign investors; between 1989 and 1994, annual foreign investment in the country rose from $8.8 million to $65.7 million. Britain, Portugal and South Africa accounted for more than three-quarters of total foreign investment in Mozambique between 1985 and 1994 (some US $272 million). The second largest investment sector in the country between 1985 and 1993 was tourism at $122 million, exceeded only by agriculture. Many South African investors perceive Mozambique to be the "paradise not yet lost" in contrast to the beaches in their own country which have succumbed to mass tourism.

New investments have ranged from a few tents or hostels to large bungalow parks and hotels. The International Finance Corporation, the private-sector lending arm of the World Bank, funded the restoration of the Polana hotel in Maputo to its colonial splendor as part of a joint venture between a South African company and the hotel's original Portuguese owners; other hotels, such as the Cardoso, owned by British multinational Lonrho, have also been renovated.

At the same time, the numbers of visitors increased. Between 1993 and 1994, the number of nights that visitors (domestic and foreign) spent in hotels almost doubled to reach 9,546. Camp sites and bungalow parks, however, continue to be the most popular places to stay; in 1971, about one quarter of tourists spent their holidays camping and the figure is probably similar today.[4]

Promoting Tourism

A 1994 World Bank study advocates tourism on the grounds that it is labor intensive, creates decentralized employment for less privileged people, contributes to political stability and brings in foreign exchange. The Economist Intelligence Unit estimates a potential tourism income for Mozambique of $80 million per year.[5]

The Bank study also acknowledges the general limitations of tourism as a development strategy: currency "leakages" to foreign airlines and hotel chains and to pay for tourist consumption of imported goods; foreigners rather than local people holding management jobs; environmental degradation; demographic distortions; and moral and cultural disintegration, typified by increases in prostitution.

Indeed, the recent rise in the number of hotel beds in Mozambique was not accompanied by an increase in employment: according to official tourism statistics, hotels employed just over 2,400 workers from 1992 to 1994, even though the number of beds rose by a third to some 9,000. An apparent wage increase of 164 percent over the same period was offset by the rapid depreciation of the local currency against the US dollar—the dollar increase was just seven percent.

Tourism's present or potential contribution to the Mozambique's national economy is difficult to assess because of limited and unreliable statistics. Moreover, the country's official economy has long been paralleled by a shadow or black economy of South African rands and US dollars resulting in additional leakages. Its impact at local levels, however, is more apparent.

For instance, a South African entrepreneur has opened a beach camp site accommodating up to 400 guests about 150 kilometers south of Maputo in Ponta do Ouro, close to the South African border, an area ravaged during the civil war. The economic benefits it provides for Mozambicans are limited because of the few jobs available.[6]

Although most recent investment in tourism is concentrated in coastal areas, there are also initiatives inland. A South African enterprise, Gaza Safari, obtained a four-year licence in 1992 to operate on 10,000 square kilometers of *Coutada* 16 for an annual fee of 750,000 Meticais (US $300) and additional taxes on any animals hunted.[7] The total including trophies amounts to $18,000 per year. Gaza Safari can raise its annual fee easily: as one of its camps accommodates 12 persons each paying US $850, two days full occupation yields US $20,400. The impact of Gaza Safari on the local economy appears to be minuscule. Depending on the season and the number of guests, the company employs 20 to 80 local inhabitants out of a population of approximately 4,200. Gaza Safari's relations with communities in the area are "not at their best" because its tight poaching control limits their access to resources such as timber, firewood, and wildlife for household consumption.[8]

While the economic returns for Mozambicans are slim, the ecological damage caused by some tourist projects is already considerable as tourists steer their four-wheel drive vehicles over the beach, litter the area, destroy corals and catch protected fish species. Destruction and pollution of reefs, beaches and sand dunes due to the illegal construction of holiday resorts is a constant concern of national policymakers.

The Threat of Eucalyptus

But tourist operators are not the only ones grabbing land in Mozambique. South African pulp and paper companies such as SAPPI, Mondi and Hunt Lenchers have all shown interest as well.

In 1987, for example, SAPPI proposed a 32,000 hectare eucalyptus plantation in the Matutuíne district, 150 kilometers south of Maputo to grow timber for its factory in Richards Bay, South Africa. The project was held up by the civil war, so that by 1994 it had planted only some 5,000 hectares.

Environmentalists have objected to the SAPPI project on several grounds. One was that the proposed plantation would be in one of the 240 biodiversity centers worldwide recognized at the 1992 UNCED "Earth Summit".[9] Planting eucalyptus in such a sensitive region could cause irreversible ecological losses inside and outside the plantation area, they claimed. The trees' high water consumption could cause the groundwater table to drop within a few years, potentially affecting species composition in the entire area and the coastal lake system and associated floodplains. Inundated to a depth of just 10 centimeters, these are important breeding and feeding grounds for several fish species.[10]

Plantation critics also pointed to negative economic effects. If the groundwater table dropped, palm trees would probably disappear, ruining the local trade in palm wine from which women, older people and other local inhabitants earn up to R500 (US $140) per month by selling it in South Africa.[11] People's access to land, could also be severely curtailed.

On the Horns of a Dilemma

Besides SAPPI's eucalyptus plantation, however, the Matutuíne district has also been targeted by US millionaire James Blanchard III for a massive "eco-tourist" complex at a cost of $800 million on the Machangulo Peninsula and Inhaca Island close by. Such multiple land grabs have confronted environmentalists with an uncomfortable dilemma: defeating one proposal seemed possible only by supporting the other.

The Peninsula's densely-forested sand dunes reach down to the Indian Ocean which are full of turtles, dolphins, dugongs, rays and sharks. Under Blanchard's proposal, some 10,000 hectares, an area covering about two-thirds of the Peninsula and including a range of saline and freshwater lakes, would be set aside as a reserve which he would stock with game such as lions, buffaloes and giraffes—even though giraffes are not endemic to the area.

A major tourist attraction would be a steam train to carry passengers down the Peninsula to watch "wildlife" on the one side and the waves of the Indian Ocean on the other. Another attraction would be the 25,000 hectare Maputo Elephant Reserve to the south of the Peninsula which Blanchard has offered to rehabilitate by repairing its fence and supporting the National Forestry and Wildlife Direction's management efforts. The project proposes to expand the Elephant Reserve to the south to link with the Tembe conservation area in Kwazulu/Natal in South Africa, covering in total about 225,000 hectares.

The eco-tourism project aims to attract eco-tourists who will spend at least $500 a day, a similar audience to that targeted by Gaza Safari in *Coutada* 16. It also aims to make profits by allocating plots of land of between 1,000 to 3,000 square meters around the lakes to wealthy people, probably foreigners, to build houses on. The rights to build will be sold under a tax-friendly construction scheme for US $200,000 each.

The project claims that it will create some 12,000 direct jobs and an estimated 8,000 spin-off jobs. If such employment materializes, the Peninsula inhabitants, some 4,000, could well be turned into a minority because of an influx of outsiders. In addition, as two-thirds of them live in the proposed reserve, they could have to abandon their farms and move out.

Blanchard has requested a 99-year concession for the Peninsula and a 50-year concession for the Maputo Elephant Reserve and the area between the reserve and the South African border. If granted, the project will be the main factor in the economic development of a significant area of Mozambique.

My Enemy's Enemy?

Blanchard is no stranger to Southern African politics. He backed Chief Buthelezi's Inkatha Movement, the apartheid regime's "black ally" in KwaZulu/Natal, and while in Mozambique, has been an important supporter of Renamo, continuing to maintain his ties through a specially-created NGO, the US-based Mozambique Foundation for Education in Health.[12]

In early 1995, the boundaries for SAPPI's eucalyptus concession were marked out, impinging on those of the tourism project. When Blanchard found out, he declared that he would cancel his investment if the SAPPI concession went ahead. He allied himself with various groups trying to stop the eucalyptus project: non-governmental organizations such as the Endangered Wildlife Foundation (Fundação Natureza em Perigo, originally South African), Mozambican biologists, the Ministry of Environment and the World Bank (which is keen on linking up conservation areas).[13]

Whereas SAPPI claimed that it would create about 1,200 jobs in the area, Blanchard's representatives stressed that their project would create ten times as many. They said that because the plantation would be so close to the South African border, Mozambique would remain simply a source of raw materials with the value added by industrial processing accruing to South Africa. Blanchard claimed that the benefits of SAPPI's investment would probably remain limited to the wages paid to locally-recruited workers and the fee paid to the government for the use of the land.[14] Almost all their arguments, however, could equally well apply to the tourism project.

The Lesser Evil?

Ironically, SAPPI's eucalyptus plantation might prove to be Blanchard's salvation. People in Mozambique are increasingly concerned with the land grabs by tourism entrepreneurs and the ecological damage caused by them.[15] But by depicting eco-tourism as a lesser evil than pulp and paper enterprises, potential opposition to his project has been defused. In February 1996, the Mozambican government "unofficially" cancelled SAPPI's concession to plant eucalyptus; it has not yet, however, confirmed Blanchard's project.[16]

Judging by past experience, the ecological damage that Blanchard's hotel and tourist complex are likely to cause may well be much larger than observers have suggested, while the economic benefits may well be much smaller than promised. Moreover, many believe that the Frelimo-dominated government favors Blanchard's project over SAPPI's so as to improve its relations with Renamo supporters and conservative powers in the US and to broaden its own political base. The consequences for political stability are uncertain.

Trapped by the current development model, many environmentalists have seemed unable to step back from the "choices" before them and oppose both projects as being environmentally and socially destructive.

Notes and References

1. Slater, M., *Guide to Mozambique: The Essential Visitor's Companion*, Struik, Cape Town, 1994.
2. de Sousa, A. J. R., Jorge, C.A.P., and Nunes, G.M.V., *Turismo. Relatórios Sectorais*, Volume V. IV Plano de Fomento, Parte III, Estado de Moçambique, Direcção dos Serviços de Planeamento. P. I., Maputo, 1973, p. 19.

3. Miles, B., *Blood on the Tracks: A Rail Journey from Angola to Mozambique,* Picador, London, 1994; Slater, M., op. cit. I.
4. de Sousa et al., op. cit. 2.
5. Economist Intellegence Unit, *Mozambique Country Report,* 2nd quarter 1995.
6. Elias, A.,"Sul-africanos estão a mamar o'ouro azul'," *Savana,* Maputo, 20 January 1995.
7. DNFFB, "GEF Transfrontier Conservation Areas and Institutional Strengthening Project, Preparatory Studies", Draft Report, Maputo, Direcção Nacional de Florestas e Fauna Bravia, 1994.
8. CARE,"Integrated Conservation and Development in the Gazaland Transfrontier Conservation Area", Maputo, mimeo, 1995, pp. 8-10.
9. *Notícias,* Maputo, 15 February 1996.
10. Austral, "Environmental Impact Assessment of the Machangulo Peninsula and Expanded Elephant Reserve Development", Maputo, mimeo, 1996 p. 101. The ecologists' concern about the water table was shared by the National Physical Planning Institute (INPF), *Savana,* 9 February 1996.
11. Ibid, p. 101.
12. Machado da Graça,"O ouro da ponta," *Savana,* 16 February 1996; Eddie Koch,"O'Paraíso na Terra' seguno Blanchard," *Savana,* 16 February 1996 (also published in the South African *Mail & Guardian).* This NGO also played a role in the proposed tax-friendly construction.
13. See Benjamin Wilson's contribution "'Natureza em Perigo' contra plantio de eucaliptos," *Domingo,* (Maputo), 13 January 1996; the reader's letter by António Almeida Matos of the Association of the Friends of Ponta do Ouro (ADAPO) in the *Notícias* of 9 February 1996; Elias, A.,"Mosa Florestal: Governo dá marcha à ré'," *Savana,* 9 February 1996.
14. Marco Basto, national consultant to James Blanchard, in an interview with António Elias, *Savana,* 9 February 1996.
15. See, for example,"Nas praias do país: Turistas estrangeiros practicam desmandos," *Notícias* 12 February 1996.
16. *Notícias.* 8 February 1996. In later statements, the government did much to obscure whether this really implied the project's cancellation. See *Savana,* 23 February 1996.
17. Gêmo, O.,"Reserva de caça de Maputo carece de atenções e investimentos," *Notícias,* 11 November 1995.

CONTRIBUTION OF CONSERVATION TO SUSTAINABLE LIVING THROUGH HEALTH PROMOTION

MANUEL CESARIO

The conservation community is faced with the need for protected areas to attract public and political support. It needs to be shown that the contribution of protected areas to society is relevant and important. This article assesses the promotion of human health facilitated by the opportunities created by the presence of a national park, and the contribution of such protected areas to the health of the community. A case study approach has been adopted. The observed improvements in the living conditions of the communities neighboring the Serra da Capivara *National Park, Brazil, have arisen from the need to conserve the local natural and cultural heritage. It is suggested that the interrelationship between health and conservation can enable local communities to sustain their living and increase their commitment to neighboring protected areas.*

Introduction

Every 10 yrs the World Conservation Union (IUCN) hosts a World Congress on National Parks and Protected Areas, during which a course of action for the following decade is charted. The IVth World Congress on National Parks and Protected Areas in Caracas, Venezuela, 1992, emphasized the challenge which the conservation community is facing: the need for protected areas to attract public and political support. It is necessary to show that the contribution of protected areas to society is relevant and important.[1] New strategies to link local communities to their neighboring protected areas are necessary if the benefits are to be shared by society and if the local people are to provide support and be brought into the protected-area management process.[2,3]

McNeely and MacKinnon[4] state that the *World Conservation Strategy,* prepared by IUCN in 1980, recognizes that under current conditions, nature is likely to prosper best when it serves the human interest. Conservation was defined as "the management of human use of the biosphere so that it may yield the greatest sustainable benefit to present generations while maintaining its potential to meet the needs and aspirations of future generations". The principles of the World Conservation Strategy have been widely accepted by governments, development agencies, and conservation organizations. The question is no longer whether conservation is a necessary part of social and economic development, but rather how conservation can be achieved. If protecting natural areas serves basic human needs, the chances of survival for natural areas will increase. Linking protected areas to human needs can support ecologically-sound development, which takes on practical meaning for governments and local people.

The United Nations Conference on Environment and Development, Rio de Janeiro, June 1992, reaffirming the Declaration of the United Nations Conference on the Human Environment, Stockholm, June 1972, adopted the *Rio Declaration on Environment and Development* which states: "Human beings are at the centre of concerns for sustainable development. They are entitled to a healthy and productive life in harmony with nature."[5] It also stated that "In order to achieve sustainable development, environmental protection shall constitute an integral part of the development process and cannot be considered in isolation from it."

The challenge facing professionals working in protected areas is to find ways of demonstrating that the conservation of nature and the sustainable use of natural resources has a fundamental relevance to the daily lives of people, including those who never visit a protected area. There is a need to emphasize the purposes of protected areas as contributing to the quality of life. If the aim is for people to be in favor of parks, then it must be demonstrated far more clearly that parks are for people.[6]

This study addresses the interrelationship between human health and conservation, and how this can enable local communities to sustain their living and increase their commitment to neighboring national parks and protected areas.

The World Health Organization's holistic definition of health states that "health is a complete state of physical, mental and social well-being and not merely the absence of disease or infirmity."[7] "Humans are ecological beings; health is a social by-product; illness is a political fact: the sanitary attitude of a people expresses its culture."[8] Therefore, domains such as conservation, leisure, culture, education, housing and the right to employment, etc., should be considered as components of mental and social well-being, and therefore regarded as fundamental to human health.

The community level of the health system is "primary health care". It comprises actions in the fields of health promotion, disease prevention, disease treatment/cure, and individual rehabilitation.[9] Health promotion and rehabilitation regard the person as a whole. On the other hand, prevention and treatment/cure consider the sick person, the patient.

As part of conservation community rethinking of the role of national parks and other protected areas in society and the economy, health promotion should be given higher priority. It is possible to integrate conservation projects for protected areas with sustainable rural development projects, i.e. health promoting projects, for those directly affected by the existence of the protected area. It is also possible to add to the definition of the role and functions of protected areas, the important issue of the promotion of human health.

Methodology

Case-study methodology in human ecology was used to allow trans- and interdisciplinary, integrated modes of enquiry, and a combination of quantitative scientific data with qualitative data from the social sciences or the humanities.[10–13] Case-study methodology deals with operational links traced over time, rather than mere frequencies or incidence. It is applicable when the investigator has little control over events and when the relevant behavior patterns cannot be manipulated. It also applies when the focus is on a contemporary phenomenon within a real-life context: and, especially when the boundaries between phenomenon and context are not clearly evident.[14] The case-study methodology allows an investigation to retain the holistic and meaningful characteristics of real-life events, such as individual life cycles, organizational and managerial processes, neighborhood chance, and international relations. It can cope with the technically distinctive situation in which there will be many more variables of interest than data points; it relies on multiple sources of evidence, such as documents, archival records, observation and interviews, with data needing to converge in a multi-triangulating fashion which incorporates multiple sources of evidence (data triangulation), of methods (methodological triangulation) and of perspectives (theory triangulation).[14,15]

Data for the case study on the *Serra da Capivara National Park* were provided by primary and secondary sources. A critical assessment of documentary evidence[16–32] and archival records (health and school files) was used in this study. The study included direct and participant observation (town of São Raimondo Nonato; villages of Sítio do Mocó, Barreirinho, Borda and Coronel José Dias; Community Support Centres at Sítio do Mocó and Barreirinho; apiculture project; archaeological sites; tourist trails in the Park: unstructured and informal interviewing with key informants, people involved in the project administration, archaeological research, tourism, apiculture, pottery, and health and education activities: and readministration of a questionnaire that was first administered to villagers in 1989, prior to the developmental project. These activities represented fieldwork conducted by the author in 1993.

Case Study: Serra da Capivara National Park

Location

Serra da Capivara National Park is located in the southeast of Piauí State, in the northeast region of Brazil: Perimeter 214 km, area 129 140 ha. The total population of the four municipalities in which the Park is located is 120 000 inhabitants. The village of Sítio do Mocó (253 inhabitants), where the population nearest to the Park is found, is some 26 km distant from the town of São Raimondo Nonato (10,000 inhabitants).

History of Creation

In 1963, the municipal governor of São Raimondo Nonato visited the archaeological exhibitions of the museum at the University of São Paulo. He asked the person in charge—Niéde Guidon, a Brazilian archaeologist—to look at some photographs of existing cave paintings in the mountains of his region. She recognized that these case paintings were different from anything already known.

In 1970, when she saw the paintings for the first time, she confirmed the importance of the discovery. At the time, the local people were acquainted with 4 or 5 sites. She asked them to locate new paintings. By 1973 they had identified 55 new sites, and in 1975, 52 more. A total of 362 sites had been recorded up to 1991; 242 presenting cave paintings; 76 are open-air sites representing hunter/gatherer settlements; 15 ceramic/agricultural settlements; 23 occupations in caves or shelters; 1 is a burial site; and 5 are archaeopalaeontological sites. There are cave paintings dated 14,000 years B.P. The diggings at *Toca do Boqueirão do Sítio da Pedra Fuerada* have established the earliest known presence of man in the Americas, c. 50 000 yrs B.P. An abundance of fossilized plants found in the Toca de Cima dos Pilão cave, and fossil bones found in the Antonião site confirms the presence of mastodons, American horses, llamas, sloths and giant armadillos up to at least 12 000 yrs B.P., when the climate began to change from tropical to arid, reducing the region's humidity as well as the chances of survival.

The archaeological research program. developed since 1970, showed the importance of the sites for understanding the colonization of Brazil; the ecological studies developed showed preserved areas of forests and caatingas (savannah) that protected endemic species of animals and plants, some of which were in danger of extinction. In 1974, Niéde Guidon's team reached an agreement with the Federal University of Piauí, which resulted, in the establishment of the French-Brazilian Mission in 1975. After three months of archaeological work, the team decided to communicate their important findings and the unique ecological features of the region to the Brazilian authorities, asking for the creation of a protected area. The *Serra da Capivara National Park* was created by Presidential Decree No. 83 548 of 5 July 1979.

Furthermore, in 1986, those involved in the São Raimondo Nonato research decided to create the Museum of American Man Foundation. The objective was not only to further interdisciplinary research, but also to lend support to the development and welfare of the local populations.

Importance of the National Park

The importance of the park was recognized in 1991, when it was declared a World Heritage Site by the United Nations Educational, Scientific and Cultural Organization (UNESCO). It is the only national park situated in the morphoclimatic dominion of Caatingas, possessing the greatest concentration of prehistoric sites in the country, and the most important in the Americas. The diggings at Toca do Boqueirão do Sítio da Pedra Furada have established the earliest known presence of man in the Americas, providing contradiction to the ancient classic theory, which had established that American man came originally from Africa, Europe and Asia, having crossed the Bering Straits, passed through North and Central America and finally arriving in Brazil about 12,000 years B.P.

Background

The 860 000 km^2 Caatinga remains one of the least populous and certainly the poorest among the five major ecosystems of the country. The long cycles of severe droughts periodically force local populations either to migrate to slums in urban areas or to exhaust the region's natural resources. Illegal and uncontrolled hunting have resulted in the extinction of animal and plant species. Tree-felling and fires, started by hunters or farmers, have destroyed extensive zones of primary vegetation.

The great drought of the period 1980–1985 caused a substantial depletion of the resources of small, local farmer, and forced many of them to become rural workers who searched daily for any kind of work and were extremely badly paid, with conditions of health and hygiene ever more difficult. One of the sources of income is the extraction of limestone to make lime: massive blocks of karstic (limestone) are crashed with marble, put into a lime-kiln and heated for three days to form lime. For the heating process up to 7 trucks of wood are used. The blocks of limestone belong, in the main part, to proprietors who employ daily workers to transform noble raw materials such as hardwood and marble into a cheap product, lime.

Depletion of natural resources was identified as an important threat to the local natural and cultural heritage and became the principle worry for the Museum of American Man Foundation, which proposed to intervene in the region's development. A questionnaire answered by the villagers in 1989, identified the main problem, i.e. the lack of local work opportunities which forced the men to migrate to bigger towns and distant cities. The aim of the project was, therefore, to provide local communities with the ability to carry out innovative activities which have a chance to succeed in the area, and which can provide a sustainable livelihood. In addition, this would prevent the destruction of the archaeological sites and cave paintings.

Findings Prior to the Development Program

Local children had obvious learning problems which affected their physical and intellectual development. Among the salient factors influencing this situation were:

1. scarcity of food, which forced the children to go to school without having had any food in the morning;
2. lack of clean water facilities, which, produced gastro-intestinal and other infections which lead to malnutrition;
3. an absence of contact with parents during the day;
4. lack of stimulation for linguistic expression, in the context of a culture strongly founded in oral expression:
5. lack of stimulation and incentives for school attendance, because of inadequate teaching methods;
6. poor physical state of the school buildings and facilities
7. absence of preschool education.

A survey carried out by the Secretariat of Education of the municipality of São Raimundo Nonato showed that 48% of matriculated students in the district rural schools, administered by the prefecture, did not complete the school year, 35.4% failed exams, and 13.1% abandoned school during the year.

Factors related to 'deschooling' included:

1. the oldest students, and even teachers, abandoned lessons during the planting and harvesting seasons;
2. lack of clothing for the school children;
3. lack of school lunches;
4. lack and inadequacy of teaching materials;
5. low level of competence of teachers, who had either scarce or no complete primary level education;
6. the low salary of teachers (50% of minimum wage in 1986 and 20% of minimum wage in 1989).

In addition, municipal rural schools were usually in a poor state of repair (cracks in the walls containing bugs-vectors of Chagas disease), no running water or toilets, no systematic cleaning system, little school furniture, and broken doors and windows.

The health conditions of the municipality do not differ from those observed in other rural areas of the majority of the municipalities of northeastern Brazil. Among the main findings were poor primary health care, low level of vaccination, high prevalence of infantile malnutrition, poor dental health and frequent diarrhea.

Development Program of Activities and Achievements

The development aim of the Museum of American Man Foundation is being achieved through the implementation of sites of community action designated as

Community Support Centres (in Portuguese: Núcleos de Apoio Comunitário—NACs), where diverse activities which aim to benefit the local communities are concentrated. The objective is to increase the number of NACs around the Serra da Capivara National Park in a manner designed to integrate the local communities with activities that avoid depredation and which can create better conditions. In the NACs, professional courses such as embroidery, cabinet-making, painting and design, ceramics, horticulture and goat-herding, can be offered to children, adolescents and adults. These courses can provide means to breaking the current cycle, in which agriculture is the only productive activity available.

Since the beginning of the project, apiculture (beekeeping) courses have been offered to those interested in learning about this productive activity. Studies on the environment agree that the development of apiculture produces highly profitable results for the preservation of the Park, because it favors the pollination of many of the Park's species. Beekeeping employs 2 to 6 people, some honey being used by the schools while the rest is sold. In the first year, production reached 2800 liters of honey of which 2500 liters were sold. This activity is already self-supporting. The beekeepers are assisting in the conservation of the vegetable gardens adjacent to the schools. These gardens are irrigated by the water used in the community lavatories, taking advantage of the slopes of the terrain. The school children also help in taking care of the gardens, which provide vegetables for their meals.

The tourist potential of the region is enormous, with its natural beauty and archaeological sites with rock paintings of great scientific and artistic value. The educational program supports the aspects of the preservation of the environment, such as the training of people to accompany tourists in the Park. A 16-room hotel (each room with air-conditioning, TV, telephone, refrigerator and shower) for up to 36 guests, began operating in December 1992. It also has two small swimming pools and a good restaurant with seating for 56 people. This hotel, located in the town of São Raimondo Nonato provides work for 20 permanent staff and 9 occasional employees. It is already self-supporting.

In addition to these activities, the educational program and the health program form the basis of the project developed by the Museum of American Man Foundation, with each NAC having a school and a health center. The educational program and the health program were designed to be interconnected, supporting the need to integrate health professionals, community health agents, school activities and parents.

In July 1993, the education program was designated by UNICEF as one of 14 model programs in the world; based on the following findings:

1. illiteracy in adults was eradicated;
2. the level of understanding of children of the NACs is superior to that of children in other schools of the region, and handwriting is noticeably better;
3. basic English and French are taught;
4. the children remain in the NACs school from 8:00 a.m. to 5:00 p.m. They have a bath every day at school because water is not available in their

homes. The practice of cleaning their teeth, a complete novelty, has also been introduced

5. they receive breakfast, lunch and a fruit juice drink before leaving school. Meals are prepared by teams of mothers;
6. if lessons are given in one session, the following time being used for light activities and sports;
7. a choir has been created;
8. environmental education is part of the curriculum such as the necessity of protecting the Park's animal life. The children transfer the environmental messages to their parents, who discuss the issues at parents' meetings.

After just four years of existence, the achievements of the health program are remarkable: primary health care is available for all; all children aged between 0–5 years participated in the official vaccination scheme and all children aged between 7–14 years were immunized against tetanus: not one case of vaccine-preventable disease has been reported over the past 4 years; infantile malnutrition has been eliminated: and dental caries is decreasing as are diarrheic diseases.

Community Participation

Community participation in the work of the NACs was initially difficult to obtain, but after 4 years changes were obvious. At the beginning of the project it was difficult to interest the mothers in the preparation of meals for the children. Today, the initial position has changed and mothers participate with interest in these activities.

Alternative methods of agricultural production first regarded with suspicion are now fully supported, in particular the beekeeping and pottery activities. Initially. vegetable gardens were regarded with suspicion by adults and only one adult grew his own vegetables. After the success of the children's garden, with the surplus being sold and generating a profit for the children, adults also started to grow their own vegetables.

Certain community works have been and are still being completed. In the NAC of Sítio do Mocó, building maintenance work is performed by the inhabitants, while the required materials are provided by the Museum of American Man Foundation. In the NAC of Barreirinho community participation has been more dynamic and enthusiastic from the beginning.

The structure of the buildings allows the members of the community to develop various recreational and school activities. In both NACS, such activities are particularly well-attended at weekends. Community meetings, held fortnightly, become better attended each time; participation is growing considerably, and involves the whole age spectrum. Conferences are developing and videos illustrating different themes are being shown. The available space in the NACs is being used more and more for community activities. Once a week a rented film from the town of São Raimondo Nonato is shown at the NACs, and attendance is massive.

Sustainability

During the first stage of the project, all activities were funded by the Museum of American Man Foundation, which in its turn raised the necessary funds from Brazilian and international funding agencies. Some activities, such as beekeeping and the hotel, have already achieved the'break-even point'and are in the process of self-sustainability, paying their own costs and making profits.

Discussion

It is accepted that protected areas"provide major direct and indirect benefits to the local and national economies and models for sustainable conservation."[33] The benefits to the local communities which were observed in the case studied include the provision of educational and recreational facilities, and the creation of employment opportunities. They are among the 16 identified benefits of protected areas for regional development (5). Although evident. human health has not yet been included as one of these benefits.

The observed improvements in the health conditions of the communities close to the Serra da Capivara National Park, Brazil, arose from the unique opportunities and abilities present in this area. In other words, the need to conserve the local natural and cultural heritage, identified by a group of scientists, persuaded the Brazilian Government to create a national park in the region. The existence of a protected area brought together a multidisciplinary team of specialists, who were interested in addressing the different problems found locally under a holistic approach. This holistic approach made possible the design and implementation of a multidisciplinary project, which was aimed at protecting the local natural and cultural heritage by improving the health conditions of the local communities. By health conditions are understood the access to primary health care and, more importantly, education, decent housing, and employment opportunities. Such improvements were not observed in communities in the same area, which were not linked to the development activities described in this case study.

In addition, integrating concepts of health promotion to environmental management is useful in understanding the role of environmental managers and conservation biologists in the larger society. It is tempting to examine substantive norms of human health as a guide to providing a definition of ecosystem health.[34]

To what extent and in what ways can health promotion be part of a strategy to give purpose to protected areas and to show their contribution to sustainability? The suggestion made here is that the connection between health and conservation could enable local communities to sustain their living and increase their commitment to the surrounding protected areas.

Further research is required to understand the findings of this case study. It is also necessary to find other representative cases and to produce general principles from which to elucidate models which are applicable elsewhere.

Manuel Cesario MD is experienced in coordinating multidisciplinary teams to work on community-based health and developmental projects with indigenous groups and local communities. He is presently pursuing doctoral research at the University of Edinburgh, Institute of Ecology and Resource Management, Centre for Human Ecology, under the auspices of the Brazilian Council for Scientific and Technological Development. His address: 88 Brunstane Road, Edinburgh, Scotland, EH15 2QR.

References and Notes

1. Caroline, M. and McNeely, J. 1992. Managing parks for the 21st Century: Advice from the Parks Congress. *Parks 3*, 13–21.
2. McNeely, J. A. 1992. Protected areas in a changing world: The management approaches that will be required to enable primates to survive into the 21st century. In: *Tropics in Primatology, Vol. 2, Behaviour, Ecology, and Conservation.* University of Tokyo Press, Tokyo, pp. 373–383.
3. Davis, S. H. 1993. Introduction. In: *The Social Challenge of Biodiversity Conservation.* Global Environmental Facility, Washington. pp. 1–4.
4. McNeely, J. A. and MacKinnon, J. R. 1989. Protected areas, development, and land use in the tropics. *Resour. Mgmt Optimization* 7, 189–206.
5. UNCED (United Nations Conference on Environment and Development), 1992. The Rio declaration on environment and development. In: *Earth Summit '92.* Quarrie, J. (ed.). The Regency Press Corporation, London, pp. 11–13.
6. IUCN (Commission on National Parks and Protected Areas). 1994. *Parks for Life: Action for Protected Areas in Europe.* IUCN, Gland, Switzerland and Cambridge, UK, 154 pp., recently launched in 31 countries across Europe as a basis for European conservation over the next decade.
7. WHO (The World Health Organization). 1978. *Report of the International Conference on Primary Health Care, Alma-Ata-USSR, 6–12 September 1978.* World Health Organization and United Nations Children's Fund.
8. Almedia Lessa, quoted by Machado, P. A., 1984. *Introdução à Ecologia Humana.* Brasilia: CNPq, Editora Cortez/Editora Autores Associados. (In Portuguese).
9. Veras, F. M. F. 1986. Atenção Primária em Saúde. In: *Epidemoiologia e Saúde,* Rouquayrol, M. Z. (eds). MEDSI—Editora Médica e Científica Ltda, Rio de Janeiro, pp. 363–370. (In Portuguese).
10. Follér. M. L. 1992. Social determinants of health and disease: the role of small-scale projects illustrated by the Koster health project in Sweden and Ametra in Peru. *Cadernos de Saúde Pública 8*, 229–239.
11. Holling, C. S. 1993. New science and new investments for a sustainable biosphere. In: *Investing in Natural Capital: Why, What and How?* ISEE Press, Solomons, MD.
12. Jungen, B. 1985. *Integration of Knowledge in Human Ecology.* Göteborg University, Göteborg, Sweden.
13. Tengström, E. 1985. *Human Ecology—A New Discipline?* Göteborg University, Göteborg, Sweden.
14. Yin, R. L. 1994. *Case Study Research: Design and Methods,* Sage Publications, Thousand Oaks, London, New Delhi.
15. Rose, H. 1993. Case studies. In: *Handbook for Research Students in the Social Sciences.* The Falmer Press, London, pp. 190–202.

16. Araújo, A. J. G. 1989. *Saúde e Desenvolvimento em São Raimundo Nonato—Sudeste do Piauí* (mimeo). (In Portuguese).

17. Araújo, A J. G. 1990 *Relatório de Atividades da Coordenação de Saúde FUMDHAM/TERRA NUOVA/FIOCRUZ—São Raimundo Nonato—Piauí* (mimeo). (In Portuguese).

18. Araújo, A., Pessis, A. M., Marasciulo, A. C. E., Guerin, C., Olmos, F., Parenti, F. and Pellerin, J. 1991. *Plano de Manejo do Parque Nacional da Serra da Capivara. Brasília: SEMA-PR/IBAMA/FUMDHAM.* (In Portuguese).

19. FUMDHAM/FIOCRUZ/TERRA NOUVA. 1989 *Impressões Preliminares do Processo Saúde/Doença nas Comunidades do Sítio do Mocó e do Barreirinho—São Raimundo Nonato—Piauí* (mimeo). (In Portuguese).

20. FUMDHAM/TERRA NOUVA. 1989. *Programa Educativo no Sítio do Mocó* (mimeo) (In Portuguese).

21. FUMDHAM/TERRA NOUVA. 1992. *Relatório Bienal sobre o Convénio FUMDHAM/TERRA NOUVA.—1991/1992—São Raimundo Nonato—Piauí* (mimeo). (In Portuguese).

22. Gomes, M. A. T. 1993. *Potencial de Transmissão da Tripanossomíase Americana nas Localidades do Sítio do Mocó e Borda, Município de São Raimundo Nonato, Sudeste do Piauí.* Tese de Mestrado, Rio de Janeiro: Escola Nacional de Saúde Pública. Fundação Oswaldo Cruz. (In Portuguese).

23. Macêdo, R. E. N. 1991. *Relatório das Atividades do Posto de Saúde do Sítio do Mocó—Agosto* (mimeo). (In Portuguese).

24. Macêdo, R. E. N. 1991. *Relatório das Atividades do Posto de Saúde do Sítio do Mocó—Dezembro* (mimeo). (In Portuguese).

25. Macêdo, R. E. N. 1992. *Relatório das Atividades do Posto de Saúde do Sítio do Mocó—Janeiro* (mimeo). (In Portuguese).

26. Macêdo, R. E. N. 1992. *Relatório das Atividades do Posto de Saúde do Sítio do Mocó—Agosto/Setembro* (mimeo). (In Portuguese).

27. Macêdo, R. E. N. 1992. *Relatório das Atividades do Posto de Saúde do Sítio do Mocó—Dezembro* (mimeo). (In Portuguese.)

28. Macêdo, R. E. N. 1993. *Análise da Situação Atual das Crianças nos Núcleos de Apoio as Comunidades Sítio do Mocó e Barreirinho* (mimeo). (In Portuguese).

29. Guidon, N. and Arnaud, B. 1991. The Chronology of the New World: Two Faces of One Reality. *World Archaeol. 23,* 167–168.

30. Parenti, F. 1993. Le Gisement Prèhistorique du Pleistocène Supérieur de Pedra Furada (Piauí, Brésil) Considérations Chronostratigraphiques et Implications Paléoanthropologiques. *Actes de la Table Ronde Europèene Paléontologie et Stratigraphie d'Amerique Latine.* Lyon 1992, Docum. Lab. Géol. Lyonn. 125. (In French).

31. Parenti, F. 1993. Le premier peuplement américain. *Archeologie 295,* 54–59. (In French).

32. Parenti, F. 1993. *Pedra Furada: le Point sur le Peuplement de l'Amérique.* Musée de l'Homme. N. 2. (In French).

33. IUCN. 1993. Parks, protected areas and the human future—The Caracas declaration. In: *Parks and Progress: Protected Areas and Economic Development in Latin America and the Caribbean.* Valerie Barzetti (ed.). IUCN—The World Conservation Union, Washington DC, pp. 217–220.

34. Norton, B. G., 1991. Ecological health and sustainable resource management. In: *Ecological Economics: The Science and Management of Sustainability.* Columbia University Press, New York, pp. 102–117.

35. The author is very grateful to Dr Rafal Serafin (Heritage Research Programme of the Estreicher Institute in Kraków. Poland) for the suggestions relating to the research subject and to Prof Dr Colin Whittemore and Prof Dr Susan Smith (Institute of Ecology and Resource Management and Department of Geography of the University of Edinburgh, respectively) for the criticism provided.

36. First submitted 6 December. Accepted for publication after revision, 26 July 1995.

SOCIO-ECONOMIC CHALLENGES FOR SUSTAINABLE TOURISM IN THE MEDITERRANEAN BASIN

ANDREAS ASHIOTIS

The Mediterranean Basin has been the womb where great civilisations of our past have been born, evolved and lost in the depth of history. In fact, one could suggest that the history of Europe in particular has been written along the coasts of the Mediterranean.

Even today, in an era of rapid technological evolution and development, when distances diminish with every year and information technology bridges the gap between the peoples of our world to form an integrated universal system without precedent in human history, the Mediterranean remains an extremely significant region of the world.

The Mediterranean has always been and remains till today a physical element which functions in contradictory ways; it links people with strong, albeit salient, bonds, but at the same time it divides an impressive variety of nations, cultures, economies and religions. Peoples living along its coasts have widely differing aspirations, expectations and lifestyles but they share uniquely a common destiny: the Mediterranean as a common life-supporting element of nature, the future of which will define both directly and indirectly their own future. Depending on the chosen course of action, the Mediterranean Basin may sustain prosperity and welfare, or alternatively it might as well provoke protracted social unrest and conflict.

During the last forty years, the contemporary generations exploit the potential of the Mediterranean without any effective concern for the irreversible damage inflicted on the overall capacity of an extremely sensitive environmental system to support life in the future. Eventually, as a result of our priorities and choices, the Mediterranean Basin is facing a state of crisis.

Tourism and tourism-related activities and development might be considered responsible for many of the severe problems affecting sustainability in the Mediterranean Basin. For this reason, discussion on sustainable development in this particular region usually focuses on tourism and its environmental impact.

The Impact of Tourism on the Mediterranean

For the most part of this century, the Mediterranean has been an extremely attractive tourism destination and millions of Europeans have visited the region for holidays. This tremendous influx of visitors over the years has contributed to the economic development of coastal areas of Europe and to the overall welfare of local communities. The significance of tourism for the Gross National Product of most Euro-Mediterranean countries proves that tourism will remain an extremely vital service-industry which is especially valuable for national economies during periods of recession and high unemployment. At the same time, tourism has helped local populations to remain in otherwise poor coastal regions of the Basin, thus contributing to the demographic stability of many areas in Portugal, Spain, France, Italy, Greece, Cyprus, etc.

However, over the last decade it is gradually becoming clear that the "blessing" of tourism has imposed on the Mediterranean environment pressures and stresses of an unprecedented scale and magnitude. Today, it is generally accepted that most Euro-Mediterranean countries are facing a paradoxical phenomenon, where an environmental asset (the Mediterranean Basin) has attracted tourism, and that simultaneously the asset itself is victimised through over-exploitation and misuse.

Environmental degradation due to over-exploitation of the primary asset is the direct result of the drive to maximise tourist development and the immediate economic benefit generated by the influx of visitors. Nevertheless, other indirect effects multiply the overall impact on the Mediterranean environment.

Ribbon tourism development in the form of hotels, apartments, resorts, and second holiday homes stretch over hundreds of kilometres along the Euro-Mediterranean coast, transforming the unspoiled countryside of coastal regions into a continuous urban area. The very fast rate of these developments has not coincided in most cases with the provision of the infrastructures, necessary to support them (ie, waste and sewage treatment systems). Highways, airports, power plants and power lines, marinas, etc. have scarred what was previously perceived as the picturesque Mediterranean environment. Small fishing villages have been lost forever to be replaced by massive tourism settlements.

Natural resources are being over-exploited beyond their capacity to regenerate in order to satisfy increasing demand for new tourism-related development. Golf courses, amongst others, add up to the excessive consumption of huge quantities of fresh water, which is a particularly scarce resource in this region. Environmentally sensitive areas of high value and their fragile ecosystems are affected by the influx of visitors and development pressures.

Adverse socio-economic phenomena are increasingly alarming in many countries. Local labour forces are unable to support tourism, even where this sector attracts a work force from agriculture and industry, weakening further their respective competitiveness. The resulting dominance of tourism in local economies is not good news for any country, especially when it has been proven that tourism is not as reliable a sector of the economy as it was perceived to be a few years ago.

In many cases massive tourism exceeds the social and cultural carrying capacity of local communities and resulting social tensions and changes in the traditional social structures and balances are warning indications which should be monitored as early as possible.

The widening of regional imbalances is another indirect effect of the rapid tourism-related development of coastal areas. In most Euro-Mediterranean countries the economic benefits from tourism tend to be confined within a relatively narrow coastal zone, while the hinterland remains mostly underdeveloped and neglected. In addition, a significant proportion of the young and productive population of these hinterland areas moves to the coastal zone to work and live.

Developments on the Southern Coast

Sustainable development in specific environmental units (to the extent that this identification is relatively feasible), and especially in "closed" and vulnerable systems like the Mediterranean Basin, is an objective which needs to be pursued irrespective of national boundaries or any other delineation. Regardless of any effort made or action taken by any state or group of states in the European coast of the Basin, sustainable development in this region will not be achievable without an equally strong commitment by countries on the southern coast of the Mediterranean.

This presumption is the foundation of the interest and particular attention that Europe should give to the conditions which affect the ability of countries on the southern rim of the Basin to adjust their policies, priorities and overall orientation towards the common objective.

This particular aspect of sustainable development in the Mediterranean Basin is not the subject of this article. However, it should be noted that the issues at stake are very significant indeed, especially when evaluated within a broad and long-term perspective. Demographic developments and migration trends, the performance of national economies and relevant problems, socio-cultural balances and the development prospects in these countries should be studied in depth in order to formulate a set of policies and incentives which will encourage them to join the common effort towards safeguarding that this development will be sustainable in the future.

Sustainable Tourism Development

Past experience should be used to define the new approach towards sustainable tourism development in the Mediterranean. Generally, tourism development should guarantee the integrity of all non-renewable resources, while remaining economically viable in an increasingly competitive international market.

Tourism development should be perceived as part of an integrated spatial and environmental system, which needs to be well balanced and interwoven to other major activities, development and the policies for protecting the natural habitat. A decisive priority should be allocated in the upgrading of existing facilities for adjusting to the changing needs of the consumer, whilst the excessive consumption of scarce land, especially in environmentally sensitive areas, should be discouraged.

The local population should be encouraged to participate in an effective manner in the planning process related to tourism development.

A permanent dialogue with local people will make them an equal partner in the implementation and management phases and will minimise social cost.

Euro-Mediterranean countries should initiate an intensive dialogue and co-operation process between themselves and with countries on the southern coast of the Basin in order to co-ordinate:

- the development of complementary tourism products:
- the transfer of experiences on the negative effects from tourism in order to avoid repeating past mistakes:
- the formulation of a common integrated tourism strategy, which should ensure the sustainability of all future developments.

Eventually, it should be realised that massive tourism and all related development along the Mediterranean coast affect permanently the viability of the Mediterranean Basin itself as a nature resource and its potential to support the peoples of the region. All future developmental choices, therefore, should consider the environmental and social carrying capacity of localities and the Basin as a whole as the one most significant and indeed crucial criterion for regional planning decisions.

A. Ashiotis
Department of Town Planning and Housing
Ministry of the Interior
Nicosia
Cyprus

OUTCOMES OF A COMMUNITY CONTROLLED WILDLIFE UTILIZATION PROGRAM IN A ZAMBEZI VALLEY COMMUNITY

GORDON EDWIN MATZKE[1] AND NONTOKOZO NABANE[2]

Zimbabwe is devolving substantial wildlife management responsibility to local government, and ultimately to local communities, through its Communal Areas Management Program for Indigenous Resources (CAMPFIRE) program. This paper's purpose is to explore the notion that CAMPFIRE offers a useful model for sustainable development discussions by examining the program's impacts in one case study location. It presents a legislative history of the CAMPFIRE Program before specifying the spatial criteria that explain ideal CAMPFIRE sites. Thereafter, it reports results from a study of the development impacts of locally controlled wildlife management in Masoka, an isolated CAMPFIRE community in the Zambezi Valley. The village has developed a land use plan, fenced its fields and settlements, reduced wild animal attacks on people and crops, provided access to primary education for both boys and girls, created local employment, and provided money for household food purchases during a severe drought. In Masoka, the implementation of CAMPFIRE seems to successfully provide incentives to protect megafauna and their habitats. For Masoka, CAMPFIRE has provided an alternative model to statist solutions emphasizing centralized control for biodiversity conservation purposes.

Key words: CAMPFIRE; Zimbabwe; sustainable development; wildlife management; biodiversity; wildlife utilization.

Introduction

The megafauna of Africa are popular subjects for media attention. There is an endless outpouring of magazine articles, television documentaries, and books which extol the beauty of the beasts while decrying threats to their survival (e.g., Bonner, 1993). One of the greatest threats is a set of anonymous indigenous people pejoratively reviled as "poachers" (Mackenzie, 1988). Poachers, usually local people, kill animals and are universally portrayed as villains. On the other hand, antipoaching paramilitary agents of the central government are portrayed as wildlife guardians who arrest, or even kill, poachers in the name of wild animal protection. If the popular media is our guide, the solution to the poaching threat is getting tough with poachers. If the experience of many independent African governments is our guide, the paramilitary model is a failure. It is a failure because it is based on the flawed premise that wildlife management exclusively directed by the central government constitutes a good model to foster long-term wildlife conservation.

Marks (1984) was one of the first writers to offer a serious challenge to the management-from-the-center *modus operandi.* He argued that professional wildlife management was developed in the "North," particularly in North America, and was transported wholesale to the "South," especially to Central Africa. The northern approach disenfranchised local people from the resources they had managed for millennia. As conservation structures developed, policies evolved to separate human settlement from animal habitat and to transform hunting into "a ritual of the elite" (Mackenzie, 1988). The allocation of legal wildlife harvest rights became the exclusive domain of central authority which provided them to urbanites and foreigners while capturing the benefits of licensing for itself. All of the "negatives" of the wildlife presence, including damage from marauding animals and arrests for illegal harvest, accrued to people living in close proximity to the megafauna (Riney, 1967) and who ultimately decide whether there is a place for wildlife in the neighborhood. This approach to management does not encourage local people to foster the presence of megafauna.

Recent African initiatives are attempting to create different approaches to wildlife management (Moyo *et al*, 1991). "The new paradigm is an advance on older confrontationist stances, now treating local communities as a means rather than an impediment to conservation" (Murphree, 1993a). Following early experiments in Kenya (Western, 1982, 1994) variations on the participatory community-based management theme developed in Zambia, Tanzania, Namibia, and Botswana. One high impact program is CAMPFIRE (Communal Areas Management Program for Indigenous Resources) in Zimbabwe.

The CAMPFIRE program has received international recognition as an innovative approach to the devolution of natural resource management to local communities. It seeks to obtain the voluntary participation of communities and to introduce a system of ownership with defined access rights to natural resources for community residents in target areas. At the same time, it enables residents to realize economic benefits from their wildlife resources (Martin, 1986).

The rationale for the CAMPFIRE program is based on the notion that people from producer communities—those controlling areas with wildlife resources—ought to gain benefits from the wildlife that their lands produce on a sustainable basis. Producer communities are thereby encouraged to make responsible decisions about wildlife and act to conserve them as a matter of enlightened self-interest.

The concept of a "producer community" introduces operational issues with strong spatial dimensions in the CAMPFIRE context. In order to operate, CAMPFIRE functionaries[3] must define a group of people upon whom they can eventually confer proprietorship and management responsibility. In the final analysis, the definition of a producer community means that "these people in this place" shall assume control. Although CAMPFIRE aims at a participatory approach to resource management, it does not provide equal participatory opportunities for the entire country and of only 22 districts 1/3 were deemed suitable for the grant of CAMPFIRE authority. Even within authorized districts, there is great areal variation in the level of possibilities for CAMPFIRE.

Murphree (1993b) has argued: "The unit of proprietorship should be as small as practicable, within ecological and sociopolitical constraints." Since ecological and socioeconomic conditions show spatial variation, we have identified a set of criteria (Figure 1) that help explain the features associated with potentially suitable sites for community-based wildlife initiatives. Using these criteria, the potential for campfire type resource management endeavors is greatest in the lightly settled and isolated marginal lands that appear as empty spaces on a map of Zimbabwe's modernization surface. Our study site meets these conditions and is located in Guruve District, one of the first two districts in the country to be given appropriate authority to initiate campfire programs.

Appropriate authority does not mean either total authority, nor ultimate authority, in the realm of management. Even in fully functioning CAMPFIRE districts, legal authority is largely retained off-site. For example, hunting quotas are still established by the Department of National Parks and Wildlife Management (DNPWM) and the District Council does financial accounting for local communities. CAMPFIRE establishes a framework whereby local level priorities are injected into the decision making process of the legal authorities. The more empowered communities are those successfully negotiating outcomes beneficial to themselves. Our study reports on CAMPFIRE outcomes in one locale after 5 years of program participation.

The Study and Methods

This paper examines the historical background of CAMPFIRE before focusing on the extent and manner of local community participation in Masoka,[4] a CAMPFIRE village in the Zambezi Valley of Zimbabwe. It ends by discussing some broader implications of this experiment. Since Masoka was an early participant in the CAMPFIRE program (Crane, 1991; Metcalfe, 1994), it is an excellent place to make seminal observations on planning, implementation, and realization of benefits.

CRITERIA USEFUL IN SELECTING CAMPFIRE SITES

BEST CONDITION	CRITERION	WORST CONDITION
SMALL	HUMAN POPULATION	LARGE
AGGLOMERATED	SETTLEMENT PATTERN	DISPERSED
LARGE	UNSETTLED HINTERLAND	SMALL
CROPLAND CONCENTRATED NEAR SETTLEMENT	CROPPING PATTERN	CROPLAND SCATTERED THROUGHOUT HINTERLAND
POOR	AGRICULTURAL POTENTIAL	GOOD
NONE	LIVESTOCK IN HINTERLAND	MANY
FEW	ALTERNATIVE INCOME OPPORTUNITIES	MANY
STATE LAND IN PROTECTED STATUS, E.G., NATIONAL PARK	LANDUSE ON AREA OUTSIDE COMMUNITY'S HINTERLAND	DENSE SETTLEMENT
LARGE	HINTERLAND WILDLIFE POPULATIONS	SMALL
HIGH MARKET VALUE DANGEROUS GAME, E.G., BUFFALO, ELEPHANT	WILDLIFE TYPES	UNMARKETABLE PEST SPECIES, E.G., BUSHPIG SPRING HARE, BABOON
WILLING TO DEVOLVE AUTHORITY AND BENEFITS TO COMMUNITY	DISTRICT GOVERNMENT	INSISTS ON CONTROL OF DECISIONS ON MANAGEMENT - BENEFITS
COHESIVE	COMMUNITY IDENTITY	FRACTIOUS

Figure 1. Criteria useful in selecting CAMPFIRE sites. The possibilities for the successful establishment of wildlife based common property development initiatives vary widely within Zimbabwe.

We explore the management and distribution of benefits from CAMPFIRE in Masoka to see whether they are adequate to support the continued existence of the program in the village. This exploration is based on a review of published research reports, discussions with key participants, observations during periods of residence, and a field study conducted from December 1992–August 1993. The field study was designed, in part, to replicate pieces of an earlier survey conducted by Cutshall (1989)[5] in order to produce data that compare the current circumstances with those prevailing before the implementation of CAMPFIRE.[6]

The elements of the CAMPFIRE approach in Masoka that will be discussed include the formation of a local wildlife authority (part of CAMPFIRE's community empowerment strategy), the authority's choice of management tools, its strategy for income generation (leasing of a safari hunting area), the community-based mechanisms for deciding on the distribution of benefits derived from wildlife management, and the *de facto* outcomes of community directed benefit distribution.

Precursors to the CAMPFIRE Program

It is important to understand the history of wildlife management in Zimbabwe as present policies are derived from it (Murphree, 1990; Hill, 1991). Under colonial rule, all wildlife was state property and legal utilization of wildlife was the province of Whites. Wildlife management was centralized and controlled by the state. At the same time, African settlement was pushed to the most agriculturally marginal dry lands under a system of "Native Reserves," now called "Communal Lands," where private land ownership was not allowed. Whites converted the best highland agricultural lands into a commercial farming region where a system of private freehold land tenure was formalized. Murphree (1990) has referred to this state of affairs as "double expropriation" in that the indigenous population was denied both suitable land for agriculture and legal access to the wildlife that survived on the marginal lands. The population commonly responded by poaching, which was centrally viewed as a threat to the wildlife resource base.

CAMPFIRE evolved as a response to numerous problems of the centralized wildlife management system. The system lacked sufficient funds and personnel to police illegal utilization of wildlife resources, particularly in communal areas. The outside policing system was costly, inefficient, and incapable of controlling high illegal offtake. Even before independence, it was clear that something had to change and alternative solutions were suggested (e.g., Dasmann, 1964).

Zimbabwe, after 1980, experienced new challenges to the centralized management model when immigrants started to move into previously war ravaged and tsetse infested areas. With the arrival of peace, the European Economic Community sponsored a large-scale and long-term tsetse control program. The presence of tsetse in the Zambezi valley had inhibited human settlement and, at the same time, created *de facto* opportunities for wildlife conservation. Lower human populations were associated with relatively low poaching activities (Peterson, 1991c). Tsetse eradication, or even the promise of it, stimulated the spontaneous movement of people into the Valley and threatened its abundant wildlife resources. This, in turn, forced the development of new approaches to conservation of wildlife.

Starting in the 1960s, a growing concern for wildlife conservation throughout Africa stimulated various attempts to solve the problems of illegal offtake of wildlife resources (Hunter *et al.*, 1990). CAMPFIRE in Zimbabwe emerged after several trials with other approaches including the 1975 Parks and Wildlife Act and the 1978 WINDFALL (Wildlife Industries New Development for All) Program. Because rural community people were not really included in decision making and implementation, these attempts were not successful in communal areas (Murindagomo, 1990).

The 1975 Parks and Wildlife Act transferred effective control over wildlife on private lands from the state to land holders (Peterson, 1991c). The new law gave the benefits of wildlife management to landowners. This provided an economic rationale to reinforce the scientific, aesthetic, and moral justification for wildlife conservation (Murindagomo, 1990). The reasoning behind this was that "local proprietorship of wildlife resources was likely to promote investment for their efficient and sustainable management" (Murindagomo, 1990). This Act empowered White commercial farmers, but did nothing for residents of the communal areas where private landholding did not exist. A tool was needed to address conservation in places that were managed under common property regimes.

The Department of National Parks and Wildlife Management launched the WINDFALL program in 1978 in an attempt to address this need. It was the "first attempt to enable rural communities to realize economic benefits from wildlife" (Murindagomo, 1990; Peterson, 1991c). WINDFALL attempted to reduce conflicts between humans and wildlife and to improve attitudes toward conservation in settled communal areas by returning revenues from wildlife in reserves to nearby people via payments through (Rural) District Councils (Murindagomo, 1990).[7]

WINDFALL had some positive results, but it fell far short of meeting most of its objectives. Its two major problems were the limited local participation in decision making, and its programmatic base in wildlife from state reserves—not communal areas (Murindagomo, 1990). Communities did not receive enough benefits to encourage them to participate meaningfully. Only a small amount of revenue belatedly reached the communities and it was rarely very directly linked to wildlife production.

CAMPFIRE emerged as a new approach to natural resource conservation through the empowerment of local people. Peterson (1991c) noted that CAMPFIRE was met with some skepticism in the beginning as fears of over exploitation in line with Hardin's (1968) *tragedy of the commons* theory were voiced. However, CAMPFIRE was not an "open-access" resource management program, but rather a defined access "common property" resource management scheme limited to specifically enfranchised communities.

CAMPFIRE is a unique approach to wildlife management in the sense that it "combines the three principles of sustainable development, community empowerment, and conservation through utilization" in one program (Peterson, 1991c). Having emerged from an awareness of problems experienced by prior attempts (e.g., through the WINDFALL program), CAMPFIRE was designed to give rural communities full control of wildlife resources. The rationale behind CAMPFIRE was that people will invest in environmental conservation if they can utilize these resources on a sustainable basis for their own benefit (Murindagomo, 1990).

Legislative Changes

The root of CAMPFIRE'S successful implementation lies in legislative changes initiated by the independence government. The government has shown a commitment to localized planning and implementation.

A first step toward "true empowerment" (Peterson, 1991c) of local people occurred immediately after independence with the new directions for self-government in communal areas.[8] This was envisaged through the 1980 District Council Amendment Act, whereby District Councils were given, in addition to the normal functions of local authorities, the responsibility to manage most resources on communal lands (Peterson, 1991c).

A 1984 Prime Minister's Directive on Development further enhanced participation by local people in planning their own development. Ward Development Committees (WADCOS) and Village Development Committees (VIDCOS) were established as units for citizen participation (Murindagomo, 1990). These subdistrict institutions are key to the implementation of the CAMPFIRE program in the communal areas (see Figure 2; Thomas, 1992).

The government incorporated the CAMPFIRE objectives into the country's First Five-Year National Development Plan 1986–1990 and further encouraged its development. This plan enlisted as one of the natural resources sector's objectives "bringing benefits of wild game parks and other natural resources to the people living in and around game parks in an organized way, which will help control poaching of protected wildlife whilst improving the standard of living of the people" (Peterson, 1991c). Consistent with national objectives of decentralized development, the CAMPFIRE concept has become increasingly articulated as part of overall national development and in some ways is a more vigorous decentralization model than is the devolution of power to local government (Roe, 1992).

Figure 2. Zimbabwe governmental hierarchy as of July 1, 1993. The CAMPFIRE program transfers appropriate authority for many aspects of wildlife management, and the benefits derived from them, to selected Rural District Councils. Councils are encouraged to further devolve this authority to the ward level.

Changes in Management Approach Toward Community Empowerment

The legislative changes set the stage for local control of wildlife management. Given the radical shift in approach that this concept implied, it was not certain that communities would really be empowered or "given power to think and act in terms of some resource over which they have control" (Peterson, 1991c).

The foundation of community empowerment lies in devolution of management decisions to the local level. Just giving the communities economic resources from wildlife is not CAMPFIRE. In CAMPFIRE, the concept of community empowerment means "actually giving the community the power to decide on the allocation of these resources" (Peterson, 1991c). This process of ensuring decision making by local communities strengthens community leadership and fosters the development of mechanisms to make collective judgments.

The ability to collectively make decisions is an incentive for collective action to preserve the benefit-producing resource base (Peterson, 1991b). A major hindrance to local community empowerment is the reluctance of district councils to devolve proprietorship to more grassroots levels (WADCOS and VIDCOS). Council control has frustrated CAMPFIRE efforts in Nyaminyami, a district that started its program about the same time as Masoka (Metcalfe, 1994). In Nyaminyami's case, "local participation in management simply does not exist" (Murombedzi, 1993). In Masoka's case, the District has taken an alternative path and allowed the local community a great deal of control of its wildlife generated revenues. If Nyaminyami anchors one end of the local participation continuum, Masoka represents the other end. Metcalfe (1994) maintains that most CAMPFIRE sites fall somewhere between these extremes.

Search for a Test Site

Changes in management philosophy, even when backed by appropriate legislation, mean little if they are not put into practice. In this context, the village of Masoka played a special role in Zimbabwe's CAMPFIRE program. In 1986, it was selected by staff of the University of Zimbabwe's Center for Applied Social Science as a place where the seed of CAMPFIRE could be planted (Murphree, 1992; Metcalfe, 1994). Once planted, the seed germinated. It provided a focal site for study by many researchers starting with a 1988 baseline socioeconomic survey by Cutshall (1989).

Masoka was an ideal CAMPFIRE site because it met all of the best site conditions outlined in Fig. 1. Situated far from the beaten track in the mid-Zambezi Valley, its lands juxtaposed extensive areas of government controlled game estate that contained considerable dangerous game (Figure 3). Only a small riverine portion of its land was suitable for agriculture, and the presence of tsetse assured the absence of livestock. The small agglomerated settlement pattern made it easy to identify a producer community that was sufficiently cohesive to define and express community attitudes while offering the possibility of social controls in

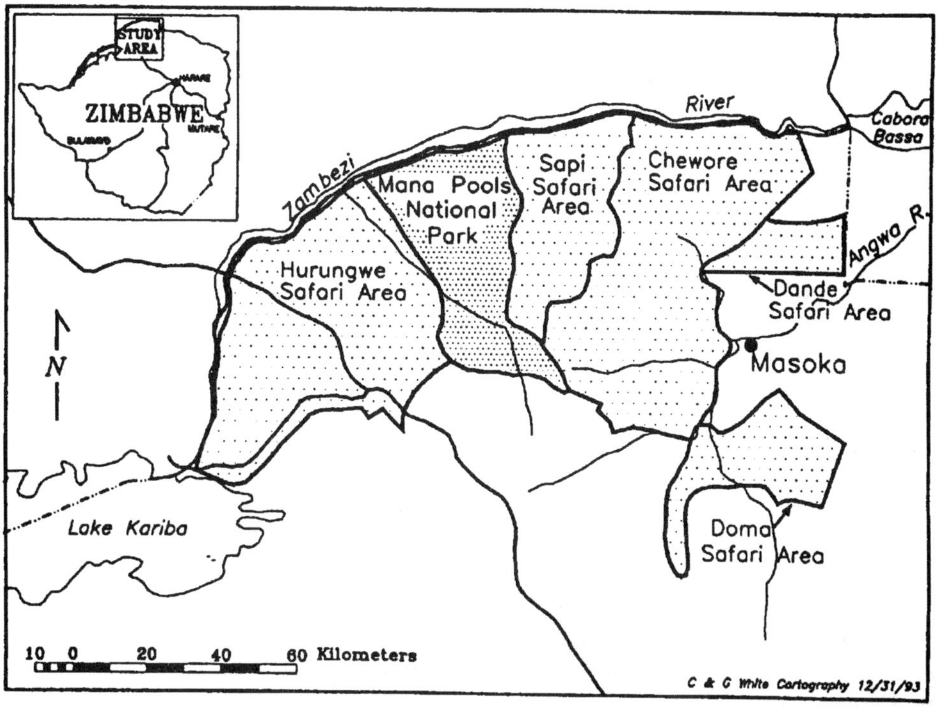

Figure 3. Masoka and surrounding land use map. Masoka's lands are juxtaposed to state protected safari areas which provide reservoirs of wildlife. This allows the village's CAMPFIRE program to capture benefits while simultaneously providng a managed and protected buffer for national game estate lands.

support of local decisions. Masoka was an archetypical "poaching community" that was the bane of old school wildlife protectors.

The people of Masoka were not immediate converts to the CAMPFIRE approach. Their skepticism was understandable given that many of the residents had lost their lands to government wildlife initiatives, which in earlier times had carved both a national park and several safari areas from their ancestral homelands.

An important shift in community perception coincided with the District Council's decision to use wildlife revenues to build a school in 1988. Prior to the school's construction, children had a dangerous 30-km walk through the bush to the nearest school. Since a daily commute by foot was impossible, only those willing and able to spend the entire week away from home (almost always boys) were schooled. When wildlife revenues were locally credited with delivering a community school, the CAMPFIRE initiative took root in Masoka (Murphree, 1992).

Elements of CAMPFIRE in Masoka

Community priorities for wildlife management were first expressed through an innovative land use planning exercise that identified about 20 km^2 of the best alluvial terrace lands along the Angwa River for agriculture and residential settlement.[9] The rest of the land, about 400 km^2 was designated as the wildlife area (Figure 3). Although the lands within the settlement area included good farmland, not all potentially good land was reserved for human occupance. An upstream extension of good alluvial soils was purposefully left unoccupied to maintain an important wildlife access corridor linking the community's wildlife area and the neighboring game estate controlled by DNPWM. Importantly, the community developed this strategy and it resulted in a larger wildlife area than recommended by a technical advisory team that operated out of the capital city with only an aerial photography information base (Murphree, 1992).

The land use plan was a means to achieve several important community hazard reduction goals related to life among dangerous beasts. Large game represented both direct and indirect threats to people's well being in Masoka. An average of 6% of the households annually suffered injuries and 2% experienced a death caused by wildlife.[10] Indirectly, their crops were threatened and each household suffered an average of nearly four crop damage incidents per year (Cutshall, 1989).

Prior to the arrival of the CAMPFIRE initiative, the only recourse for Masoka villagers was to kill the offending animals after the fact. When the community took charge, they consolidated their settlement and decided to enclose the settled area within a solar powered electric fence. The results were dramatic.

In the 3 years after the fence was completed, only one villager lost his life to animals, and that incident took place outside the fence along the route formerly traveled by school children on their weekly 30-km walking commute to school. Animal-caused injuries have disappeared (Figure 4), and people, especially men, report a greater sense of personal safety with the protective fence in place.

The reported incidents of crop damage also declined (Figure 5), and people were in near universal agreement that the amount of time spent guarding crops from wildlife was down (Figure 6).[11] A noticeable shift in the type of animals

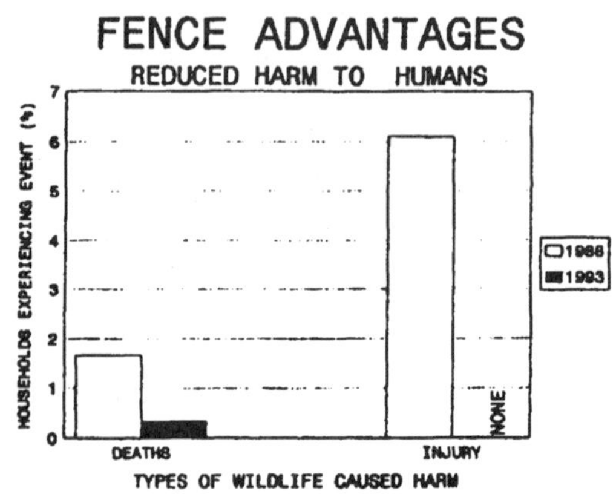

Figure 4. Reduction in wildlife-caused death and injuries. In the 3 years since the construction of the protective fence, wildlife attacks have disappeared. (One villager died from buffalo attack outside of the fence. Also, a professional hunter and client were killed by a buffalo in the safari area.)

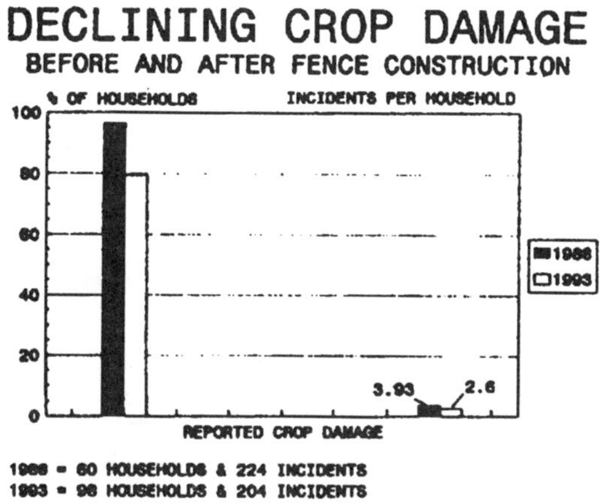

Figure 5. Declining crop damage. Reported crop damage decreased substantially after fence construction.

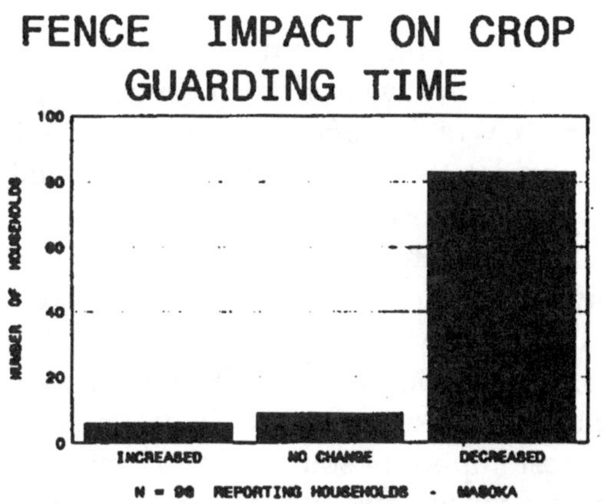

Figure 6. Changes in time spent guarding crops. Residents reported protective fence decreased the time spent guarding crops.

causing crop damage occurred as most dangerous game was excluded by the fence. Proportionately more damage resulted from smaller animals. These either resided within the enclosure, or could pass through it with relative ease (Figure 7).

Masoka's experience inside the fence suggests that the community's approach to resource management created a win-win situation. People gained personal safety and crop protection while reducing a major irritant that formerly led to demands for the destruction of both the offending animals and those found guilty by virtue of their species' affiliation. These outcomes were made possible by the transfer of greater management responsibility, and revenues generated from safari leasing, to the local level.[12]

Masoka's land use planning exercise created a community controlled safari area suitable for leasing to big game hunting safari operators. With this in place, the village created both an important revenue source to fund community development

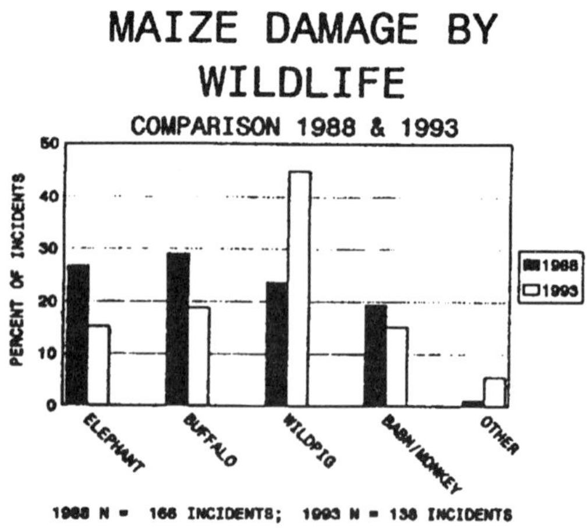

Figure 7. Shift in animals causing crop damage. The relative importance of smaller animal pests increased as the fence reduced problems caused by larger animals.

(Table 1) and a built-in monitoring system to ensure only controlled levels hunting off-take occur. Lease-holders have a strong incentive to prevent outsider hunting and since many employees of the safari operator are from Masoka any attempt by an operator to circumvent the quotas would be immediately reported to the community where the employees are important stake-holders and CAMPFIRE beneficiaries with a vested interest in maintaining community dividends.

Job opportunities from the CAMPFIRE enterprise are important in Masoka, where few alternative employment opportunities exist apart from subsistence agriculture. In 1993, 78% of households with a paid labor opportunity reported it to be the result of the CAMPFIRE enterprise (Figure 8). Although the safari camp operation was the most important source of work opportunities, the community also hired fence maintenance personnel and game scouts from funds generated by the safari area leasing arrangement (Figure 9).

The community's willingness to employ its own game scouts to prevent illegal hunting in the safari area is evidence of changing community attitudes toward the value of wildlife. Under the centralized management model Masoka was the source of poachers; its empowerment through CAMPFIRE has transformed it into a bastion of support for wildlife protection and enhancement. The evidence goes beyond the local employment and training of game protection

Table 1 Masoka Safari Leasing Revenue Growth[a]

1989	$ 47,310
1990	$ 76,170
1991	$ 89,293
1992	$276,745
1993	$459,851
1994	$639,290

[a]Source: Guruve District Files and Marshall Murphree (Personal communication).

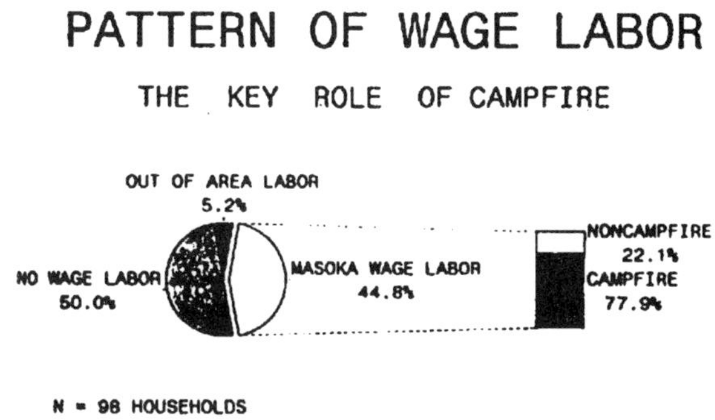

Figure 8. The pattern of wage labor in Masoka. The CAMPFIRE program plays a key role in Masoka's employment pattern.

personnel, including a new willingness to tolerate the actions of problem animals when they occasionally break through the fence. Rather than destroy the animals, the community organizes to drive their "cattle" back to safari area in a nonlethal manner. This changing view of wildlife seems to be the result of attitude shifts tied closely to the development and empowerment of the locally based resource management culture and associated emerging institutions.

The germination of CAMPFIRE in Masoka was aided by a number of personnel from government agencies, nongovernmental organizations (NGOs), and the University of Zimbabwe. The fact that CAMPFIRE has sunk deep roots reflects the devolution of substantial decision making authority to the village level. The devolution was not simply a reversion of authority to preexisting "traditional structures" that had been temporarily displaced by the confiscatory powers of central government. Preexisting authority structures including the chief,

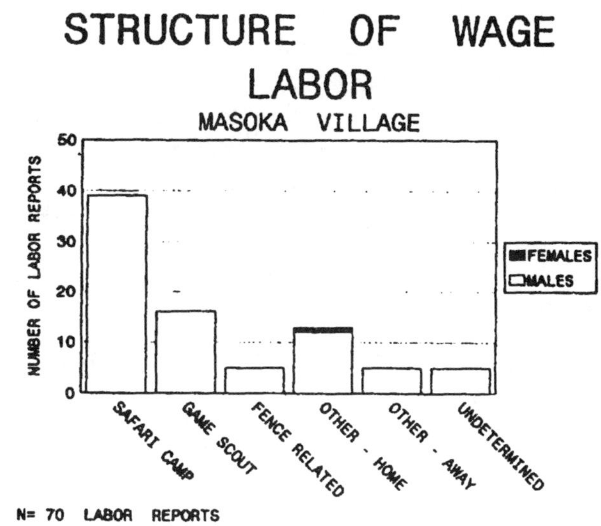

Figure 9. The structure of wage labor. Masoka's safari camp generates most direct employment. In addition, leasing fees allow the community to hire game scouts and fence-maintenance personnel.

headmen, spirit medium, and elected member of the district council continue to exercise their functional responsibilities. The community has added an elected wildlife committee to oversee CAMPFIRE activities while simultaneously developing a tradition of village assemblies to discuss and reach consensus on pending major decisions. One consequence of these new developments is the emergence in leadership positions of younger and better educated village members in positions of substantial influence. By virtue of their literacy and English language skills, these people have come to the fore in a quasimeritocracy where leaders who can operate in the extra-village context to negotiate with NGOS, government officials and private safari operators are favored by the electorate.

The speed of village institutional development and the insight demonstrated by their decisions is instructive. Only 5 years after the Masoka community's first tentative steps toward locally controlled wildlife-based development, many illustrations of sophisticated decision making process are available. Two examples are cited here; they both revolve around the disposition of wildlife generated revenues.

True local decision making allows village members to distribute safari-generated revenues in any way they see fit. In practice, they have varied the disposition of their funds to fit the changing circumstances. After drought-induced crop failure during the 1991–1992 cropping year, most CAMPFIRE money was dispersed in payments of $400 (Zim) to each household to allow families to purchase food. In the following year with even greater CAMPFIRE revenues available and with good crop yields making food plentiful, they decided to reduce household cash payments to $76 and reserve the bulk of the money for the construction of a health clinic. After the health clinic was under construction, the purchase of a community tractor was raised as the next priority.

News of the benefits of CAMPFIRE could serve as a major attraction to people wanting to cash in on Masoka's success. Realizing this, the wildlife committee developed a set of rules that prevented individuals from either migrating, or forming households, just to gain CAMPFIRE benefits. In-migrant families must establish a 5-year period of residence to be eligible for household payments. In addition, marriages must endure more than 1 year before the new household can receive cash payments. Additional rules consider the cases of polygynous marriages, divorces, widows, widowers, and multiple generation households.

Lessons from CAMPFIRE in Masoka

Community empowerment provides the people an opportunity to manage and make decisions about their wildlife, while at the same time realizing wildlife related benefits. To succeed, community-based wildlife management requires people to accept wildlife maintenance as a viable land use option. The success of such programs also depends partly on the creation of adequate local wildlife-derived incentives which are not present under direct state management (Child and Peterson, 1991). These conditions appear to be satisfied in Masoka's experience.

The traditionally dominant top-down approach to planning has been criticized as overly rational, reductionist, elitist, and essentially supportive of existing power relationships in societies (Daley and Angulo, 1990). If this approach has negative wildlife conservation impacts, or retards the achievement of more general development goals, participatory common property approaches offer an alternative model. Participation may occur at any or all levels of project activity, including design, implementation, management, and realization of benefits (Kiss, 1990). Masoka's experience mirrors these conditions and suggests their replication may have wider consequences for the conservation and development planning.

"The 1992 Earth Summit succeeded in alerting the conscience of the world to the urgency of achieving environmentally sustainable development" (Serageldin, 1993). The discourse on this subject is often global in perspective and framed in the language of ecology which focuses on subjects such as ecosystem integrity, biodiversity, and carrying capacity. On occasion, the economic objectives of growth, equity, and efficiency are introduced (Munasinghe, 1993). Less frequently, the social components of sustainable development are elevated for consideration (Cernea, 1993). The CAMPFIRE program in Zimbabwe is a bold approach that simultaneously addresses all three aspects of sustainability in the context of ecosystems with megafaunal components. The Masoka case offers some tentative measures of success as global sustainability concerns are operationalized at the very local level. It also challenges researchers to go beyond Masoka and to examine the implications of the CAMPFIRE experiment in the broader context of sustainable development.

CAMPFIRE management in Masoka addresses ecosystem concerns by concentrating human settlement and by extending many aspects of protected area status beyond the boundaries of the officially designated game estate shown on maps of national parks and safari areas. As practiced in Masoka, CAMPFIRE preserves habitat and creates substantial spatial buffers suitable for large animal movement and occupance.

CAMPFIRE developments in Masoka suggest the economic sustainability criteria of equity, growth, and efficiency are addressed by the devolution of wildlife management responsibility to the village level. Under the centralized management model, most benefits were captured and retained outside the environs of the wildlife-producing community, even while the substantial penalties of the wildlife presence were paid by local residents in the form of crop losses, animal attacks, and poaching arrests. And, while the people suffered these indignities, they had no school, clinic, or other development benefits. CAMPFIRE changed all of that as both responsibility and major portions of revenue streams were shifted to the village level. The CAMPFIRE approach produced efficient feedback loops that reinforced the merit of ecosystem protection in the minds of the system's most important actors—the local citizens with axes, hoes, and weapons.

Factors favoring the social sustainability of CAMPFIRE in Masoka have emerged over the short span of 5 years as the local citizenry developed institutions for both decision making and management. "Evidence from many countries suggests that the statist solution applied exclusively, promises more failure than success" (Cernea, 1993). If the Zimbabwe government and its local district representatives continue to foster locally-based management approaches, Masoka will serve as a benchmark locale for monitoring social sustainability of biodiversity initiatives over longer periods of time.

CAMPFIRE in Masoka also reinforces the call for new frameworks for the old *tragedy of the commons* debate (Roe, 1994). Many authors have demonstrated that the commons need not equal tragedy where systems of community control are in place. Masoka demonstrates that new forms of community control can be powerful tools in the practice of sustainable development. In this time and in this place, CAMPFIRE seems to provide an alternative model to biodiversity preservation strategies that emphasize central control.

References

Bonner, R. (1993). *At the Hand of Man: Peril and Hope for Africa's Wildlife.* Alfred A. Knopf, New York.

Cernea, M. M. (1993). The sociologist's approach to sustainable development. *Finance and Development* 30(4):11–13.

Child, G. S., and Peterson, J. H. (1991). CAMPFIRE in Rural Development: The Beitbridge Experience. DNPWLM/CASS Working Paper No. 1/91, Centre for Applied Social Science, University of Zimbabwe.

Crane, M. D. (1991) *CAMPFIRE, From Theory to Practice in the Village of Masoka.* Center for Applied Social Science, University of Zimbabwe, Harare.

Cutshall, C. R. (1989). *Masoka/Kanyurira Ward: A Socioeconomic Baseline Survey of Community Households.* Center for Applied Social Science, University of Zimbabwe, Harare.

Daley, J. M., and Angulo, J. (1990). People centered community planning. *Journal of The Community Development Society* 21(2):88–101.

Dasmann, R. (1964). *African Game Ranching.* MacMillan, New York.

Hardin, G. (1968). Tragedy of the commons. *Science* 162:1243–1248.

Hill, K. A. (1991). Zimbabwe's wildlife conservation regime: Rural farmers and the state. *Human Ecology* 19(1): 19–34.

Hunter, M., Hitchcock, R., and Wyckoff-Baird, B. (1990). Women and wildlife in southern Africa. *Conservation Biology* 4(4):448–451.

Kiss, A. (ed.) (1990). Introduction: Living with Wildlife: Wildlife Resource Management with Local Participation in Africa. World Bank Technical Paper No. 130, Washington, D.C., pp. 1–15.

Mackenzie, J. M. (1988). The natural world and the popular consciousness in southern Africa. In Kaarsholm, P. (ed.), *Cultural Struggle and Development in Southern Africa.* Baobab Books, Harare.

Marks, S. (1984). *The Imperial Lion: Wildlife Management in Central Africa.* Westview Press, Boulder.

Martin, R. B. (1986). *Communal Areas Management Programme for Indigenous Resources.* Terrestrial Ecology Branch, Department of National Parks & Wildlife, Harare.

Metcalfe, S. (1994). The Zimbabwe communal areas management program for indigenous resources CAMPFIRE. In Western, D., and Wright, R. M. (eds.), *Natural Connections: Perspectives in Community-based Conservation.* Island Press, Washington, D.C., pp. 161–192.

Moyo, S., Robinson, P., Katerere, Y., Stevenson, S., and Gumbo, D. (1991). *Zimbabwe's Environmental Dilemma: Balancing Resource Inequalities.* ZERO Publishers, Harare.

Munasinghe, M. (1993). The economist's approach to sustainable development. *Finance and Development* 30(4):16–19.

Murindagomo, F. (1990). Zimbabwe: WINDFALL and CAMPFIRE. In Kiss, A. (ed.), Living with Wildlife: Wildlife Resource Management with Local Participation in Africa. World Bank Technical Paper No. 130, Washington, D.C., pp. 123–139.

Murombedzi, J. (1993). *Decentralization or Recentralization? Implementing CAMPFIRE in the Omay Communal Lands of the Nyaminyami District.* Center for Applied Social Science, University of Zimbabwe, Harare.

Murphree, M. W. (1990). *Decentralizing the Proprietorship of Wildlife Resources in Zimbabwe's Communal Lands.* Center for Applied Social Science, University of Zimbabwe, Harare.

Murphree, M. W. (1992). Personal communication.

Murphree, M. W. (1993a). The Role of Institutions. Paper presented at a Workshop on Community-Based Conservation; in *Proceedings.* Island Press/Earth Scan (forthcoming).

Murphree, M. W. (1993b). *Communities as Resource Management Institutions.* International Institute for Environment and Development, London.

Peterson, J. (1991b). *A Proto-Campfire Initiative in Mahenye Ward, Chipinge District: Development of a Wildlife Utilization Programme in Response to Community Needs.* Centre for Applied Social Science, University of Zimbabwe, Harare.

Peterson, J. (1991c). *CAMPFIRE: A Zimbabwean Approach to Sustainable Development and Community Empowerment through Wildlife Utilization.* Centre for Applied Social Science, University of Zimbabwe, Harare.

Riney, T. (1967). *Conservation and Management of African Wildlife.* FAO/IUCN, Rome.

Roe, E. M. (1992). *Report on the Amalgamation of District Councils and Rural Councils.* Center for Applied Social Science, University of Zimbabwe, Harare.

Roe, E. M. (1994). Very new frameworks for a very old tragedy of the commons and an aging common property resource management. *Agriculture and Human Values* 11(1):29–36.

Serageldin, I. (1993). Making development sustainable. *Finance and Development* 30(4):6–12.

Thomas, S. J. (1992). *The Legacy of Dualism and Decision-Making: The Prospects for Local Institutional Development in "CAMPFIRE."* Centre for Applied Social Science, University of Zimbabwe, Harare.

Western, D. (1982). Amboseli National Park: Enlisting landowners to conserve migratory wildlife. *Ambio* 11(5):302–308.

Western, D. (1994). Ecosystem conservation and rural development: The case of Amboseli. In Western, D., and Wright, R. M. (eds.), *Natural Connections: Perspectives in Community-Based Conservation.* Washington, D.C., Island Press, pp. 15–52.

Endnotes

1. Geosciences Department, Oregon State University, Corvallis, Oregon 97331–5506.
2. Centre for Applied Social Science, University of Zimbabwe, Harare, Zimbabwe.
3. This term is purposefully vague as CAMPFIRE has a wide range of actors participating in its establishment in any particular site. Some actors foster community participation through insight/persuasion, others exercise political influence, and some have legal authority to make decisions. CAMPFIRE functionaries may include private sector people (e.g., safari operators, commercial farmers, community residents), central government personnel, academics, district councils, development groups, and environmental organizations.
4. Masoka is a village located in Kanyurira Ward in Guruve District. Some publications refer the location as Kanyurira, but the term Masoka is used here because it refers to the focal

study location which was enclosed within a protective wildlife fence and is the site of the community school, stores, and clinic.

5. Cutshall surveyed all Masoka households (60) in 1988. His results are reported as means without estimates of variance. Therefore, our study also reports mean responses for comparison purposes as tests of significance could not be applied.
6. The survey instrument was administered to all 98 households resident within the village wildlife exclusion fence. Men and women were interviewed separately by interviewers matched to the sex of the respondents. In all, the 98 households yielded interviews with 54 matched husband/wife pairs, seven women living as second wives in matched pair households, ten men only household interviews, and 34 women only household contacts. A total of 148 interviews were completed. Except where differences are noted, household data as reported in this paper reflect an agreement between householders of both sexes.
7. Subsequently, District Councils were amalgamated with the Rural Councils of commercial farming regions. In theory, all of Zimbabwe now has a unified local government structure under newly formed Rural District Councils (Roe, 1992).
8. A more complete review history of CAMPFIRE's evolution in Zimbabwe is available in Metcalfe (1994). Metcalfe's discussion includes a review of the timing, issues, people, legislation, and seminal project associated with CAMPFIRE's genesis in Zimbabwe.
9. This exercise was undertaken at the request of the community. Although personnel from the World Wildlife Fund (WWF) and the University of Zimbabwe's Centre for Applied Social Science (CASS) provided assistance, the community took ownership and amended the efforts of the "experts." Technical and financial assistance for construction of the fence was provided by WWF.
10. Cutshall reported almost 1/4 of the 60-households in Masoka recounted an injury or death from wild animals over a 3-year period. There was a total of 14 incidents that resulted in three deaths.
11. Evaluation of the decline in crop damage was problematic because of the limited information gathered in Cutshall's (1989) original survey. Cutshall only recorded whether a particular crop type had been damaged by a named species—for example, "Did you suffer any maize damage by elephants?" It took no measure of the extent of damage, or of multiple incident by the same species. For comparison purposes, the same procedure was repeated in our survey. Villagers are not obligated to live, or cultivate, within the fence, and some choose not to do so. Hence, their crops cannot be protected and contribute disproportionately to damage reports.
12. Not all management decisions are localized. DNPWM decides on off-take quotas while the District Council supervises the leasing process and maintains accounting responsibilities. In many ways, the community negotiates with other entities to achieve its desired outcomes. CAMPFIRE provides a context for negotiation rather than specific programmatic prescriptions. To assist in this process, ZimTrust (an NGO) provides leadership training and CASS provides advice and evaluative research support.